# APPLIED MATHEMATICAL MODELS AND EXPERIMENTAL APPROACHES IN CHEMICAL SCIENCE

*Innovations in Chemical Physics and Mesoscopy*

# APPLIED MATHEMATICAL MODELS AND EXPERIMENTAL APPROACHES IN CHEMICAL SCIENCE

*Edited by*
**Vladimir I. Kodolov, DSc**
**Mikhail A. Korepanov, DSc**

Apple Academic Press Inc. | Apple Academic Press Inc.
3333 Mistwell Crescent | 9 Spinnaker Way
Oakville, ON L6L 0A2 | Waretown, NJ 08758
Canada | USA

©2017 by Apple Academic Press, Inc.

First issued in paperback 2021

*Exclusive worldwide distribution by CRC Press, a member of Taylor & Francis Group*
No claim to original U.S. Government works

ISBN 13: 978-1-77463-711-1 (pbk)
ISBN 13: 978-1-77188-382-5 (hbk)

### Library and Archives Canada Cataloguing in Publication

Applied mathematical models and experimental approaches in chemical science / edited by Vladimir I. Kodolov, DSc, Mikhail A. Korepanov, DSc.

(Innovations in chemical physics and mesoscopy)
Includes bibliographical references and index.
Issued in print and electronic formats.
ISBN 978-1-77188-382-5 (hardcover).--ISBN 978-1-77188-383-2 (pdf)

1. Chemistry--Mathematical models. I. Kodolov, Vladimir Ivanovich, author, editor II. Korepanov, Mikhail. A., author, editor III. Series: Innovations in chemical physics and mesoscopy

QD39.3.M3A66 2016          540.1'51          C2016-905452-7          C2016-905453-5

### Library of Congress Cataloging-in-Publication Data

Names: Kodolov, Vladimir I. (Vladimir Ivanovich), editor. | Korepanov, Mikhail A., editor.
Title: Applied mathematical models and experimental approaches in chemical science / editors, Vladimir I. Kodolov, DSc, Mikhail A. Korepanov, DSc.
Description: Oakville, ON ; Waretown, NJ : Apple Academic Press, 2017. | Series: Innovations in chemical physics and mesoscopy | Includes bibliographical references and index.
Identifiers: LCCN 2016035475 (print) | LCCN 2016043006 (ebook) | ISBN 9781771883825 (hardcover : alk. paper) | ISBN 9781771883832 (ebook)
Subjects: LCSH: Photoelectrochemistry. | Solar cells. | Nanostructured materials--Research.
Classification: LCC QD578 .A67 2017 (print) | LCC QD578 (ebook) | DDC 540.1/51--dc23
LC record available at https://lccn.loc.gov/2016035475

Apple Academic Press also publishes its books in a variety of electronic formats. Some content that appears in print may not be available in electronic format. For information about Apple Academic Press products, visit our website at **www.appleacademicpress.com** and the CRC Press website at **www.crc-press.com**

# CONTENTS

# LIST OF CONTRIBUTORS

**Hassan Abdoos**
Department of Nano Technology, Nano Materials Engineering Group, Semnan University, Semnan, Iran, E-mail: Hassan.abdoos@gmail.com

**Bahram Azad**
Department of Material Science and Engineering, Semnan University, Semnan, Iran

**Yu. V. Berestneva**
L.M. Litvinenko Institute of Physical Organic and Coal Chemistry, Donetsk 83114, Ukraine

**A. N. Blyablyas**
Institute of Mechanics of the Ural Branch of the Russian Academy of Sciences (IM UB RAS), Izhevsk, Russia

**Ehsan Borhani**
Department of Nano Technology, Nano Materials Engineering Group, Semnan University, Semnan, Iran, E-mail: Ehsan.borhani@profs.semnan.ac.ir

**E. G. Butolin**
Izhevsk State Medical Academy, Izhevsk, Russia

**I. Cherenkov**
Udmurt State University, Izhevsk, Russia

**N. N. Chuchkova**
Izhevsk State Medical Academy, Izhevsk, Russia

**A. B. Eresko**
L.M. Litvinenko Institute of Physical Organic and Coal Chemistry, Donetsk 83114, Ukraine

**A. Yu. Fedotov**
Institute of Mechanics of the Ural Branch of the Russian Academy of Sciences (IM UB RAS), Izhevsk, Russia

**F. Z. Gilmutdinov**
Physic-Technical Institute of the Ural Branch of the Russian Academy of Sciences (PhTI UB RAS), Izhevsk, Russia

**O. Yu. Goncharov**
Institute of Mechanics of the Ural Branch of the Russian Academy of Sciences (IM UB RAS), Izhevsk, Russia

**N. V. Goncharova**
Institute of Mechanics of the Ural Branch of the Russian Academy of Sciences (IM UB RAS), Izhevsk, Russia

**S. A. Gruzd**
Kalashnikov Izhevsk State Technical University, Izhevsk, Russia

**A. A. Gurov**
Perm National Research Polytechnic University, Perm, Russia

**E. Ilina**
National Research Tomsk Polytechnic University, Tomsk, Russia

**V. F. Kablov**
Volzhsky Polytechnic Institute, Volgograd State Technical University, Volgograd, 400131, Russia

**M. M. Kanunnikov**
Izhevsk State Medical Academy, Izhevsk, Russia

**O. M. Kanunnikova**
Physic-Technical Institute of the Ural Branch of the Russian Academy of Sciences (PhTI UB RAS), Izhevsk, Russia

**O. V. Karban**
Physic-Technical Institute of the Ural Branch of the Russian Academy of Sciences (PhTI UB RAS), Izhevsk, Russia

**A. Yu. Karpova**
Institute of Mechanics of the Ural Branch of the Russian Academy of Sciences (IM UB RAS), Izhevsk, Russia

**V. G. Kochetkov**
Volzhsky Polytechnic Institute, Volgograd State Technical University, Volgograd, 400131, Russia

**V. I. Kodolov**
Kalashnikov Izhevsk State Technical University, Izhevsk, Russia

**V. Komissarov**
Izhevsk State Medical Academy, Izhevsk, Russia

**M. A. Korepanov**
Institute of Mechanics of the Ural Branch of the Russian Academy of Sciences (IM UB RAS), Izhevsk, Russia

**N. V. Kostenko**
Volzhsky Polytechnic Institute, Volgograd State Technical University, Volgograd, 400131, Russia

**P. A. Kuryleva**
Samara State Aerospace University, Moskovskoye sh., 34, Samara, Samara Oblast, Russia

**V. I. Ladyanov**
Institute of Mechanics of the Ural Branch of the Russian Academy of Sciences (IM UB RAS), Izhevsk, Russia

**L. Larionova**
National Research Tomsk Polytechnic University, Tomsk, Russia

**V. P. Lebedev**
Elekond Open Joint-Stock Company, Sarapul, Russia

**L. A. Lenartovich**
Belarusian State Technological University, Minsk, Sverdlov St., 13-a, 220006, Belarus, Tel. 3275738

**V. S. Lifanov**
Volzhsky Polytechnic Institute, Volgograd State Technical University, Volgograd, 400131, Russia

**A. M. Lipanov**
Institute of Mechanics of the Ural Branch of the Russian Academy of Sciences (IM UB RAS), Izhevsk, Russia

**M. M. Lylina**
Research Institute of Chemical Reactants and Ultrapure Compounds (IREA), Moscow, Russia

**Sh. Maghsoodlou**
University of Guilan, Rasht, Iran

**T. M. Makhneva**
Institute of Mechanics of the Ural Branch of the Russian Academy of Sciences (IM UB RAS), Izhevsk, Russia

**S. S. Mikhaylova**
Physic-Technical Institute of the Ural Branch of the Russian Academy of Sciences (PhTI UB RAS), Izhevsk, Russia

**V. V. Mukhgalin**
Physic-Technical Institute of the Ural Branch of the Russian Academy of Sciences (PhTI UB RAS), Izhevsk, Russia

**A. V. Muratov**
L.M. Litvinenko Institute of Physical Organic and Coal Chemistry, Donetsk 83114, Ukraine

**O. S. Nabokova**
Institute of Mechanics of the Ural Branch of the Russian Academy of Sciences (IM UB RAS), Izhevsk, Russia

**A. M. Nemeryuk**
Research Institute of Chemical Reactants and Ultrapure Compounds (IREA), Moscow, Russia

**E. Nesterov**
National Research Tomsk Polytechnic University, Tomsk, Russia

**O. M. Novopoltseva**
Volzhsky Polytechnic Institute, Volgograd State Technical University, Volgograd, 400131, Russia

**V. G. Petrov**
Institute of Mechanics of the Ural Branch of the Russian Academy of Sciences (IM UB RAS), Izhevsk, Russia

**S. Poreskandar**
University of Guilan, Rasht, Iran

**S. E. Porozova**
Perm National Research Polytechnic University, Perm, Russia

**V. Pozdnjakov**
Udmurt State University, Izhevsk, Russia

**N. R. Prokopchuk**
Belarusian State Technological University, Minsk, Sverdlov St., 13-a, 220006, Belarus, Tel. 3275738, E-mail: prok_nr@mail.ru

**B. E. Pushkarev**
Physic-Technical Institute of the Ural Branch of the Russian Academy of Sciences (PhTI UB RAS), Izhevsk, Russia

**E. V. Raksha**
L.M. Litvinenko Institute of Physical Organic and Coal Chemistry, Donetsk 83114, Ukraine, E-mail: elenaraksha411@gmail.com

**G. V. Sapozhnikov**
Physic-Technical Institute of the Ural Branch of the Russian Academy of Sciences (PhTI UB RAS), Izhevsk, Russia

**N. V. Savinova**
Izhevsk State Medical Academy, Izhevsk, Russia

**V. Scuridin**
National Research Tomsk Polytechnic University, Tomsk, Russia

**V. Sergeev**
Udmurt State University, Izhevsk, Russia

**A. V. Severyukhin**
Institute of Mechanics of the Ural Branch of the Russian Academy of Sciences (IM UB RAS), Izhevsk, Russia

**I. N. Shabanova**
Physic-Technical Institute of the Ural Branch of the Russian Academy of Sciences (PhTI UB RAS), Izhevsk, Russia

**A. A. Shaklein**
Institute of Mechanics of the Ural Branch of the Russian Academy of Sciences (IM UB RAS), Izhevsk, Russia

**M. A. Shumilova**
Institute of Mechanics of the Ural Branch of the Russian Academy of Sciences (IM UB RAS), Izhevsk, Russia

**B. V. Skvortsov**
Samara State Aerospace University, Moskovskoye sh., 34, Samara, Samara Oblast, Russia

**A. A. Smetkin**
Perm National Research Polytechnic University, Perm, Russia

**M. V. Sobennikova**
Physic-Technical Institute of the Ural Branch of the Russian Academy of Sciences (PhTI UB RAS), Izhevsk, Russia

**S. P. Starostin**
Elekond Open Joint-Stock Company, Sarapul, Russia

**E. Stasuk**
National Research Tomsk Polytechnic University, Tomsk, Russia

**N. S. Terebova**
Physic-Technical Institute of the Ural Branch of the Russian Academy of Sciences (PhTI UB RAS), Izhevsk, Russia

**S. Yu. Treschev**
Physic-Technical Institute of the Ural Branch of the Russian Academy of Sciences (PhTI UB RAS), Izhevsk, Russia

**N. A. Turovskij**
Donetsk National University, Universitetskaya Street, 24, Donetsk, 83001, Ukraine

**A. V. Vakhrouchev**
Institute of Mechanics of the Ural Branch of the Russian Academy of Sciences (IM UB RAS), Izhevsk, Russia

**N. Varlamova**
National Research Tomsk Polytechnic University, Tomsk, Russia

**D. S. Vokhmyanin**
Perm National Research Polytechnic University, Perm, Russia

**G. E. Zaikov**
Institute of Biochemical Physics, Russian Academy of Sciences, Kosygin Street, 4, Moscow, 117 334, Russian Federation, E-mail: chembio@sky.chph.ras.ru

**R. Zelchan**
National Research Tomsk Polytechnic University, Tomsk, Russia

**D. K. Zhirov**
Institute of Mechanics of the Ural Branch of the Russian Academy of Sciences (IM UB RAS), Izhevsk, Russia

**D. M. Zhivonosnovskaya**
Samara State Aerospace University, Moskovskoye sh., 34, Samara, Samara Oblast, Russia

# LIST OF ABBREVIATIONS

| | |
|---|---|
| AA | aluminum alloys |
| AFM | atomic force microscopy |
| AO | atomic orbitals |
| APM | area of protective measures |
| ARB | accumulative roll bonding |
| BQ | benzoquinone |
| BSA | bovine serum albumin |
| CARB | cross accumulative roll bonding |
| CEAL | Central Ecological and Analytical Laboratory |
| COP | coefficient of performance |
| CV | cyclic voltamperometry |
| CVD | chemical vapor deposition |
| DFT | density functional theory |
| DMFA | dimethylformamide |
| DMSO | dimethylsulfoxide |
| DRS | dielectric relaxation spectroscopy |
| DSC | differential scanning calorimetry |
| DTA | differential thermal analysis |
| EAM | embedded atom method |
| ECAE | equal channel angular extrusion |
| ECAP | equal channel angular pressing |
| ECP | electroconductive polymers |
| EDX | energy dispersive X-ray |
| ES | emeraldine salt |
| FG | fiberglass |
| FTIR | Fourier-transform infrared spectral |
| GS | galvanostatic method |
| HAGBs | high angle grain boundaries |
| HCF | high cycle fatigue |
| HOMO | highest occupied molecular orbital |
| HPT | torsion under high pressure |

| HQ | hydroquinone |
|---|---|
| IM UB RAS | Institute of Mechanics of the Ural Branch of the Russian Academy of Sciences |
| IPD | intensive plastic deformation |
| JEFF | Journal of Engineered Fabrics & Fibers |
| LCF | low cycle fatigue |
| MD | molecular dynamics |
| MEAM | modified embedded-atom method |
| ND | normal direction |
| PAn | polyaniline |
| PCM | polymer composite materials |
| PD | potentiodynamic method |
| PEC | photoelectric converters |
| PLm | polyluminol |
| PS | potentiostatic method |
| RCS | repetitive corrugation and straightening process |
| RD | rolling direction |
| REU | repetitive extrusion and upsetting process |
| RS | Raman scattering |
| SEM | scanning electron microscopy |
| SLN | sentinel lymph nodes |
| SPD | severe plastic deformation |
| SPZ | sanitary protection zone |
| ST | solution treated |
| TCEC | tube cyclic extrusion-compression process |
| TD | transverse direction |
| THF | tetrahydrofuran |
| TMS | tetramethylsilane |
| UFG | ultrafine grained |
| UHMWPE | ultrahigh molecular weight polyethylene |
| WE | working electrodes |
| XPA | X-ray phase analysis |
| XPS | X-ray photoelectron spectra |
| XRD | X-ray diffraction |

# LIST OF SYMBOLS

| | |
|---|---|
| $a$ | fiber radius |
| $a_{AV}$ | the mean of fiber radius |
| $d$ | down direction index |
| d | fiber diameter |
| $e$ | electrical charge |
| $F'$, $Z'$ | dimensionless direction |
| $F_c$ | Columb force |
| $F_{cap}$ | surface tension force |
| $F_{EI}$ | electrical field force |
| $G$ | elastic modulus |
| $H$ | distance between the capillary exit and the ground |
| $h$ | spinning distance |
| $i, j$ | index |
| $i, j, k$ | axis coordinate |
| $K_i$ | curvature of the jet |
| $L$ | distance between two charges |
| $l$ | length a part of jet |
| $l'$ | dimensionless length a part of jet |
| $m$ | mass |
| $Q$ | dimensionless electrical charge |
| $R_{ij}$ | vector distance |
| $V$ | velocity |
| $V_0$ | electric potential |
| $V_c$ | critical voltage |
| $X_i$, $Y_i$, $Z_i$ | Cartesian coordinates |
| $\alpha$ | surface tension coefficient |
| $\mu$ | viscosity |
| $\sigma$ | stress |

# PREFACE

This volume is the collection of topics in three parts to reflect the diversity of recent advances in *chemical physics and mesoscopy* with a broad perspective, which may be useful for scientists as well as for graduate students and engineers. This new book presents leading-edge research from around the world in this dynamic field. It focuses on concepts above formal experimental techniques and theoretical methods of chemical physics for micro and nanotechnologies.

The first part of this book offers scope for academics, researchers, and engineering professionals to present their research and development works that have potential for applications in several disciplines of engineering and nano-science. Contributions range from new methods to novel applications of existing methods to gain understanding of the nano-material and/or structural behavior of new and advanced nano-systems.

Contributions in part 2 of the volume were sought from many areas of polymer science and engineering in which advanced methods are used to formulate (model) and/or analyze the problem. In view of the different backgrounds of the expected audience, readers are requested to focus on the main ideas and to focus as much as possible on the specific advantages that arise from applying modern ideas. A chapter may, therefore, be motivated by the specific problem, but just as well by the advanced method used which may be more generally applicable.

Part 3 is a collection of articles that highlight some important areas of current interest in recent advances in chemistry and physics of engineering materials. It gives an up-to-date and thorough exposition of the present state of the art of chemical physics. It describes models and techniques now available to the chemist and discusses their capabilities, limitations, and applications. It provides a balance between chemical and material engineering, and basic and applied research.

We would like to express our deep appreciation to all the authors for their outstanding contributions to this book and to express our sincere gratitude for their generosity. All the authors eagerly shared their experiences and expertise in this new book. Special thanks go to the referees for their valuable work.

*—Vladimir I. Kodolov, DSc*
*Mikhail A. Korepanov, DSc*

# ABOUT THE EDITORS

**Vladimir I. Kodolov, DSc**

*Professor and Head, Department of Chemistry and Chemical Technology, M. I. Kalashnikov Izhevsk State Technical University, Izhevsk, Russia; Chief of Basic Research, High Educational Center of Chemical Physics and Mesoscopy, Udmurt Scientific Center, Russia*

Vladimir I. Kodolov, DSc., is Professor and Head of the Department of Chemistry and Chemical Technology at M. T. Kalashnikov Izhevsk State Technical University in Izhevsk, Russia, as well as Chief of Basic Research at the High Educational Center of Chemical Physics and Mesoscopy at the Udmurt Scientific Center, Ural Division at the Russian Academy of Sciences. He is also the Scientific Head of Innovation Center at the Izhevsk Electromechanical Plant in Izhevsk, Russia.

He is Vice Editor-in-Chief of the Russian journal *Chemical Physics and Mesoscopy* and also is a member of the editorial boards of several Russian journals. He is the Honorable Professor of the M. T. Kalashnikov Izhevsk State Technical University, Honored Scientist of the Udmurt Republic, Honored Scientific Worker of the Russian Federation, Honorary Worker of Russian Education, and also Honorable Academician of the International Academic Society.

**Mikhail A. Korepanov, DSc**

*Professor, Udmurt State University, Izhevsk, Russia*

M. A. Korepanov, DSc, is Professor at Udmurt State University, Izhevsk, Russia, where he teaches thermodynamics, theory of combustion, and physical-chemical fluid mechanics. He is editor-in-chief of the journal *Chemical Physics and Mesoscopy* (Russia) and has been named an Honorable Scientist of the Udmurt Republic. His research interests are hydrodynamics, chemical thermodynamics, numerical simulation of processes in technical systems and natural phenomena.

# ABOUT THE SERIES
# INNOVATIONS IN CHEMICAL
# PHYSICS AND MESOSCOPY

The Innovations in Chemical Physics and Mesoscopy book series publishes books containing original papers and reviews as well as monographs. These books and monographs will report on research developments in the following fields: nanochemistry, mesoscopic physics, computer modeling, and technical engineering, including chemical engineering. The books in this series will prove very useful for academic institutes and industrial sectors round the world interested in advanced research.

## BOOKS IN THE SERIES

# PART I

# NANOTECHNOLOGY

# CHAPTER 1

# SELF-ORGANIZATION OF NANOSCALE STRUCTURES BY EPITAXIAL DEPOSITION

A. V. VAKHROUCHEV, A. V. SEVERYUKHIN, and
A. YU. FEDOTOV

*Institute of Mechanics of the Ural Branch of the Russian Academy of Sciences (IM UB RAS), Izhevsk, Russia*

## CONTENTS

## 1.1  INTRODUCTION

In recent literature there are more information on the use of nanostructured elements to improve the energy and mass characteristics of thin photoelectric converters (PEC). Scientists and engineers of Oregon State University, the US proposed a new technology that allows to cover the thin film surface nanostructures of different materials [1]. The technology can be used to change the optical properties of these materials, in particular, to reduce the reflectance of the material by applying to the surface of the nanoparticles. The deposition process is carried out in a special chemical microreactor, wherein the material deposited on the surface of the thin film nano-elements in the form of tiny pyramids. A reduction in reflectance of the material is achieved due to this. Such material is capable of absorbing a greater amount of incident light than the non-treated. Such materials can be used in various fields.

The main application of materials coated nanofilms is to provide photoelectron converters having increased efficiency. In addition, new materials will find application in the production of optics and lenses for eyeglasses. Significant progress has been received in area of solar cells from complex semiconductor compounds [2–5], which has the coefficient of performance (COP) superior to simple theoretical efficiency of silicon solar cells (up to 23%).

The latest development is a solar cell based on GaP, GaAs, and InGaAsN. The solar cell has a four layers: the upper is composed of indium gallium phosphide, the second is composed of gallium arsenide, the third is composed of 2% nitrogen "in the indium gallium arsenide" and the fourth is composed of germanium. Each layer absorbs and converts electricity into a certain range of light wavelengths. First layer is absorbed green and yellow, the second layer is absorbed from green to deep red, the third layer is absorbed between the deep red and infrared, and fourth layer is absorbed from infrared and beyond. Currently, the best efficiency of 4-layer solar cells is 40.7%. But this is a special prototype of solar cells. Distribution is only obtained silicon solar cells, in which the real efficiency lies in the range 12–18%.

Low efficiency silicon solar cell due to the fact that he works in a narrow range of the solar spectrum, in the orange-red light. The remainder of the spectrum is not involved in the production of electricity, but only causes undesired heating of the device. In the latest solar cell efficiency improvement is achieved by expanding the range of "efficient" solar radiation by complicating the structure and 4-layer cake. But in this case only action involved in green, yellow, orange, red, and infrared rays of sunlight. The rest of the solar spectrum, from ultraviolet to green, including purple, blue and blue, completely excluded from the "useful" photoelectric conversion.

As for the ultraviolet solar radiation, the nanofilm of nanomagnesia improves solar cell efficiency by converting ultraviolet rays into visible orange-red band in the case of silicon solar cells. With the help of an additional layer of nanofilms of nanomagnesia you can improve efficiency by 5–7%. If we could use in photoelectric conversion of silicon solar cells purple, blue, cyan, green, yellow and infrared part of the solar spectrum, the real efficiency could be raised by 15–20%. Then the solar energy conversion efficiency reaches 32–50%.

The aim is to describe the methodology of modeling processes of special nanostructured layers in epitaxial structures for sophisticated photovoltaic cells. It is interesting to conduct computational experiments on creation and use of nanostructured and nanoscale elements that can be used in solar cells and other photovoltaic devices.

## 1.2  STATEMENT OF THE PROBLEM AND THE THEORY

The problem of modeling processes of special nanostructured layers in the structures for the PEC has been solved by molecular dynamics (MD). MD method is widely used in modeling the behavior of nanosystems due to the ease of implementation, satisfactory accuracy and low cost of computing resources. The basis of this method is the solution of differential equations of motion of Newton for each particle.

Depending on the type of building and external forces in the system, the problem of modeling processes of special nanostructured layers in the structures for the PEC will vary in accuracy and different thermodynamic parameters. Question preparation and research potential parameters are complex and time-consuming. As sources of data can serve as a first-principles calculation methods or experimental data.

Sets fairly well-matched to the parameters of the same type of molecules are combined in a special database and library-force fields. View capabilities and the potential energy of the photoelectron nanostructured systems make a decisive contribution in kind, the nature and value of interactions between objects nanosystems. The potentials are divided into many-and guys and spatial axially symmetric. In many ways, kind of potential in solving the problem of parameters determined by the presence in libraries and force fields of databases for modeling of nanostructure photoelectric systems.

Lennard-Jones [6, 7] widely used for calculating the properties of gases, liquids and solids, intermolecular forces are manifested. Potential describes the pair interaction of spherical non-polar molecules, and includes the dependence of the energy of interaction between two particles of the distance between them – a short-range repulsion and attraction at large. Lennard-Jones potential is as follows:

$$U^{LD}(r) = 4\varepsilon\left(\left(\frac{\sigma}{r}\right)^{12} - \left(\frac{\sigma}{r}\right)^{6}\right) \tag{1}$$

where $\varepsilon$ – depth of a potential hole, $r$ – distance between centers of particles, $\sigma$ – distance at which energy of interaction becomes equal to zero. Sizes $\varepsilon$ and $\sigma$ are also defined by properties of material. The type of potential is shown in Figure 1.1.

This potential is attractive at large distances, has a minimum at the point $r_{min} = \sqrt[6]{2}\sigma$ is repulsive at short distances. When $r > r_{min}$ the attractive forces prevail over the forces of repulsion, which corresponds to the term formula $(\sigma/r)^6$. The nature of the forces of attraction is due to dipole-dipole induced interaction [8] (van der Waals force). When the distance between the centers of the particles repulsive force due to the exchange

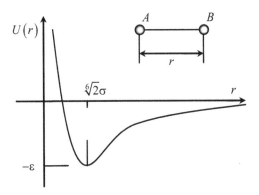

**FIGURE 1.1**    Potential of Lennard-Jones.

interaction prevail over the forces of gravity, which corresponds to the terms of $(\sigma/r)^{12}$ (the overlap of the electron clouds of the molecule begins to push off). The disadvantage of such a representation of the interaction potential is that the strength of the exchange interaction responsible for the repulsion of the particles at short distances, only approximately described by a power law. Physically, it was necessary to be more correct to choose the repulsive part in exponential form. However, the representation of the potential convenient for computer calculations. For this reason, Lennard-Jones potential is widely used in numerical modeling of the behavior of matter.

Lennard-Jones potential is also often written in the form:

$$U^{LD}\left(r\right) = \varepsilon\left(\left(\frac{r_{min}}{r}\right)^{12} - 2\left(\frac{r_{min}}{r}\right)^{6}\right) \tag{2}$$

where $r_{min} = \sqrt[6]{2}\sigma$ – the minimum point of the potential.

To speed up the calculations Lennard-Jones potential is usually cut off in the distance $r_c = 2.5\sigma - 3.5\sigma$. This choice $r_c = 2.5\sigma$ is due to the fact that at this distance the value of the interaction energy is only 1.63% of the well depth $\varepsilon$ [9]. To avoid abrupt changes in capacity and, hence, the shock energy of the system, with its potential for breakage smoothly reduced to zero. Lennard-Jones potential is a two-parameter, so it is not suitable for other types of molecules (non-spherical or have permanent dipole moments). The potential is very accurately describes the properties of a number of substances (for example, crystalline inert gases) and strength of the van der Waals forces. The advantages of the Lennard-Jones also applies its computational simplicity, does not require the calculation of irrational and transcendental functions. Lennard-Jones potential is used as a classic model potential, whose main task is to describe the general physical laws rather than accurate quantitative results [10, 11].

If only the pair interatomic interaction in the mathematical modeling of metal and/or semiconductor systems raises a number of problems. In Ref. [12], it was shown that using only a pair potential of interaction in metal and/or semiconductor done nonphysical Cauchy correlation coefficients ($C_{12} = C_{44}$).

Pairwise potentials cannot provide realistic values of the physical characteristics of the material [13]. The correct description of the properties of solids must be used many-particle potentials. It is known that none of the current capabilities is not capable of reproducing the full set of characteristics of the solids. Thus, the selection of the potential for mathematical modeling – complex problems. Most empirical potential well describe the bulk properties of the materials, but, nevertheless, some successfully used to describe and surface properties.

The actual ratio of the elastic constants of metals and semiconductors can only be obtained with the pair and many-particle interactions. In the simulation of metal and semiconductor systems, the most widely used the following approach, taking into account the many-particle interaction:

-    potential Stillinger-Weber [14];
-    potential Abel-Tersoff [15];
-    embedded atom method [12, 16, 17];
-    modified embedded atom method [18].

The embedded atom method (EAM) the binding energy of the atoms represented in the form

$$E = \sum_i F_i \left[ \sum_{j \neq i} \rho_j \left( R_{ij} \right) \right] + \frac{1}{2} \sum_{j \neq i} \varphi_{ij} \left( R_{ij} \right) \qquad (3)$$

where $\sum_i F_i \left[ \sum_{j \neq i} \rho_j \left( R_{ij} \right) \right]$ – feature embedded atom $i$, depending on the total electron density in the immersion $i$-th atom; $\varphi_{ij} \left( R_{ij} \right)$ – the pair interaction energy. Output (3) using density functional theory (DFT) can be found in [17].

Each atom system is considered as a particle immersed in the electron gas generated other atoms of the simulated system. The energy required for immersion is dependent on electron density at the point of immersion. Introduced way immersion function to determine the exchange and correlation energy of the electron gas system.

The sense of immersion function can be defined as the energy required for immersion of a single atom in a homogeneous electron gas with density.

However, there are other transformations [19] allow you to change the function (3) with the proviso that the resulting energy and interatomic forces remain unchanged.

The EAM uses the following approach:

1) The function of the electron density of one atom is spherically symmetric function that depends only on the distance between the atoms. This approach significantly limits the scope of the EAM and allows us to consider a system in which the direction of the covalent component of communication can be neglected.

2) The electron density in the embedded atom $i$ is defined as a linear superposition of the electron densities of the other atoms $\sum_{j \neq i} \rho_j \left( R_{ij} \right)$. This approach greatly simplifies the calculation of the electron density.

3) The quantity $\sum_{j \neq i} \rho_j \left( R_{ij} \right)$ in metal systems in the region of the atom $i$ varies slightly compared with the electron density of the atom $\rho_i$. Thus, $\sum_{j \neq i} \rho_j \left( R_{ij} \right)$ in the arrangement of atoms $i$ is replaced by a constant $\bar{\rho}$ [18]. The energy of the electron gas is approximated by a function that depends only on the value of the average value of the electron density in the immersion, not complicated functionals as in the method of DFT.

Currently displayed EAM potentials for most metals and some binary systems. Also calculated potentials for ternary systems [20]. However, such "triple" potentials qualitatively reproduce the physical properties of materials.

In the 90 years it was proposed by the semi-empirical approach, combining the benefits of many potentials and embedded atom method. The theory of modified embedded atom method (MEAM – modified embedded-atom method) is derived by using density functional theory (DFT) [21]. DFT method is currently considered the most recognized approach to the description of the electronic properties of solids. In the method of EAM complete electron density is a linear superposition of spherically averaged functions. This disadvantage is eliminated in the modified embedded atom method.

The method MEAM energy of the system is written in the form:

$$E = \sum_i \left( F_i \left( \frac{\bar{\rho}_i}{Z_i} \right) + \frac{1}{2} \sum_{j \neq i} \varphi_{ij} \left( R_{ij} \right) \right)$$  (4)

where $E$ – atomic energy $i$; $F_i$ – function for atom $i$, embedding in the electron density $\bar{\rho}_i$; $Z_i$ – the number of nearest neighbor atoms $i$ reference in its crystalline structure; $\varphi_{ij}$ – pair potential between atoms $i, j$, located at a distance $R_{ij}$.

In MEAM $F(\rho)$ determined as a function of embedding

$$F(\rho) = A E_c \rho \ln \rho$$  (5)

where $A$ – controlled variable; $E_c$ – binding energy.

Pair potential between atoms $i, j$ is determined by

$$\varphi_{ij}(R) = \frac{2}{Z_i} \left\{ E_i^u(R) - F_i \left( \frac{\bar{\rho}_i^0(R)}{Z_i} \right) \right\}$$  (6)

where $\bar{\rho}_i^0$ – electron density.

The total electron density at the dive includes angular dependence and written in the form:

$$\bar{\rho} = \rho^{(0)} G(\Gamma)$$  (7)

There are many types of functions $G(\Gamma)$ [74]:

$$G(\Gamma) = \sqrt{1 + \Gamma}$$  (8)

$$G(\Gamma) = e^{\frac{\Gamma}{2}}$$  (9)

$$G(\Gamma) = \frac{2}{1 + e^{-\Gamma}}$$  (10)

$$G(\Gamma) = \pm \sqrt{|1 + \Gamma|}$$  (11)

The most widespread form of writing in the form:

$$G(\Gamma) = \sqrt{1 + \Gamma} \tag{12}$$

The function $\Gamma$ is calculated according to the formula:

$$\Gamma = \sum_{h=1}^{3} t^{(h)} \left( \frac{\rho^{(h)}}{\rho^{(0)}} \right)^2 \tag{13}$$

where $h = 0 - 3$, correspond to $s, p, d, f$ symmetry; $t^{(h)}$ – weight multipliers; $\rho^{(h)}$ – values, determining the deviation of the distribution of the electron density distribution of a perfect crystal of the cubic system $\rho^{(0)}$:

$$s(h = 0) : \rho^{(0)} = \sum_i \rho^{a(0)}(r^i) \tag{14}$$

$$p(h = 1) : \left( \rho^{(1)} \right)^2 = \sum_\alpha \left[ \sum_i \rho^{a(1)}(r^i) \frac{r_\alpha^i}{r^i} \right]^2 \tag{15}$$

$$d(h = 2) : \left( \rho^{(2)} \right)^2 = \sum_{\alpha,\beta} \left[ \sum_i \rho^{a(2)}(r^i) \frac{r_\alpha^i r_\beta^i}{r^{2i}} \right]^2 - \frac{1}{3} \sum_i \left[ \sum_i \rho^{a(2)}(r^i) \right]^2 \tag{16}$$

$$f(h = 3) : \left( \rho^{(3)} \right)^2 = \sum_{\alpha,\beta,\gamma} \left[ \sum_i \rho^{a(3)}(r^i) \frac{r_\alpha^i r_\beta^i r_\gamma^i}{r^{3i}} \right]^2 \tag{17}$$

Here $\rho^{a(h)}$ – radial functions that represent a decrease in the contribution of distances $r^i$, superscript $i$ indicates the nearest atoms, $\alpha, \beta, \gamma$ – summing the codes for each of the three possible directions. Finally, the individual contribution is calculated according to the formula:

$$\rho^{a(h)}(r) = \rho_0 e^{-\beta^{(h)}\left( \frac{r}{r_e} - 1 \right)} \tag{18}$$

Consider the process of formation of special nanostructures, including quantum dots, a sophisticated photovoltaic cells on a silicon substrate. Usually nano it assumed that the properties of hard metals and semiconductors do not depend on their size. However, such an omission is valid only for systems containing an infinite number of atoms, i.e., the number of atoms characteristic of the macrocosm. Recent studies have shown that the particle sizes of less than 10 nm, many physical and chemical properties differ significantly from those properties of macroscopic objects. The unusual physical and chemical properties of nanoparticles due to three main reasons:

- nanoparticle size comparable to the Bohr radius of hydrogen-like quasiparticles – excitons (optical properties of nanoparticles);
- nanoparticles fraction of the atoms trapped on the surface, is a significant part of the total number of atoms of the nanoparticles (thermodynamic properties of nanoparticles);
- own nanoparticle size comparable to the size of the molecules (the kinetics of chemical processes at the surface of the nanoparticles).

When modeling used silicon substrate with the orientation 100 and 111 having a minimum dimension: length – 11 nm, width – 11 nm, height – 3 nm. The substrate is not rigidly fixed, it means that the atoms of the substrate can be moved freely in any direction.

By selecting the appropriate step of integration often depends on the convergence of the numerical solution of the problem. Step should be small enough to correctly display the behavior of the system. By using the methods of molecular dynamics on the value of the integration step affects the value of the mass of the simulated agents. It is selected in the range from 0.5 to 2 fs. In this case, step time integration was 1 fs. Total time calculation process was about 2 ns.

Forming of heterostructures can occur by several mechanisms: mechanism Volmer-Weber (islet growth model), the mechanism of the Frank-van der Merwe (layered growth) and mixed-Krastanov Stransky. Figure 1.2 shows a schematic of the simulated system. Along the edges of the system, along the $x$ and $y$ axis, using periodic boundary conditions. The borders along the $z$ axis is rigidly fixed.

Physical characteristics are compounds [22] used in the simulations of the growth process, are presented in the Table 1.1.

The formation of nanoscale photovoltaic structures on the surface of the substrate is greatly influenced by mismatch lattices deposited material and the substrate. Depending on the parameter mismatch have different mechanisms of growth. The experimental and theoretical studies [23] it is known that the formation of three-dimensional islands requires parameter lattice mismatch in the deposited material/substrate was large enough ($\varepsilon_0 > 2\%$). Moreover, the more the value of the lattice mismatch, the sooner the formation of coherent islands [23].

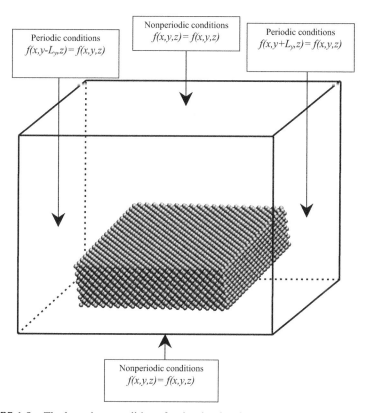

**FIGURE 1.2**    The boundary conditions for the simulated system.

**TABLE 1.1**    The Physical Properties of the Substances

| Features of substances | Si | Au | Ga | In |
|---|---|---|---|---|
| Type crystal lattice | | | | |
| The lattice parameter, $10^{-10}$ M | $a = 5,430$ | $a = 4,080$ | $a = 4,519$ $b = 7,658$ $c = 4,526$ | $a = 3,252$ $c = 4,946$ |

For example, the formation of quantum dots is observed in the InAs/GaAs system, where $\varepsilon_0 > 7\%$. In this case, the growth occurs Stransky-Krastanov initially formed on the surface of elastic-busy wetting layer having the same lattice parameter as the substrate material. When you reach a certain critical thickness of the wetting layer begin to form misfit dislocations. After the formation of dislocations epitaxial layer increases with the lattice constant of the deposited material. For this mechanism the critical thickness of the wetting layer is greater than one monolayer.

Quantum dots are formed ($\varepsilon_0 > 2\%$), and the growth of heterostructures made of layer by layer mechanism (layer-by-layer growth). For very large values of the error made by the mechanism of the growth of the Volmer-Weber. Then the critical thickness of the wetting layer is less than one monolayer and three-dimensional islands formation occurs directly on the substrate surface. An example of such a system is the system InAs/Si, where $\varepsilon_0 > 10.6\%$.

To solve the problem of modeling the processes of formation of special nanostructures in sophisticated photovoltaic cells on a substrate of material deposited atoms of a different material (Figure 1.3), followed by molecular dynamics calculations at constant temperature. To perform the test calculations are dealt with separately in the airless environment of nanostructure and its dynamics during the relaxation of self-organization of atoms.

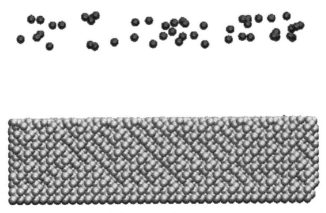

**FIGURE 1.3**   The initial state of the system for modeling the processes of formation of epitaxial PEC.

Simulation algorithm in this case is as follows. The silicon substrate is heated to a fixed temperature $T_0$. This process is described by the equation of motion in the form of Newton, with the initial conditions where the rate set in accordance with the Maxwell distribution. On the resulting system of atoms are deposited this simulated system. The equations for this phase will be similar to the description of the first stage, but other than that specified velocity for each of the deposited atoms and their direction is opposite to the direction of the axis $z$, then there is the speed directed towards the substrate. In the next step the resulting system simulation epitaxially deposited atoms of the second of this type. The equations for this phase will be similar to the description of the first stage. Figure 1.3 shows the state of the system at the first stage of precipitation.

The size and number of epitaxial islands formed depend on the size of the substrate, and the layer thickness of the deposited materials. For more details and parameters of the simulation technique used force fields are shown in published studies [24–32].

## 1.3   RESULTS

As test the calculations in this paper was reviewed by the use of a mathematical model that described a system consisting of atoms of Au, In, Ga, Si. Short-range order – ordering in the mutual arrangement of the atoms or molecules in a substance, which (unlike the long-range order) is repeated only at distances comparable to the distance between the atoms, that is, short-range order – is the presence of patterns in the distribution of neighboring atoms or molecules. Short-range order in the arrangement of the atoms or molecules exhibit, along with the crystals, and the amorphous body and the fluid. The concept of short-range order is entered through a pair distribution function. For an ideal gas radial function is unity, that is, short-range order is absent, since the location of each particle in space is independent of the location of other particles.

The pair correlation function $g(r)$ measures the correlation between the independent particles. That is, $g(r)$ – the probability that a volume element in the vicinity of particle found in the simultaneous presence in the origin of other particle [22]:

$$g(r) = \frac{Wn(r)}{4\pi r^2 \Delta r N} \qquad (19)$$

where $n(r)$ – the average number of particles located at distances $r,...,r + \Delta r$ from the given particle; $W$ – the volume of the system; $N$ – total particles.

For each element discussed above undertook crystal size of about $10 \times 10 \times 10$ lattice periods. The boundary conditions in all areas used intermittently. The temperature in the process of calculating the constantly maintained and amounted to 300 K. The radial distribution function for gold is shown in Figure 1.4.

The autocorrelation function over time should tend to zero. This dependence for nanosystems of gold atoms can be seen in Figure 1.5.

Similar plots were obtained by test calculations crystal silicon, indium and gallium. The temperature for these calculations was set at the level of the normal – T = 300 K temperature graph for all types of simulated nano showed that the thermostat maintains the temperature at the set level. Small variations in temperature fluctuations in the system explained atoms and nanoparticles.

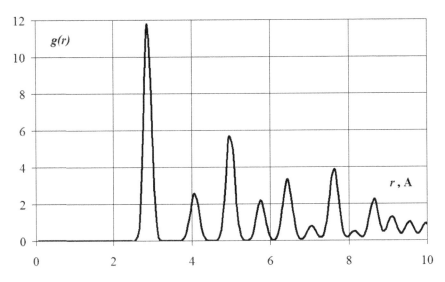

**FIGURE 1.4**    The radial distribution function for gold at T = 300 K.

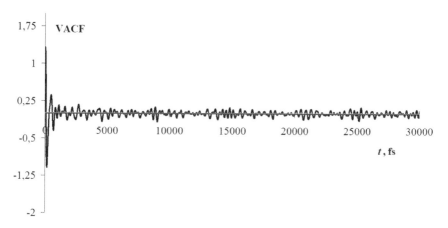

**FIGURE 1.5**    The autocorrelation function for test problem with crystal gold.

For modeling of epitaxial processes used silicon substrate with (100) orientation. Deposited on a substrate gold atoms in an amount of 1000 pieces. Then, the resulting system of deposited silicon atoms in an amount of 3000 pieces. The temperature of the simulated system was maintained constant at 800 K. As shown in Figures 1.6–1.7, on a substrate formed of gold drops of different diameters.

The size distribution of the nanoparticles shown and numbered in Figure 1.7 are shown in Figure 1.8. The size of the formed nanostructures varies from 0.7 to 2.7 nm. The shape of nanostructures prepared by a variety. Diffusion of gold atoms into the substrate is not observed (Figure 1.9). Interesting dynamics of physical education process of photovoltaic cells made of gold and silicon. First, the gold atoms deposited on a silicon substrate, and then begin to organize themselves, going to drop.

To simulate the processes of special nanostructured layers for solar cells in structures Si-Au-Ga used silicon substrate (100). The dimensions of the substrate were set as follows: width – 16 nm, length – 16 nm, the height – 2 nm. Deposited on the substrate in an amount of gold atoms 4500. Then, the resulting system of deposited gallium atoms in an amount of 33,000 pieces. The temperature of the simulated system was kept constant by a thermostat and was equal to 300 K.

As can be seen from Figure 1.11, on a substrate formed of gold drops of different diameters and shapes. There is a partial Silting nanostructures

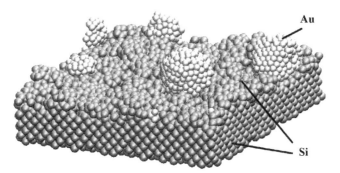

**FIGURE 1.6**    The simulation results of deposition of gold atoms to the substrate Si (100).

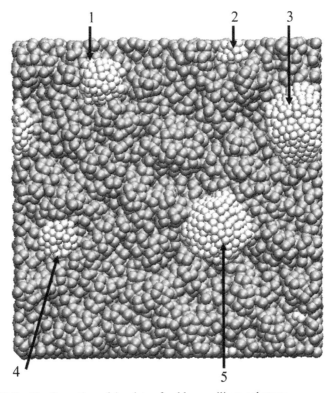

**FIGURE 1.7**    The formation of droplets of gold on a silicon substrate.

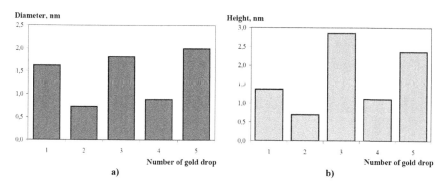

**FIGURE 1.8** Diameter distribution (a) and the height (b) droplets of gold obtained in the simulation.

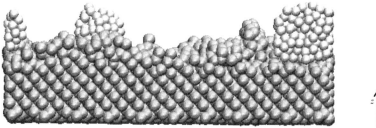

**FIGURE 1.9** Section nanosystems of gold and silicon atoms along the $y$-axis.

of special purpose gallium atoms. Gallium atoms do not form pronounced quantum dots, the distribution of the silicon surface is uneven. Diffusion of gold atoms and gallium in the silicon substrate is observed (Figure 1.11). The surface structure of the atoms of gallium and gold deposition on the substrate after forming the relief, there are fluctuations in the heights.

Using periodic boundary conditions allows us to extend the results of modeling the behavior of the system to a certain area. Figure 1.12 is a plan view of the substrate with quantum dots, broadcast 5 times $x$ and $y$ axis. Frequency repetition gold quantum dots in Figures 1.10 and 1.12 makes it possible to speak about the suitability of their use for the creation of the photoelectric effect.

To simulate the processes of special nanostructured layers for solar cells in structures Si-In-Ga used silicon substrate (100). The dimensional

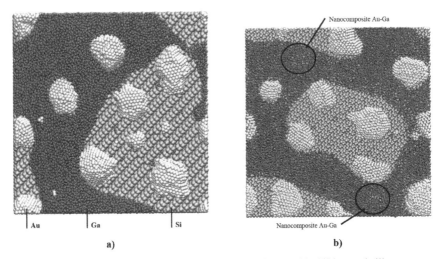

Au          Ga          Si

a)                              b)

**FIGURE 1.10**  Dripping gold and gallium silicon substrate (a). Silting and silicon atoms of gold islands (b).

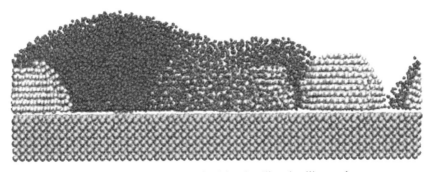

**FIGURE 1.11**  The diffusion of atoms of gold and gallium in silicon substrate.

parameter values of the substrate were as follows: width – 16 nm, length – 16 nm, height – 2 nm. Deposited on a substrate of indium atoms in an amount of 4500 pieces. Then, the resulting system of deposited gallium atoms in an amount of 33,000 pieces. The temperature of the simulated system was kept constant by a thermostat and was equal to 300 K.

The process of formation and deposition of nanostructures of indium atoms different from a similar process of gold atoms. If the gold atoms deposited on the substrate first, and then going into the quantum dots, the clustering of indium atoms began and continued preferably in a gaseous

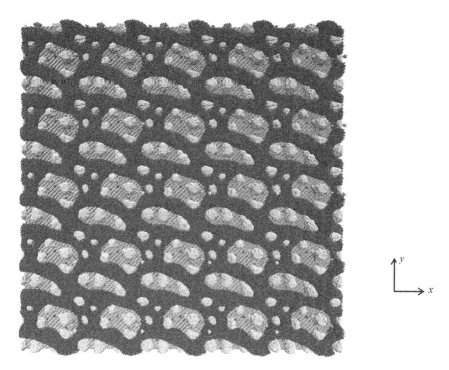

**FIGURE 1.12**    The periodic system of quantum dots of gold and gallium atoms on the silicon substrate.

medium. Deposited on a substrate already formed nanostructures India. India is trying to form conglomerates spherical. In the next stage of modeling deposited gallium atoms. The resulting gas vapor from gallium and indium nanoparticles was attracted silicon substrate. As can be seen from Figures 1.13 and 1.14 are formed on a substrate of indium drops of different diameters and shapes.

As a result, two steps deposition modeling atoms of indium and gallium get the picture shown in Figure 1.15. indium atoms form a quantum of education expressed a special form in the form of hemispheres. Gallium is distributed over the entire substrate, sometimes zaraschivaya nanoparticles of indium (indium overgrown nanostructures are shown in Figure 1.15a). The diffusion of atoms of indium and gallium in the silicon substrate, similarly to the case of gold quantum dots are not observed.

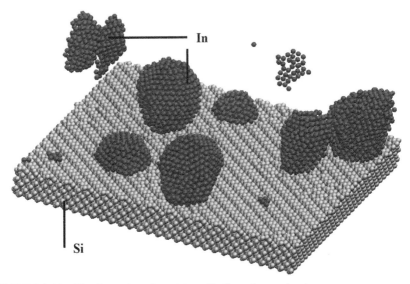

**FIGURE 1.13**   The formation of particles of indium, isometric view.

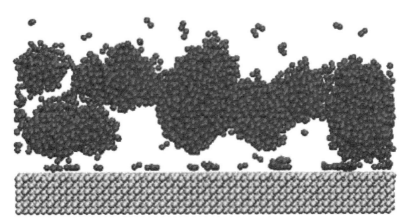

**FIGURE 1.14**   The formation of particles of indium, side view.

For indium nanostructures, numbered and presented in Figure 1.15b, was constructed distribution heights and sizes. These graphs are shown in Figure 1.16. The average size of nanostructures indium is 3–3.5 nm and larger than the quantum dots of gold obtained in the simulation of such a system.

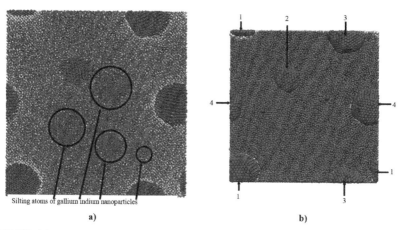

**FIGURE 1.15**   Various forms of quantum dots of indium and gallium on the surface of the silicon substrate.

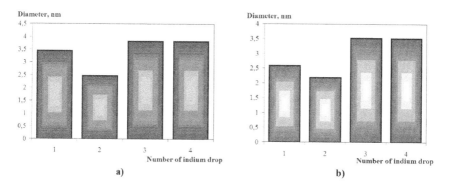

**FIGURE 1.16**   Diameter distribution (a) and the height (b) droplets of indium obtained in the simulation.

## 1.4   CONCLUSION

This chapter describes the theoretical foundations of computer modeling of processes of special nanostructured layers of epitaxial structures for sophisticated photovoltaic cells, comprising a mathematical model and setting goals.

Methods of simulation specialized nanostructures contains basic equations of motion of nano-objects, algorithms maintain the temperature and pressure in the system, the mechanisms of formation of the initial velocities and coordinates related to the thermodynamic parameters of nanostructured objects. The different types of potential and force fields that can be used for solving problems of modeling photovoltaic nanoelements. The advantages and disadvantages of different types of potential, showing the benefits of using the potential of many in the study of nanostructures of special purpose.

The initial forming elements considered silicon atoms, gallium, indium and gold. Showcased different mechanisms of special nanostructures epitaxial structures for sophisticated photovoltaic cells. Epitaxial and self-organization processes have shown that disruption of the internal order of nanostructures and substrate not observed. When the deposition of individual atoms was recorded in part silting nanostructures on the substrate.

The physical process of solar cells depends on the material to be deposited. The simulation showed that gold is characterized by forming a hemisphere on the surface of silicon. Indium atoms begin to assemble into nanoparticles is still in a gaseous medium into contact with the silicon substrate. Gold nanoparticles are elongated hemisphere and education. For indium nanoparticles are characterized as a spherical shape, and a hemisphere.

The resulting system overgrown with gallium atoms. In the case of indium is characterized by an almost uniform distribution of the gallium atoms on the substrate. There silting nanoparticles of indium gallium, that is, the formation of photovoltaic cells. In the case of gold, gallium peculiar non-uniform distribution of the substrate is formed pronounced quantum dots. Diffusion of gold atoms of indium, gallium into the upper layers of the substrate is observed.

The work was supported by the Russian Foundation for Basic Research (grant 13-08-01072) and by state assignment of Kalashnikov ISTU № 201445-1239.

## KEYWORDS

- epitaxy
- molecular dynamics
- nanostructures
- photoelectric converters
- simulation

# REFERENCES

1. Oregon State University. Nanotechnology Provides Advances in Eyeglasses, Solar Energy. Oregon State University News. September 16, 2009. URL: http://oregon-otato.odu/ua/nes/archives/2009/sep/newnanostructure-tech–nology-provides-advances-eyeglass-solarenergy-performance (date of access 05.05.2014).

2. Kim, S. G., & Mambeterzina, G. K. Nanomagnezija dlja stabilizacii jekologii [Nanomagnesia to stabilize the ecology]. *Vestnik KASU. Ust'-Kamenogorsk: Izdatel'stvo Kazahstansko-Amerikanskogo svobodnogo universiteta – Bulletin Kafu. Ust-Kamenogorsk: Publishing House of the Kazakh-American Free University*, 2007, no. 3, p. 196.

3. Mambeterzina, G. K., & Kim, S. G. Povyshenie dolgovechnosti i svetovoj otdachi razrjadnyh istochnikov sveta [Increasing longevity and light output bit light sources]. *Svetotehnika – Lighting*, 2005, no. 3, pp. 60–61.

4. Istochniki jenergii. *Fakty, problemy, reshenija. Serija "Informacionnoe izdanie"* [Energy sources. Facts, problems and solutions. A series of "Information Booklet"]. Ed. V. S. Larus, SPb., Science and Technology Research Center, 1997, no. 3. p. 110.

5. Kim, S. G., & Mambeterzina, G. K. Al'ternativnaja jenergetika na vozobnovljaemom istochnike jenergii i poluprovodnikovyh toplivnyh jelementah [Alternative energy renewable energy and semiconductor fuel cell]. *Industrija Kazahstana. Karaganda – Industry of Kazakhstan. Karaganda*, 2006, vol. 4, no. 48, pp. 39–42.

6. Lennard-Jones, J. E. On the Determination of Molecular Fields. II. From the Equation of State of a Gas. Proc. Roy. Soc. A, 1924, vol. 106, pp. 463–477.

7. Lennard-Jones, J. E. Wave Functions of Many-Electron Atoms. Proc. Camb. Phil. Soc., 1931, vol. 27, pp. 469–480.

8. Zinenko, V. I., Sorokin, B. P., & Turchin, P. P. *Osnovy fiziki tverdogo tela* [Fundamentals of Solid State Physics]. Moscow, "Fizmatlit" publ., 2001, 336 p.

9. Stoddard, S. D., & Ford J. Numerical Experiments on the Stochastic Behavior of a Lennard-Jones Gas System. Phys. Rev. A, 1973, vol. 8, no. 3, pp. 1504–1512.

10. Krivcov, A. M., & Krivcova, N. V. Metod chastic i ego ispol'zovanie v mehanike deformiruemogo tverdogo tela [Particle method and its use in solid mechanics]. *Dal'nevostochnyj matematicheskij zhurnal – Far Eastern Mathematical Journal*, 2002, vol. 3, no. 2, pp. 254–276.

11. Wilson, N. T. The structure and dynamics of noble metal clusters. PhD Thesis. University of Birmingham, Sept. 2000.

12. Daw, M. S., & Baskes, M. I. Embedded-atom method: derivation and application to impurities, surfaces, and other defects in metals. Phys. Rev. B, 1984, vol. 29, no. 12, pp. 6443–6453.

13. Ercolessi, F. A. Molecular dynamics primer. Spring College in Computational Physics, ICTP, Trieste, June 1997. URL: http://www.fisica.uniud.it/~ercolessi/ (date of access 10.03.2014).

14. Stillinger, F. H., & Weber, T. A. Computer simulation of local order in condensed phases of silicon. Phys. Rev. B, 1985, vol. 31, pp. 5262–5271.

15. Tersoff, J. New empirical approach for the structure and energy of covalent systems. Phys. Rev. B, 1988, vol. 37, no. 12, pp. 6991–7000.

16. Daw, M. S., & Baskes, M. I. Semiempirical, Quantum Mechanical Calculations of Hydrogen Embrittlement in Metals. Phys. Rev. Letters, 1983, vol. 50, no. 17, pp. 1285–1288.

17. Daw, M. S. Model of metallic cohesion: The embedded-atom method. Phys. Rev. B, 1989, vol. 39, no. 11, pp. 7441–7452.
18. Baskes, M. I. Modified embedded-atom potentials for cubic materials and impurities. Phys. Rev. B, 1992, vol. 46, no. 5, pp. 2727–2742.
19. Ruda, M., Farkas, D., & Abriata, J. Interatomic potentials for carbon interstitials in metals and intermetallics. Scripta Materialia, 2002, vol. 46, pp. 349–355.
20. Tomar, V., & Zhou, M. Classical molecular-dynamics potential for the mechanical strength of nanocrystalline composite fcc Al+$\alpha$-Fe$_2$O$_3$. Phys. Rev. B, 2006, vol. 73, no. 17, pp. 174116.1–174116.16.
21. Hohenberg, P., & Kohn, W. Inhomogeneous Electron Gas. Phys. Rev. B, 1964, vol. 136, no. 3, pp. 864–871.
22. Kitel, Ch. *Vvedenie v fiziku tverdogo tela* [Introduction to Solid State Physics]. Moscow, "Science" publ., 1978, 789 p.
23. Dubrovskij, V. G. *Teorija formirovanija jepitaksial'nyh nanostruktur* [The theory of the formation of epitaxial nanostructures]. Moscow, "Fizmatlit" publ., 2009, 352 p.
24. Severyukhina, O. Y., Vakhrouchev, A. V., Galkin, N. G., & Severyukhin, A. V. Modelirovanie processov formirovanija mnogoslojnyh nanogeterostruktur s peremennymi himicheskimi svjazjami [Modeling of formation of multilayer nanoheterostructures with variable chemical bonds]. *Himicheskaja fizika i mezoskopija – Chemical Physics and Mesoscopics.*, 2011, vol. 13, no. 1, pp. 53–58.
25. Vakhrouchev, A. V., Severyukhin, A. V., & Severyukhina, O. Y. Modeling the initial stage of formation of nanowhiskers on an activated substrate. Part 1. Theory foundations. International Journal of Nanomechanics Science and Technology, 2012, vol. 3, no. 3, pp. 193–209.
26. Vakhrouchev, A. V., Severyukhin, A. V., & Severyukhina, O. Y. Modeling the initial stage of formation of nanowhiskers on an activated substrate. Part 2. Numerical investigation of the structure and properties of Au–Si nanowhiskers on a silicon substrate. International Journal of Nanomechanics Science and Technology, 2012, vol. 3, no. 3, pp. 211–237.
27. Vakhrouchev, A. V., Severyukhina, O. Y., Severyukhin, A. V., Vakhrushev, A. A., & Galkin, N. G. Simulation of the processes of formation of quantum dots on the basis of silicides of transition metals. International Journal of Nanomechanics Science and Technology, 2012, vol. 3, no. 1, pp. 51–75.
28. Vakhrushev, A. V., & Fedotov, A. Y. Issledovanie processov formirovanija kompozicionnyh nanochastic iz gazovoj fazy metodom matematicheskogo modelirovanija [The study of the formation of composite nanoparticles from the gas phase by mathematical modeling]. *Himicheskaja fizika i mezoskopija – Chemical Physics and Mesoscopics*, 2007, vol. 9, no. 4, pp. 333–347.
29. Vakhrushev, A. V., Fedotov, A. Y., Vakhrushev, A. A., Golubchikov, V. B., & Givotkov, A. V. Multilevel simulation of the processes of nanoaerosol formation. Part 1. Theory foundations. International Journal of Nanomechanics Science and Technology, 2011, vol. 2, no. 2, pp. 105–132.
30. Vakhrushev, A. V., Fedotov, A. Y., Vakhrushev, A. A., Golubchikov, V. B., & Givotkov, A. V. Multilevel simulation of the processes of nanoaerosol formation. Part 2. Numerical investigation of the processes of nanoaerosol formation for suppression of fires. International Journal of Nanomechanics Science and Technology, 2011, vol. 2, no. 3, pp. 205–216.

31. Vakhrushev, A. V., & Fedotov, A. Y. Issledovanie verojatnostnyh zakonov raspredele-
nija strukturnyh harakteristik nanochastic, modeliruemyh metodom molekuljarnoj
dinamiki [Investigation of probability distribution laws of the structural characteristics of nanoparticles, simulated by molecular dynamics]. *Vychislitel'naja mehanika
sploshnyh sred – Computational continuum mechanics*, 2009, vol. 2, no. 2, pp. 14–21.
32. Vakhrushev, A. V., Golubchikov, V. B., Fedotov, A. Y., & Zhivotkov, A. V. Mnogou-
rovnevoe modelirovanie processov kondensacii molekuljarnoj smesi v ajerozol'nyh
ognetushiteljah [Multilevel simulation of molecular mixture condensation aero-
sol fire extinguishers]. *Himicheskaja fizika i mezoskopija – Chemical Physics and
Mesoscopics*, 2011, vol. 13, no. 3, pp. 340–350.

# CHAPTER 2

# THEORETICAL BASES OF NON-CONTACT MEASUREMENTS OF ELECTROMAGNETIC PARAMETERS OF NANOMATERIALS

B. V. SKVORTSOV, D. M. ZHIVONOSNOVSKAYA, and P. A. KURYLEVA

*Samara State Aerospace University, Moskovskoye sh., 34, Samara, Samara Oblast, Russia*

## CONTENTS

## ABSTRACT

The chapter theoretically proves non-invasive on-line measurement method for electromagnetic parameters of thin films and nanomaterials, which based on the following: the controlled surface is sounded with electromagnetic signal, with reflected signal being analyzed afterwards.

Dielectric and magnetic inductivity as well as electrical conductivity specify electromagnetic properties of thin films and nanomaterials. Mathematical description of measurement procedure is supplied. Based on that, parameters in question are being calculated. This includes measurement of reflected signal amplitude and phase at different frequencies. Then, combined nonlinear equations are solved.

## 2.1   INTRODUCTION

Modern methods of monitoring electromagnetic parameters of thin films and nanomaterials, are quite varied [1–4], but sparse. The electromagnetic properties of thin films and nanomaterials are determined permittivity ($\varepsilon$) and ($\mu$) permeability, as well as electrical conductivity ($\sigma$). Currently, as a rule, used contact methods are determined separately and the dielectric and magnetic properties of materials. Studies are made on a variety of plants, require a fundamentally different designs of measuring devices. In the modern theory there is no uniform mathematically-based method of contactless operative measuring of these electromagnetic parameters of thin films and nanomaterials, which would allow to realize the measurement procedure in a single integrated device. The paper theoretically justified method of contactless operative measurement of electromagnetic parameters of thin films and nanomaterials, including their conductivity ($\sigma$), permittivity ($\varepsilon$) and permeability ($\mu$), at frequencies of gigahertz to optical $\omega = 10^9 \div 10^{14}$ Гц. The purpose of research is theoretical basis and mathematical modeling of procedure for measuring electromagnetic parameters of nanomaterials by sensing of their surface. The parameters for carbon and semiconductor materials and their values are given in Table 2.1 [2–5], that define the technical requirements for appliances for conversion range.

**TABLE 2.1**    The Electromagnetic Parameters of Nanomaterials

| Material | $\mu_{rel}$ | $\varepsilon_{rel}$ | $\sigma$ |
|---|---|---|---|
| Pure metals | 0.999912÷1.000253 | $\to \infty$ | $10^6 \div 10^8$ |
| Ferromagnetic materials and liquids | 16÷80,000 | $2.2 \div 70$ | $10^{-15} \div 10^{-5}$ |
| Carbon and semiconductor structures, graphite | 0.999895÷1.0000001 | 3.3–15 | $(0.1 \div 0.2) \cdot 10^6$ |
| Glass, quartz, ceramics | 0.999987÷0.999985 | 4÷11 | $10^{-14} \div 10^{-10}$ |

## 2.2  MATHEMATICAL MODEL

The procedure for non-contact measurement of these parameters of materials based on sensing of test surface by electromagnetic signal and is illustrated in Figure 2.1.

The emitter 3, being in an environment with known electromagnetic parameters $\mu_1$, $\sigma_1$, $\varepsilon_1$ (usually a gas medium), generates an electromagnetic signal $\Phi_1$ with a spatial angle $2\alpha$, falling on the material 1 with controlled electromagnetic parameters $\mu_x$, $\sigma_x$, $\varepsilon_x$ angle $\theta_1$. Reflected from the interface and the refracted at an angle $\theta_2$ energy flows $\Phi_2$ and $\Phi_3$ are marked accordingly, with $\theta_2 < \theta_1$. Hereinafter, the angles of incidence, reflection and refraction are measured from the axis of the beam propagating. Reflected from the surface an electromagnetic signal $\Phi_2$ entering the receiver 4 carries information about the required electromagnetic parameters $\mu_x$, $\sigma_x$, $\varepsilon_x$. Also reflected from the surface of the signal $\Phi_2$ in the receiver falls as the flow $\Phi_5$, which is part of the flow $\Phi_4$ reflected from the bottom of the interface. The signal $\Phi_5$ interferes with $\Phi_2$ and must be taken into account

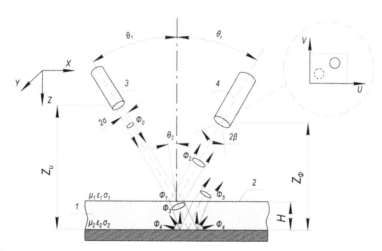

**FIGURE 2.1**  Illustration for the mathematical modeling of non-contact measurement of electromagnetic parameters of nanomaterials ($\mu$, $\sigma$, $\varepsilon$ – electrodynamic parameters appropriate environment and material, $\theta_1$ – angle of incidence, $\theta_2$ – angle of refraction, $\Phi$ – electromagnetic flux, $H$ – material thickness; $z_U$, $z_\Phi$ – structural parameters, $\alpha$ – spatial angle pattern radiator, $\beta$ – the angle of the receiver capture the reflected signal, 1 – controlled environment (material), 2 – the boundary between the media, 3 – emitter, 4 – receiver).

or eliminated during the measurement. The probe signal can generally be continuous or pulsed. As the emitter can be used not only electromagnetic directional antenna, but laser or LED with polychromatic spectrum signal. Depending on the method of signal processing in a receiver antenna may be used, or a photodetector, including matrix. The transmitter and receiver can be constructively combined.

The reflection and refraction of the signal may be polarized if the source gave no polarized radiation. The desired electromagnetic parameters may occur in amplitude, phase, propagation time, the spectral density of the reflected signal, and in the diffraction pattern or position of the reflected signal at the photodetector array (using optical signals). Use of each of these parameters of the reflected signal to estimate the electromagnetic proper- ties of nanomaterials define different methods of measurement. Thus, the measuring procedure is associated with the spread of the directional elec- tromagnetic signal in layered media. When the mathematical description of the measurement procedures assume that contact homogeneous isotropic medium, stationary and linear in terms of the electromagnetic properties.

In general, the problem is three-dimensional. We choose a rectangular coordinate system so that the plane $x0y$ parallel to the boundary between the media and plane $x0z$ coincides with the plane of propagation of the signal. In this case, the problem can be considered as flat (i.e., $y = 0$). Consider the sensing surface of the continuous wave with frequency $\omega$. The intensity (power density [W/m$^2$]) of the electromagnetic beam which has come down to the test surface is determined by the formula [6, 7].

$$\Phi_1 = \Phi_0 e^{-2\alpha_1 R_1} e^{\left[ j \left( 2\omega \left( t - \frac{R_1}{V_{\phi 1}} \right) + \varphi_1 \right) \right]} \tag{1}$$

where $\Phi_0 = \left| \overline{E}_m \times \overline{H}_m \right|$ the module of the Poynting vector of a probing signal, the electric field E and magnetic intensity H are related by [6]:

$$\overline{E} = Z \overline{H} \tag{2}$$

$$Z = \sqrt{\frac{j\omega\mu}{\sigma + j\omega\varepsilon}} = \sqrt[4]{\frac{\mu^2 \omega^2}{\sigma^2 + \varepsilon^2 \omega^2}} e^{j \, arctg \frac{\sigma}{\omega\varepsilon}} = \sqrt[4]{\frac{\mu^2 \omega^2}{\sigma^2 + \varepsilon^2 \omega^2}} e^{j\varphi} \tag{3}$$

where $\varphi_1$, $\alpha_1$, $V_{\Phi 1}$, $R_1$ – phase shift, the absorption coefficient, the phase velocity and the distance the signal has passed to the test surface (Figure 2.1) in the medium above the interface are determined, respectively, by the formulas [6]

$$\varphi = -\frac{1}{2} arctg \frac{\sigma}{\omega \varepsilon} \tag{4}$$

$$\alpha = \sqrt{\frac{\mu \omega (\sqrt{\sigma^2 + \varepsilon^2 \omega^2} - \varepsilon \omega)}{2}} \tag{5}$$

$$V_{\Phi} = \sqrt{\frac{2\omega}{\mu(\varepsilon \omega + \sqrt{\sigma^2 + \varepsilon^2 \omega^2})}} \tag{6}$$

$$R_1 = \frac{z_U - H}{\cos \theta_1} \tag{7}$$

At the interface wave is partially reflected, partially penetrates the controlled environment, and partly dissipated, which is determined according to the reflection $G$, transmission $T$ and scattering $Q$ coefficients, defined mainly roughness and cleanliness controlled surface [6]: $G + T + Q = 1$. The intensity of the reflected electromagnetic signal arriving at the receiver is determined by the following formula, which follows from (1).

$$\Phi_2 = G\Phi_1 e^{-2 \pm_1 R_2} e^{j \left[ 2\omega \left( t - \frac{R_2}{V_{\Phi 1}} \right) + \varphi_1 \right]} = G\Phi_0 e^{-2 \pm_1 (R_1 + R_2)} e^{j \left[ 4\omega \left( t - \frac{R_1 + R_2}{V_{\Phi 1}} \right) + 2\varphi_1 \right]} \tag{8}$$

The intensities of the signals transmitted inside the controlled environment $\Phi_3$, and come down to the lower limit of the material $\Phi_4$, determined by the formulas:

$$\Phi_2 = G\Phi_1 e^{-2\alpha_1 R_2} e^{j \left[ 2\omega \left( t - \frac{R_2}{V_{S1}} \right) + \varphi_1 \right]} = G\Phi_0 e^{-2\alpha_1 (R_1 + R_2)} e^{j \left[ 4\omega \left( t - \frac{R_1 + R_2}{V_{D1}} \right) + 2\varphi_1 \right]} \tag{9}$$

$$\Phi_2 = G\Phi_1 e^{-2 \pm_1 R_2} e^{j \left[ 2\omega \left( t - \frac{R_2}{V_{S1}} \right) + \varphi_1 \right]} = G\Phi_0 e^{-2 \pm_1 (R_1 + R_2)} e^{j \left[ 4\omega \left( t - \frac{R_1 + R_2}{V_{D1}} \right) + 2\varphi_1 \right]} \tag{10}$$

$\alpha_2$, $V_{\Phi2}$, $R_2$ – phase shift, the absorption coefficient, the phase velocity and electromagnetic phase shift in the medium below the interface,

$$R_3 = \frac{H}{\cos\theta_2}.$$

Equations (8)–(10) provide a mathematical basis for the study of electromagnetic properties of the materials. Consider the method for determining the parameters controlled by the reflected signals. Expression (8) to the
signal intensity can be written as

$$R_3 = \frac{H}{\cos\theta_2} \tag{11}$$

$$B = \Phi_0 e^{-2\alpha(R_1+R_2)} = \Phi_0 e^{-\frac{2\alpha(z_\phi+z_H-2H)}{\cos\theta_1}} \tag{12}$$

$$B = \Phi_0 e^{-2\alpha(R_1+R_2)} = \Phi_0 e^{-\frac{2\alpha(z_\phi+z_H-2H)}{\cos\theta_1}} \tag{13}$$

The reflection coefficient depends on the angle of incidence of the
wave resistance and contacting media, is a complex value, and generally
for unpolarized probing waves is determined by the formula [7].

$$B = \Phi_0 e^{-2\alpha(R_1+R_2)} = \Phi_0 e^{-\frac{2\alpha(z_\phi+z_H-2H)}{\cos\theta_1}} \tag{14}$$

$Z_1$, $Z_2$ – complex impedances of the contacting media, defined by
the formula (3). $A_G$, $\psi_G$ – modulus and phase of the complex reflection
coefficient.

The angle of refraction $\theta_2$ related to the angle of incidence $\theta_1$ [7]

$$\frac{\sin\theta_1}{\sin\theta_2} = \frac{n_2}{n_1} = \frac{V_{\Phi1}}{V_{\Phi2}} = n_{21} \tag{15}$$

$n_1$, $n_2$, $V_{\Phi1}$, $V_{\Phi2}$ – refractive indices and phase velocity, respectively in
contiguous media.

From (15) follows

$$\theta_2 = \arcsin(\frac{n_1}{n_2}\sin\theta_1) = \arcsin(n_{12}\sin\theta_1) \tag{16}$$

It is known [7] that the angles of incidence $(\theta_1 < 15°)$ reflection coefficient is practically independent of the polarization of the electromagnetic wave and the angle of incidence and can be approximately defined by the formula deriving from Eqs. (15) and (16) at normal incidence $(\theta_1 = 0)$:

$$G = \frac{Z_2 - Z_1}{Z_2 + Z_1} = \frac{[\sqrt{\mu_1(\sigma_2 + j\omega\varepsilon_2)} - \sqrt{\mu_2(\sigma_1 + j\omega\varepsilon_1)}]^2}{\mu_1(\sigma_2 + j\omega\varepsilon_2) + \mu_2(\sigma_1 + j\omega\varepsilon_1)} = A_G e^{j\varphi_G} \quad (17)$$

According to the Eq. (17) are calculated modulus and phase of the reflection coefficient of the signal for the materials listed in Table 2.1 at a frequency $\omega = 10^9$ Hz, where the environment is air with parameters $\sigma_1$, $\mu_1$, $\varepsilon_1$ (Table 2.2).

Note that the angle of incidence $\theta_1 < 15°$ is consistent with the requirements of the implementation of the structural unit. The values in Table 2.2 show that the amplitude and phase of the reflection coefficient, depend substantially on the monitored parameters, especially for carbon and semiconductor structures, which are the main object of our measurements. Given that the electromagnetic parameters $\mu_1$, $\sigma_1$, $\varepsilon_1$ of environment medium assumed to be known, and the material parameters $\mu_2$, $\sigma_2$, $\varepsilon_2$ are sought, the reflection coefficient can be written as

$$G = A_G(\mu_2, \sigma_2, \varepsilon_2)e^{j\psi_G(\mu_2, \sigma_2, \varepsilon_2)} \quad (18)$$

$A_G$, $\psi_G$ – modulus and phase of the reflection coefficient, are not amenable to direct analytical description, but may be defined numerically by the Eqs. (14)–(17). After substitutions expression (11) arrived at the receiver signal is written

**TABLE 2.2** Values of the Complex Reflection Coefficient for Different Materials

| Material | | Re + iIm | $A_G$ | $\Psi_G$ |
|---|---|---|---|---|
| Pure metals | $\varepsilon \to \infty$ | $1 + 1.129i \times 10^{-299}$ | 1 | $1.129 \times 10^{-299}$ |
| | $\varepsilon = 10^{10}$ | $0.992 - 4.22i \times 10^{-4}$ | 0.992 | $4.22 \times 10^{-4}$ |
| Ferromagnetic materials and liquids (by $\varepsilon = 5$) | | $0.998 + 1.119i \times 10^{-13}$ | 0.998 | $2.123 \times 10^{-13}$ |
| Carbon and semiconductor structures, graphite | | $0.869 - 0.128i$ | 0.878 | $-0.146$ |
| Glass, quartz, ceramics | | $0.984 + 9.716i \times 10^{-14}$ | 0.984 | $9,881 \times 10^{-14}$ |

$$\Phi_2 = BA_G(\mu_2,\sigma_2,\varepsilon_2)e^{j\omega t}e^{j[\psi+\psi_G(\mu_2,\sigma_2,\varepsilon_2)]} \tag{19}$$

The amplitude and phase of the signal coming into the receiver carries information on the monitored parameters that is used to solve this problem.

We estimate how much weight contributes to the signal propagation medium. If the probe signal is distributed in a gaseous medium, the absorption coefficient at frequencies from microwave to optical by reference data is $\alpha = 10^{-3} \div 10^{-2}$ m$^{-1}$. With dimensions of the device, the distance from a controlled environment to the transmitter and receiver, $z_U = z_\Phi = 0.1$ m and angles $\theta_1 \leq 15°$, from (12) $B \approx \Phi$. Thus, the gaseous medium has virtually no effect on the amplitude of the reflected signal.

Phase signal has a different impact. Equation (13) shows that the phase of the reflected signal directly and substantially independent of frequency and constructional parameters and the frequency range $\omega = 10^8 - 10^{13}$ Hz and selected geometric parameters previously be $\psi = 0.3 \div 0.3 \times 10^5$ [rad], $\varphi_1 = -(10^{-9} \div 10^{-10})$ [rad], $V_\Phi \approx 3.0 \times 10^8$ [m/s].

Given that each of the monitored parameters depends on the frequency and for the ratio of the probe arrived at the receiver signals can be written

$$F = \frac{\Phi_2}{B} = A_G[\mu_2(\omega),\sigma_2(\omega),\varepsilon_2(\omega)]e^{j\omega t}e^{j\{[\psi(\acute{E})+\psi_G(\mu_2(\omega),\sigma_2(\omega),\varepsilon_2(\omega))]\}} \tag{20}$$

where

$$A_G[\mu_2(\omega),\sigma_2(\omega),\varepsilon_2(\omega)]=|G|, \psi_G[\mu_2(\omega),\sigma_2(\omega),\varepsilon_2(\omega)]=argG$$

These expressions are mathematical basis for determining the electromagnetic parameters of the controlled environment by detecting the reflected signal at different frequency of probing signal. This measurement can be used as the amplitude and phase of the reflected signal, and use them simultaneously. In the latter case, the general principle of the measurement consists in the fact that to measure the amplitude and phase of the reflected signal at different frequencies, the obtained readout creates a system of nonlinear equations, which calculate the required values of the electromagnetic parameters of reflecting material.

On the basis of Eq. (20) the system of equations will have the form

$$
\begin{cases}
\dotfill \\
A_0[\mu_2(\omega_i),\sigma_2(\omega_i),\varepsilon_2(\omega_i)] - |G(\omega_i)| - A(\omega_i) \\
\psi_G[\mu_2(\omega_i),\sigma_2(\omega_i),\varepsilon_2(\omega_i)] + \psi(\omega_i) = \arg G(\omega_i) + \psi(\omega_i) \\
\qquad\qquad = \varphi(\omega_i) \\
\dotfill
\end{cases}
\tag{21}
$$

$A(\omega_i)$, $\varphi(\omega_i)$ – numerical values of the relative amplitude and phase of the reflected signal at a frequency $\omega_i$, $i=1\dots n$. Signal $\psi(\omega_i)$ determines the phase shift in the medium of propagation of the signal is determined by the design parameters of the device and can be determined by direct calculations by Eq. (13).

The system of Eq. (21) illustrates the general approach to the problem and determine the average values of the electromagnetic parameters in the test frequency range, which is directly controlled parameters do not greatly depend on the frequency and accepted assumption that $\varepsilon_2(\omega_i) \approx \varepsilon_2(\omega_{i+1})$, $\sigma_2(\omega_i) \approx \sigma_2(\omega_{i+1})$, $\mu_2(\omega_i) \approx \mu_2(\omega_{i+1})$, but the change signals are only due to changes in the phase velocity of the wave resistance of the absorption coefficient on the frequency in accordance with Eqs. (3)–(6). In general, all the electromagnetic parameters are frequency dependent and the Eq. (21) is insufficient to determine the unknown parameter values for each of the frequencies. However, if one of the parameters is known in advance and only slightly dependent on the frequency, for finding the remaining two parameters are sufficient reference amplitude and phase of the reflected signal at the same frequency. Typically, semiconductor and carbon nanomaterials have $\mu = 1.0$, which facilitates the solution. In some cases it is possible to obtain analytical expressions for the direct calculation of the required parameters. From Eq. (17):

$$
\begin{aligned}
Z_2(\omega_i) &= \frac{Z_1(\omega_i)[1+G(\omega_i)]}{1-G(\omega_i)} = \frac{Z_1(\omega_i)[1+A_G e^{j\varphi_G}]}{1-A_G e^{j\varphi_G}} \\
&= \frac{Z_1(\omega_i)[1+A(\omega_i)e^{j[\varphi(\omega_i)-\psi(\omega_i)]}]}{1-A(\omega_i)e^{j[\varphi(\omega_i)-\psi(\omega_i)]}} = B_i e^{\phi_i}
\end{aligned}
\tag{22}
$$

$B_i$, $\varphi_i$ – determined according to the measuring procedure by the formulas.

$$B_i = \left| \frac{Z_1(\omega_i)[1 + A(\omega_i)e^{j[\varphi(\omega_i)-\psi(\omega_i)]}]}{1 - A(\omega_i)e^{j[\varphi(\omega_i)-\psi(\omega_i)]}} \right|,$$

$$\phi_i = \arg \frac{Z_1(\omega_i)[1 + A(\omega_i)e^{j[\varphi(\omega_i)-\psi(\omega_i)]}]}{1 - A(\omega_i)e^{j[\varphi(\omega_i)-\psi(\omega_i)]}} \tag{23}$$

Using Eq. (3) and comparing the module and phases, after transformation obtain

$$\begin{cases} \mu_2^2(\omega_i)\omega_i^2 = B_i^4[\sigma_2^2(\omega_i) + \varepsilon_2^2(\omega_i)\omega_i^2] \\ \sigma_2(\omega_i) = \omega_i \varepsilon_2(\omega_i)tg\phi_i \end{cases} \tag{24}$$

From the system (24) are obtained relations between each other required parameters

$$\varepsilon_2(\omega_i) = \frac{\mu_2(\omega_i)}{B_i^2}\cos\phi_i, \ \sigma_2(\omega_i) = \omega_i \frac{\mu_2(\omega_i)}{B_i^2}\sin\phi_i, \ \mu_2(\omega_i)$$

$$= \frac{B_i^2 \varepsilon(\omega_i)}{\cos\phi_i} \tag{25}$$

Expressions (25) allow you to define the electromagnetic parameters of the reflecting medium, if any one of them is known. For example, for carbon, silicon and germanium compounds magnetic permeability close to unity in the frequency range under consideration $\mu_2(\omega) \approx 1$. In this case, the dielectric permittivity and conductivity at a selected frequency determined by the first two equations of Eq. (25). If all three of the electromagnetic parameters in the system (24) is unknown, the system can be supplemented by the relation obtained in Ref. [8], the desired binding parameters at different frequencies

$$\frac{\sigma(\omega_{i+1})}{\sigma(\omega_i)} = \frac{\varepsilon(\omega_{i+1}) - \varepsilon_0}{\varepsilon(\omega_i) - \varepsilon_0} = \frac{\mu^2(\omega_{i+1}) - \mu_0^2}{\mu^2(\omega_i) - \mu_0^2} \tag{26}$$

This allows you to get additional two equations in Eq. (21), which gives an opportunity to solve the problem of determining the electromagnetic parameters of the reflecting medium in any case, at different frequencies, when there is no a priori preliminary data on the unknown values

## 2.3   CONCLUSION

The following material is the theoretical basis of the method express diagnostics of complex electromagnetic parameters of nanomaterials on which can be built different measurement procedures, specific methods and algorithms for determining the electromagnetic parameters of liquid and solid films, including nanomaterials. Contactless and efficiency of the proposed method constitute its main advantage over the known methods of measurement.

## KEYWORDS

- **conductivity**
- **control**
- **nanomaterials**
- **permeability**
- **permittivity**
- **pulse**

## REFERENCES

1. Afonskij, A. A., & D'jakonov, V. P. Jelektronnye izmerenija v nanotehnologijah i mikrojelektronike. Pod red. Prof. V. P. D'jakonova. – M.: DMK Press, 2011, 688 p.
2. Shabatina, T. I., & Golubev, A. M. Nanohimija i nanomaterialy: ucheb. Posobie. M.: Izd-vo MGTU im. N. Je. Baumana, 2014, 63 p.
3. Alferov, Zh. I., Aseev, A. L., Gaponov, S. V., Kop'ev, P. S., Panov, V. I., Poltorackij, Je. A., Sibel'din, N. N., & Suris, R. A. Nanomaterialy i nanotehnologii. Mikrosistemnaja tehnika. 2003, No. 8, 3–13.

4. Tkachev, A. G., & Mishhenko, S.V. Uglerodnye nanomaterialy. proizvodstvo, svojstva, primenenie. M.: Mashinostroenie, 2008, 320 p.

5. Kikoina, I. K. Tablicy fizicheskih velichin. Spravochnik pod red. Akademika. M, Atomizdat, 1976.

6. Rjazanov, M. I. Jelektrodinamika kondensirovannogo veshhestva. M.: Nauka, 1982, 304s.

7. Brehovskih, L. M. Volny v sloistyh sredah. M.: Nauka, 1973. 344 p.

8. Skvortsov, B. V., Zabojnikov, E. A., & Vasil'ev, R. L. Opredelenie jelektrodin-amicheskih parametrov materialov v shirokom diapazone chastot. "Izmeritel'naja tehnika" 1997, № 9, 24–26.

# CHAPTER 3

# STRUCTURE, PHYSICAL, CHEMICAL, AND BIOLOGICAL PROPERTIES OF MAGNEROT NANOSCALE DRUG

O. V. KARBAN,[1] M. M. KANUNNIKOV,[2] N. N. CHUCHKOVA,[2]
N. V. SAVINOVA,[2] V. V. MUKHGALIN,[1] F. Z. GILMUTDINOV,[1]
V. KOMISSAROV,[2] and E. G. BUTOLIN[2]

[1]*Physic-Technical Institute of the Ural Branch of the Russian Academy of Sciences (PhTI UB RAS), Izhevsk, Russia*

[2]*Izhevsk State Medical Academy, Izhevsk, Russia*

## CONTENTS

## 3.1   INTRODUCTION

Heart diseases are the first cause of death in most countries of the world. In the ex-USSR countries, this death rate is twice as high as in Europe. It evidences that the drugs intended to prevent and cure heart diseases are either inefficient or too expensive for most patients. The common approach to finding effective drugs is based on development of new chemical compounds showing improved properties. The whole cycle, including synthesis of the drug, study of the physical and chemical properties, bioactivity testing on animals, and clinical trials, takes over 10 years and costs approximately 1 billion USD. This approach needs much time and a lot of funds.

There is another way currently developing: improvement of the efficiency of the existing drugs by discovering their most bioactive spatial structure.

The connection between the spatial structure and the activity is a fundamental characteristic of the effect of medical drugs. It is their molecular structure that determines how the substance enters the organism (or the cell), how the substance is transported to the target area, how it interacts with different receptors. Therefore, the structure determines the degree and directionality of the biological effect [1]. In this connection, it is relevant to demonstrate the "structure-property" interconnection making it possible to find the most bioactive structure of some drug.

Magnesium orotate is a well-known drug. Its advantage among other simple-magnesium-salt-based drugs is that its pharmacological effect is produced by both magnesium-cation and orotate-anion. There are three isomers (tautomers) of orotate-anion: the oxo-form with two carboxylic groups and one double bond in heterocycle; the hydroxy-form with a single carboxyl group and two double bonds in heterocycle; the dihydroxy-form with two hydroxylic-groups and an aromatic heterocycle [2]. In solid state, orotates exist in oxo-form; by now, all tautomeric transformations have been observed only in solutions with varying pH-value. The active agent of the wide-spread drug "Magnerot" is magnesium orotate, mostly in oxo-form; however, we do not know if this form is preferable to hydroxy- and dihydroxy-forms.

This study is to explore the possibility of producing various tautomeric forms of solid-state magnesium orotate by means of ball milling activation, as well as to compare the physical-chemical and biological properties of the obtained tautomers.

## 3.2   EXPERIMENTAL RESEARCHES

The object of this study is Magnerot (a drug by Mauermann-Arzneimittel KG, Germany). Besides magnesium orotate, each pill contains additive agents: amylum, microcrystalline cellulose, lactose, carmellose, and talc. The content of the additive agents is about 30% per pill. To compare, we studied the electronic structure of the "magnesium orotate" chemical, "chemically pure" grade.

Magnerot was mechanotreatmented in a planetary ball mill AGO-2 over 1 h, 3 h, and 6 h. According to the emission spectroscopy analysis with inductively coupled (argon) plasma (a Spectroflame spectrometer used), the mechanoactivated powders contain no inorganic additions coming from the balls or walls of the grinding containers.

The morphology of the magnesium orotate after mechanical activation was examined by atomic force microscopy (AFM) with an Integra Prima (NT-MDT) scanning probe laboratory by semi-contact mode in the air. The powder had been preliminarily attached to a polystirol film produced by evaporating ethil-acetate from polystirol dissolved in ethil-acetrate. The film was attached to glass ceramics with subsequent fixation of the powder under ultraviolet emission.

The x-ray phase analysis (XRD) of the powders was performed with a Bruker D8 Advance x-ray diffraction meter in CuKα – radiation. The detector was represented by a Sol-XE (Bruker) solid-state Si(Li) spectrometer.

The real density of the powder was determined under picnometer method (a picnometer of 5 cm$^2$).

The X-ray absorption spectra (NEXAFS) were examined with the equipment of the Russian-German channel of BESSY II electronic storage device.

The X-ray photoelectron spectra (XPS) were excited with MgKα-emission at a modified ES-2401 [3] spectrometer. The mathematical treatment of the spectra relied on the method [9] based on Fourier transformation with improved convergence procedure.

The biological properties of magnesium orotate were investigated under microelectrophoresis method with the use of buccal cells and erythrocytes, with random-bred laboratory rats (with modeled hypomagnesaemia and steroid osteoporosis).

The microelectrophoresis method relies on measuring the amplitude of the cells oscillating within the microscope field. In the electrophoresis chamber, the cells make forced alternating motions upon the reversal of the electrode voltage sign (10 V, 0.1 Hz). The frequency of the cell oscillations is equal to the frequency of the electrode sign reversals; still, the oscillation amplitude may vary depending on the charge of the cell surface indicating the physiological condition of the cell [4].

With the use of hypomagnesemia, we studied the effect of magnesium orotate on the restoration of the magnesium level in the blood serum, the cytological parameters of the red and white cells, the bioelectric activity of the population of erythrocytes and lymphocytes. To provide hypomagnesemia, Furosemidum diuretic (*Furosemidi* 1%) had been abdominally injected to the rats over 14 days; dosage of 30 mg/kg.

Steroid osteoporosis had been modeled through subcutaneous injection of prednisolone; dosage of 50 mg/kg of the rat's weight, over 14 days [5]. The collagen metabolism condition was assessed by the content in the body homogenate of the 2nd and 3rd loin vertebrae (sponge bone tissue): overall collagen by the amount of hydroxyproline [6]; free hydroxyproline [6]; neutral salt-soluble collagen fraction [7]. The experiment used 32 random-bred rats of 80–130 g that were divided into groups:

1. animals with steroid osteoporosis modeled by subcutaneous injection of prednisolone;
2. animals with steroid osteoporosis that were taking the initial drug, Magnerot;
3. animals with steroid osteoporosis that were taking the drug mechanoactivated over 1 h.

The experiments observed the provisions of the Declaration of Helsinki concerning humane treatment of animals.

## 3.3  RESULTS AND DISCUSSION

### 3.3.1  *STRUCTURE OF MECHANOACTIVATED MAGNESIUM OROTATE*

Figure 3.1 shows an AFM-image of the magnesium orotate powder after 1-hour ball milling. The powder particles are spheroid-shaped, 100–200 nm (Figure 3.1); single particles of 40 nm are also observed. The crystallite in the particles are of 40 nm. The particles are united in loosely-bound aggregates of up to 5 um. The powder structure analysis with the use of a wide range of materials has demonstrated that such hierarchy is relevant for organic and inorganic dielectrics.

When the ball milling is extended up to 3 hours, the aggregate dimensions are almost similar (Figure 3.2a), while the density and shape change. The particles within the aggregates are bounded with both weak adhesion and strong diffusion forces. The aggregates get laminated with layers of about 70–180 nm (Figure 3.2b). The average particle size is 162 nm;

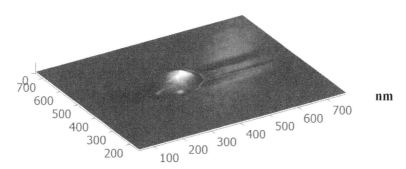

**FIGURE 3.1**  AFM-images of magnesium orotate particles after 1-hour mechanic activation.

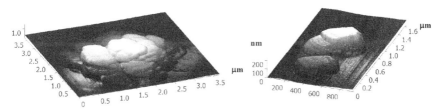

**FIGURE 3.2**  AFM-images of magnesium orotate particles after 3-hour (a) and 6-hour (b) ball milling.

meanwhile, the particles do not demonstrate grain structure; thus, amor-phization of the powder may be concluded.

Upon further mechanical treatment (over 6 hours), the aggregation pro-cess prevails over grinding. The particles are of 300–1000 nm. Particles and dense small agglomerates of about 1 µm cannot be distinguished. The particles with dense flat contacts compose laminated agglomerates with particles of about 50 nm. Being scanned with an atomic force microscope, such agglomerate behaves like a whole.

Figure 3.3 shows diffractograms of the initial and mechanoactivated orotates. Obviously, magnesium orotate in initial conditions has crystal-line structure. Amylum in the drug structure demonstrates reflexes around the angles 2 theta 20° [8]. After 1-hour ball milling activation, the structure becomes amorphic-crystalline: it is evidenced by smearing of the X-ray diffractogram reflexes; after 6 hours of mechanical activation, magnesium orotate gets completely amorphized.

The amorphization results in decrease of the real density of the mechanoactivated Magnerot and amounts to: 1,802 g/cm³; 1,341 g/cm³; 1,338 g/cm³ for the initial, 1-hour mechanoactivated, and 6-hour mecha-notreatmented drug.

### 3.3.2 CHEMICAL STATE OF THE INITIAL AND MECHANOACTIVATED MAGNESIUM OROTATE

NEXAFS spectra on K-edge of C-pyrimidine bases are of very similar thin structure. There are four well-resolved π*-resonances, two σ*-resonances

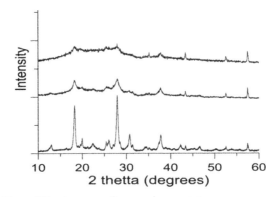

**FIGURE 3.3**   X-ray diffractograms of magnesium orotate.

near 296 and 304 eV. The peaks of π*-resonances correspond to the electron transitions from unequal core levels Cs1 of different carbon electrons
in the molecules to the delocalized π*-nonbonding orbitals. In the uracil
spectra, there are peaks at 284.6 eV, 286.0 eV, 288.0 eV, and 289.5 eV. In
the thymine spectrum, the low-energy peak shifts towards higher energies
at 0.4 eV (285 eV). The reason is appearance of the methyl substitute [9].
Still, it should be noted that the figure demonstrating the uracil chemical
composition shows the structures corresponding to oxo-forms and does
not give any data about the isomeric state of uracil and thymine. The bond
of the nitrogen-containing heterocycles is similar with aromatic heterocycles, but the explicit aromatic nature of the bonds is common only for
the dihydroxy-forms of chemical compounds.

    Figure 3.4 shows NEXAFS spectra of Cs1-edge of absorption of pure
magnesium orotate and Magnerot. The isomeric state of magnesium orotate
was not investigated. In both spectra, there is an intensive peak at 285.0 eV
corresponding with the electron transition C1s→π* (C=C) and evidencing
presence of an aromatic cycle. The low-energy peak with $E_{CB}$ = 284.2 eV
in the spectrum of magnesium orotate considerably reduces the intensity
in Magnerot's spectrum and manifests as an arm. The peak at 286.0 eV in
the magnesium orotate spectrum is more intense than the peaks at 285.0
eV and 284.2 eV; in Magnerot's spectrum, its intense is notably lower.
The peak at 286.0 eV may correspond to the electron transition C1s→π*
C=C=O in polysaccharide molecules or C=O within the orotate-anion.
The peaks 288.5 eV and 290.5 eV contribute into the electron transitions
C1s→π* in C=O and C=OH. The correlation of the intensity of these peaks

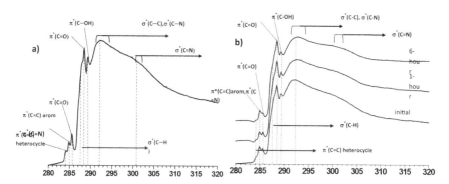

**FIGURE 3.4**   NEXAFS spectra of Cs1-edge of pure magnesium orotate (a) and Magnerot (b).

is almost equal in the spectra of magnesium orotate and Magnerot. It may be suggested that magnesium orotate – both as a pure product "magnesium orotate" and within Magnerot – has several tautomeric forms. 30%wt of additive agents per sample does not seriously influence NEXAFS spectra of Cs1-edge of absorption. After 1-hour and 6-hour ball milling, it is obvious that the influence does not seriously change Magnerot's NEXAFS spectra of Cs1-edge of absorption.

XPS C1s-, N1s-, O1s- spectra are given in Figure 3.5. In C1s-spectra, the dominant component has $E_{CB}$= 285.0±0.2 eV from carbon atoms within the adsorbed layer of hydrocarbons on the surface of the particles of $(CH_2)_n$-group powders. This component has been used to calibrate the XPS spectra. The component with $E_{CB}$= 286.2±0.2 eV corresponds with carbon atoms in composition with carboxyl groups (C-OH). The hydroxyl groups are present in hydroxy- and dihydroxy-forms of Magnerot and within the polysaccharides among Magnerot's additive agents. The component 287.0±0.2 eV refers to the carbon atoms in the bond (-C-O-C-) within the polysaccharide molecules. The component 289.0±0.2 eV refers to the carbon atoms bonded with oxygen and nitrogen (O=C-N) in the oxo- and hydroxy-forms of the orotate-anions. In the area 290.0–292.0 ±0.2 eV, there are shake-up satellites special for organic compounds with aromatic bonds [10].

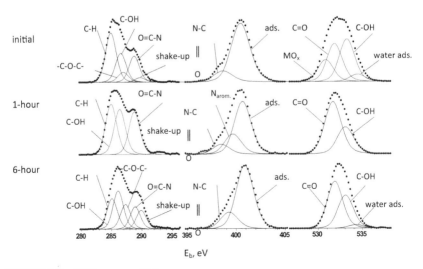

**FIGURE 3.5**   XPS spectra of Magnerot.

In O1s-spectra, there are components of oxygen atoms within the carbonyl (C=O, $E_{cB}$=531.8±0.2 eV) and hydroxyl (C-OH, $E_{cB}$=533.4±0.2 eV) groups. The component with $E_{cB}$=534.2±0.2 eV refers to the oxygen atoms in the adsorbed-water molecules; the low-energy component with $E_{cB}$=531.2±0.2 eV refers to the oxygen atoms bonded with the sample's metal base. In N1s-spectra, the intensive component 400.5±0.2 eV may be attributed to the nitrogen in NHC=O bond. The components with $E_{cB}$=398.4 ±0.2 eV and $E_{cB}$=399.6 ±0.2 eV refer to the nitrogen atoms from the aromatic heterocycle and from the nonaromatic heterocycle respectively.

The most informative were N1s-spectra, as they reflect the changes in the orotate-anion. In C1s- and O1s-spectra, considerable contribution from the carbon and oxygen atoms in the additive agents is observed. Comparison of N1s-spectra of Magnerot in initial state and after ball milling concludes that magnesium orotate in the initial drug exists in oxo-form with a minor addition of aromatic dihydroxy-form. Upon 1-hour mechanical activation, a hydroxy-from of magnesium orotate emerges; upon 6-hour mechanical activation, the hydroxy-form disintegrates and the share of the dihydroxy-form increases.

### 3.3.2.1   Physical and Chemical Properties

The organism's biological response to a drug depends firstly on its solubility, as this parameter determines the distribution of the substance in this organism and mainly conditions the pharmacokinetic properties of the drug. The solubility seriously influences the drug penetration from the intestines into the blood, i.e., such processes as absorption, filtration, diffusion etc.

Table 3.1 shows the solubility values for initial and mechanotreatmented orotates in correlation with the medium acidity (pH). The acidity is selected according to the medium acidity in different intestine and stomach parts. The mechanotreatmented Magnerot and magnesium orotate are more soluble that the initial ones; the solubility of pure magnesium orotate depends on the pH. The water-dissolution rate of mechanotreatmented Magnerot (pH 6.5) is 1.5 times higher than that of initial Magnerot.

The pH values of Magnerot's water solutions (0.5%wt) are alike: 5.76 and 5.80 for the initial and 1-hour mechanotreatmented respectively. The density of the initial Magnerot's solution is slightly higher than that of

**TABLE 3.1**　Drug Solubility

| | Mechanical activation time, h | Concentration, g/100 g of $H_2O$ | | | |
|---|---|---|---|---|---|
| | | pH 2 | pH 4 | pH 6.5 | pH 8.6 |
| Magnesium orotate | 0 | 0.140 | 0.23 | 0.230 | 0.200 |
| | 1 | 0.160 | 0.23 | 0.230 | 0.200 |
| | 6 | 0.210 | 0.29 | 0.270 | 0.400 |
| Magnerot | 0 | 0.060 | – | 0.075 | 0.072 |
| | 1 | 0.060 | – | 0.078 | 0.075 |
| | 6 | 0.630 | – | 0.078 | 0.075 |

the mechanotreatmented Magnerot. The volume expansivity of the initial Magnerot's solution is lower than that of the mechanotreatmented one. Thus, it evidences higher water-repellency of the mechanotreatmented drug. The increase of the water-repellency goes along with growing lipophily of the drug and, consequently, growing bioavailability. The analysis of the amount of Magnerot turning from a water solution (pH 7.4) into octanol (in octanol-water solution system simulating the operation of the gastrointestinal tract) proves the higher lipophily of the mechanotreatmented drug as compared to the initial one. Thus, after 1.5 hours, 11% of the initial Magnerot in the water solution 0.5%wt turns into octanol; and 20% turns in case of the mechanotreatmented drug.

**TABLE 3.2**　Temperature Dependence of Density for Water Solutions of Initial and 1-Hour-Mechanoactivated Magnerot (0.5%wt)

| Mechanical activation time, h | Density g/cm³ | | | | | Volume expansivity, $(10°–50°C)×10^{-4}$ |
|---|---|---|---|---|---|---|
| | 10°C | 20°C | 30°C | 40°C | 50°C | |
| 0 | 1.0013 | 0.9995 | 0.9980 | 0.9951 | 0.9899 | 2.4 |
| 1 | 1.0019 | 1.0005 | 0.990 | 0.9960 | 0.9905 | 2.8 |

**TABLE 3.3**　Temperature Dependence of Density for Water Solutions of Initial and 1-Hour-Mechanotreatmented Magnerot (0.5%wt)

| Solution, Mechanical activation time, h | Viscosity | | | | |
|---|---|---|---|---|---|
| | 10°C | 20°C | 30°C | 40°C | 50°C |
| 0 | 1.3308 | 1.0384 | 0.8339 | 0.6800 | 0.5661 |
| 1 | 1.3262 | 1.0184 | 0.8154 | 0.6677 | 0.5600 |

As a summary of the above research findings, it may be concluded that:

- in initial condition, magnesium orotate in Magnerot is available in two isomeric states: oxo- and dihydroxy-; ball milling activation for 1 hour results in the hydroxy form of magnesium orotate; after 6 hours of ball milling, the hydroxy-form disintegrates and the share of the dihydroxy-orotate form grows; the changing structural state of the orotate-anion and proportion of different isomeric forms of the orotate-anion may change the biological properties of the drug;
- the drug containing the hydroxy-form of magnesium orotate differs from the initial drug due to its higher lipophily; therefore, better bio-availability may be expected;
- ball milling causes reduction of the drug particle dimensions and amorphization; apparently, this is the reason of faster dissolution of the mechanotreatmented drug in water; meanwhile, the solubility is almost constant.

## 3.3.3 BIOLOGICAL PROPERTIES OF THE INITIAL AND MECHANOTREATMENTED MAGNEROT

### 3.3.3.1 Biolelectric Activity of the Cells

The experiment investigating the electrokinetic properties of the blood cells in the solutions of the initial and mechanotreatmented Magnerot has revealed that the maximum share of active cells, as well as the maximum oscillation amplitude of the cells, nucleus, plasmalemma is present in Magnerot's water solutions after 1-hour ball milling, i.e., containing the hydroxy-form of magnesium orotate. The electrokinetic potential of the cells is provided through the structural condition of the cell surface; meanwhile, it is connected with the rate of the metabolic processes in the cell, ion-exchange properties of the particles etc. Growing amount of the extracellular magnesium influences the membrane charge by changing its negative characteristics. Therefore, the mobility of plasmalemma and karyolemma changes. The disintegration of the hydroxy-form reduces the cell activity in the solution of the drug mechanotreatmented over 6 hours.

The obtained results demonstrate that it is the hydroxy-form that is the most active isomer of magnesium orotate. The oxo- and dihydroxy-forms are far less bioactive. Therefore, the tests on laboratory animals have used the initial Magnerot and the 1-hour mechanotreatmented Magnerot.

**TABLE 3.4**   Biolelectric Activity of the Buccal Epithelium Cells

| Substance, Ball milling time, h | | Biolelectric activity of the buccal epithelium cells | | | |
|---|---|---|---|---|---|
| | | Share of active cells, % | Oscillation amplitude | | |
| | | | Nucleus oscillation | Plasmalemma | Cells |
| Magnerot | 0 | 28±3.6 | 0 | 1.0±0.2 | 1.0±0.3 |
| | 1 | 100±12.0 | 6±0.4 | 1.9±0.3 | 2.7±0.3 |
| | 6 | 100±2.1 | 1±0.2 | 0.6±0.1 | 1.5±0.2 |

**TABLE 3.5**   Biolelectric Activity of the Blood Cells

| Substance, ball milling time, h | | Biolelectric activity values for blood cells | |
|---|---|---|---|
| | | Share of active cells, % | Oscillation amplitude |
| Magnesium | 0 | 83±2.6 | 7±1.7 |
| orotate | 1 | 96±2.5 | 16±3.2 |
| | 6 | 83±3.2 | 10±2.8 |

### 3.3.4   INVESTIGATION ON HYPOMAGNESAEMIA MODEL

The hydroxy-form of magnesium orotate (in the 1-hour mechanotreatmented drug) have prominent activating effect on the immune system cells (lymphocytes). It correlates with the homeostatic function of this magnesium-orotate form, when injected into the laboratory animal' organism.

Upon injection of the hydroxy-form, the amount of magnesium in the blood plasma increases by 64%; in contrast, for the comparison group (oxo-form), the amount of magnesium in the blood plasma does not increase (Table 3.6). The organism's quicker response to the hydroxy-form correlates with a higher solubility and a higher dissolution rate of this form, as compared to oxo-form.

After an injection of mechanotreatmented magnesium orotate, the cytological parameters of the white blood return to the normal figures, unlike the red cell. The hemoglobin and the erythrocytes keep reducing, but more considerably – in the group of the initial drug. A similar trend is observed for the thrombocytes: the amount of these cells is decreased in

**TABLE 3.6** Macroelements in the Laboratory Animals' Blood Serum in Case of Formation of Hypomagnesaemia and Treatment with Magnesium Orotate (mmol/L)

| | Reference (intact animals) | GME (comparison group) | 6-day treatment | |
|---|---|---|---|---|
| | | | 1-hour | Initial Magnorot |
| Magnesium | 1.75±0.08 | 0.902±0.18 | 1.12±0.10 | 0.855±0.05 |
| Sodium | 143.95±0.21 | 142.825±0.08 | 141.0±0.40 | 142.700±0.18 |
| Potassium | 4.86±0.90 | 4.390±1.52 | 5.495±0.14 | 4.405±1.56 |

**TABLE 3.7** Values of White and Red Blood of Laboratory Animals with Modeled Hypomagnesaemia and Treatment with Magnesium Orotate

| Value | Reference (intact animals) | GME (comparison group) | Hydroxy-form of magnesium orotate | Oxo-form of magnesium orotate |
|---|---|---|---|---|
| Leukocytes | 11.200±1.80 | 16.960±5.00 | 11.58±1.30 | 12.30±3.90 |
| Lymphocytes | 5.900±1.30 | 10.020±1.30 | 6.45±1.20 | 7.03±1.90 |
| Granulocytes | 3.550±0.08 | 5.400±0.07 | 4.10±0.04 | 4.17±0.11 |
| Haemoglobin | 153.500±7.20 | 140.200±4.80 | 125.00±4.08 | 110.33±12.28 |
| Erythrocytes | 8.165±0.39 | 6.984±0.27 | 6.81±0.18 | 5.85±0.31 |
| Mean cell volume | 54.620±2.60 | 59.080±2.28 | 58.33±3.51 | 60.20±4.55 |
| Thrombocytes | 1184.500±243.70 | 947.000±127.57 | 936.00±87.22 | 843.00±68.37 |

both groups after 6-day treatment. No trend to normalization is observed in the group with the target drug or in the official drug group. However, the latter group demonstrates greater reduction.

## 3.3.5 INVESTIGATION ON STEROID-OSTEOPOROSIS MODEL

The condition of collagen exchange depends on the interaction of the processes of biosynthesis and biodegradation of this biopolymer. Faster synthesis is indicated through a greater amount of the overall collagen and its neutral-salt-soluble fraction. The catabolic processes result in a greater content of the free hydroxyproline and a lower content of the overall collagen in the tissues.

For the animals with steroid osteoporosis, the concentration of the neutral-salt-soluble fraction of collagen reduced by 54.7% as compared to the reference. Both forms of magnesium orotate stimulated the synthesis activity of the osteoblasts in the bone tissue of the rats with osteoporosis. However, upon injection of the mechanotreatmented Magnerot, the content of the target collagen fraction was 130.4% (p<0.01) higher than in case of oxo-form.

The level of free hydroxyproline in the bone tissue of the animals with steroid diabetes surpassed the reference values by 400.8% (p<0.01). Using the magnesium drugs reduced these changes. Thus, upon injection of the initial and mechanotreatmented Magnerot to the rats with osteoporosis, the amount of free hydroxyproline reduced by 60% and 53.7% (p<0.05), respectively. For the animals with osteoporosis, the content of the overall collagen was lower than for the reference group by 42.8% (p<0.01). A magnesium orotate injection balanced these changes. The usage of oxo-form of the drug made it possible to reach the level of the target parameter for the reference animals. Meanwhile, the usage of hydroxy-form of magnesium orotate increased the overall collagen by 65.3 (p<0.001) as compared to the reference.

Thus, the injection of Magnerot, initial or mechanotreatmented, rectifies disorder in the bone-tissue collagen exchange. The intensification of the anabolic processes in collagen exchange is more explicit in case of the drug containing the magnesium orotate hydroxy-form

## 3.4  CONCLUSION

Magnerot contains magnesium orotate in two isomeric forms: oxo- and dihydroxy-. The ball milling of the drug causes changes in the structural state and tautomeric transformations of magnesium orotate. Upon 1-hour ball milling, the hydroxy-form of the orotate-anion is generated. Upon 6-hour ball milling, the hydroxy-form disintegrates; meanwhile, the share of the orotate-anion dihydroxy-form grows.

The ball milling does not seriously influence the solubility of the drug; the dissolution rate of the mechanotreatmented drug is 1.5 times higher than that of the initial Magnerot. The solubility of the initial and mechanotreatmented drugs unevenly depends on the solution pH: the minimum solubility is observed at pH=2; the maximum is at pH=4. The structural difference of the magnesium orotate isomers persists in the water solution. It is demonstrated that the drug containing the orotate-anion hydroxy-form is more bioactive that the drug containing oxo- or hydroxy-form.

## ACKNOWLEDGMENTS

The research was supported by the Program of the RAS Presidium, Grant N 12-P-2–1065.

The authors thank Sobennikova M.V. for her analysis of the elemental composition of the mechanoactivated magnesium orotate samples and finding of the dissolution rate and solubility in octanol. The authors also thank Dr. of Medical Sciences A.A. Solovyova for his microelectrophoresis analysis.

## KEYWORDS

- atomic force microscopy
- dihydroxy-form
- hydroxy-form
- hypomagnesemia
- magnesium orotate
- mechanic activation
- NEXAFS
- oxo-form
- steroid osteoporosis
- structure
- tautometry
- X-ray diffraction
- XPS

## REFERENCES

1. Lomovskiy, O. I. Prikladnaya mechanokyimiya: pharmatsevnika I meditsinskaya promyshlennost. Obrabotka dispersnykh vaterialov I sred. 2001, V.11, pp. 81–100.
2. Tjukavkina, N. A., & Baukov, Ju. I. Bioorganicheskaja himija:uchebnik dlja vuzov. M.: Drofa, 2005, 542 p.
3. Kanunnikova, O. M., Gil'mutdinov, F. Z., Kozhevnikov, V. I., & Trapeznikov, V. A. Metody fotojelektronnyh issledovanij neorganicheskih materialov. Uchebnoe posobie. Izhevsk: Izd-vo Udmurtskogo universiteta, 1995, 393 p.
4. Haramonenko, S. S., & Rakitjanskaja, A. A. Jelektroforez kletok krovi v norme i patologii. Minsk, 1974, p.33–35
5. Ziganshina, L. E., Burnasheva, Z. A., & Valeeva, I. H. Sravnitel'noe izuchenie jeffektivnosti dimefosfona i ksidofona pri steroidnom osteoporoze u krys. Jeksperimental'naja i klinicheskaja farmakologija. 2002, №6. pp. 55–56.
6. Proshina, L. Ja., Privalenko, M. N., & Proshina, L. Ja. Issledovanie frakcionnogo sostava kollagena v tkani pecheni. Voprosy med himii. 1982, №1, pp. 115–119.
7. Sharaev, P. N., Butolin, E. G., & Ivanov, V. G. Pokazateli obmena biopolimerov soedinitel'noj tkani pri mnogokrat-nom stresse. Ukrainskij biohimicheskij zhurnal. 1987, T.59, №3, p. 85.
8. Vinokurov, A. Ju. Issledovanie zakonomernostej i sovershenstvovanie tehnologii kationirovanija krahmala v vodnoj suspenzii. Avtoreferat diss.k.t.n., M. 2013, 25 p.
9. Zubavichus, Ja. V. Mjagkaja rentgenovskaja sinhrotronnaja spektroskopija bioorganicheskih materialov, vody i vod-nyh rastovorov. Avtoreferat diss. d. phys.math.. nauk. M. 2012, 45 p.
10. Beamson, G., & Briggs D. High resolution XPS of organic polymers: the scienta ESCA 300 Database, Wileq Intercience, 1992, 306 p.

# CHAPTER 4

# X-RAY PHOTOELECTRON STUDY OF THE MECHANISM OF THE FUNCTIONALIZATION OF THE CARBON COPPER-CONTAINING NANOSTRUCTURE SURFACE WITH SP-ELEMENTS

I. N. SHABANOVA,[1] V. I. KODOLOV,[2] N. S. TEREBOVA,[1] and G. V. SAPOZHNIKOV[1]

[1]*Physic-Technical Institute of the Ural Branch of the Russian Academy of Sciences (PhTI UB RAS), Izhevsk, Russia*

[2]*Kalashnikov Izhevsk State Technical University, Izhevsk, Russia*

## CONTENTS

## 4.1   INTRODUCTION

Unique properties of carbon metal-containing nanostructures provide various possibilities of their application. For improving the properties of materials, it is necessary to modify them with nanostructures. Since the surface of nanostructures has low reactivity, the binding between the nanostructure surface and the molecules of the ambient medium can be provided by the functionalization of the nanostructure surface, i.e., by adding sp-elements which form covalent bonds with the atoms on the nanosystems surface. In this case, the disperse ability and solubility of nanosystems improve and their coagulation into bunches is prevented due to the repulsion of sp-element atoms on the side areas. The functionalization provides the conditions for improving the nanostructure properties.

The goal of the present investigations is the study of the mechanism of the functionalization of the surface of carbon copper-containing nanostructures and the influence of sp-elements of functional groups on the formation of the chemical bond with the atoms on the surface of nanostructures and the change of their properties.

## 4.2   EXPERIMENTAL RESEARCHES

The X-ray photoelectron spectroscopy investigations were conducted on an x-ray electron magnetic spectrometer with the resolution of $10^{-4}$, luminosity of 0.085% and the excitation by AlK$\alpha$-line 1486.5 eV in vacuum $10^{-8}$–$10^{-10}$ Pa. In comparison with electrostatic spectrometers, a magnetic spectrometer has a number of advantages such as the constancy of luminosity and resolution which are independent of the energy of electrons, high contrast of spectra and the possibility of acting upon a sample during measurements [1].

The investigation objects were carbon copper-containing nanostructures prepared by low-temperature synthesis (lower than 400°C) in nanoreactors in the form of stretched cavities which are formed by macromolecules in gels of polymer materials [2]. The structure obtained was functionalized with chemical groups containing sp-elements (Si, P, S, N, F, I).

For studying the formation of the covalent (hybridized) bond between the atoms of carbon copper-containing nanostructures and sp-elements of the functional groups, the x-ray photoelectron spectra C1s and Me3s were investigated. The information about the metal atomic magnetic moment was obtained by studying the multiplet splitting of the Me3s-spectra. The relative intensity of the maxima of the 3s-spectra multiplets correlates with the number of unpaired d-electrons of the atoms in the systems of 3d-metals, and the distance between the multiplet maxima gives the information about the exchange interaction of 3s-3d shells. The changes in the 3d-shell (localization or hybridization) give the information about the changes in the chemical bond between the neighboring atoms and in the structure of the nearest surroundings of the atoms of 3d-metals.

The developed model was used for determining the atomic magnetic moment in carbon metal-containing nanotubes in comparison with that in massive metal samples [3]. It is shown that in contrast to massive samples, in the nanostructures there is a change in the relative intensity of the multiplet splitting maxima and in the distance between them. The results obtained indicate an increase in the number of noncompensated d-electrons in the atoms of the transition metals and the appearance of them on the Cu atoms in the nanostructures.

The increase of the number of noncompensated d-electrons in the carbon metal-containing nanostructures is explained by the participation of the d-electrons of the metal atoms in the hybridized chemical bond with p-electrons of the carbon atoms. Thus, in comparison with pure metal, in the carbon metal-containing nanostructures the atomic magnetic moment increases. The model was used for investigating the changes of the atomic magnetic moment on the Cu-atoms of the functionalized carbon copper-containing nanostructures. The chemical groups containing sp-elements were used for the functionalization.

## 4.3   RESULTS AND DISCUSSION

The mechanism of the functionalization of the carbon copper-containing nanostructure surface with sp-elements (Si, P, S, N, F, I) is studied. The mechanism of the formation of the chemical bond between the atoms of carbon, silicon, sulfur, nitrogen, fluorine and iodine in the investigated systems is studied on the basis of the C1s, O1s, Cu3s, Si2p, P2p, N1s, F1s and I3d core level spectra investigation.

Figure 4.1 shows the Cu3s-spectra of the nanostructures functionalized by chemical groups containing sp-elements (Si, P, S).

Table 4.1 presents the parameters of the Me3s-spectra and atomic magnetic moments on the copper atoms in the studied nanostructures.

In the earlier works [3], it is shown that in carbon copper-containing nanostructures, a copper magnetic moment appears.

As shown in Table 4.1, the nearest surrounding of Cu atoms and their chemical bond change at the functionalization. The Cu atomic magnetic moment increases in comparison to that in unfunctionalized nanostructures. The largest value of the atomic magnetic moment is observed when carbon copper-containing nanostructures are functionalized with Si atoms and when they are functionalized with P and S atoms, it decreases. It is connected with an increase in the filling of the p-shell of the sp-element atoms at going from Si to S and a decrease in the degree of the covalence or the hybridization of the d-electrons of the transition metal atoms with the p-electrons of the sp-element atoms [4, 5].

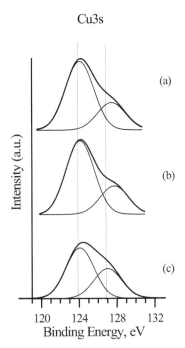

**FIGURE 4.1**   X-ray photoelectron Cu3s-spectra of nanostructures functionalized with sp-elements. (a) Cu/C+S; (b) Cu/C+P; (c) Cu/C+Si).

**TABLE 4.1** Parameters of the Multiplet Splitting of the Copper 3s-Spectra in the Nanostructures

| Sample | $I_2/I_1$ | $\Delta$, eV | $\mu_{Cu}$, $\mu_Б$ |
|---|---|---|---|
| Cu3s$_{nano}$ unfunctionalized | 0.20 | 3.6 | 1.3 |
| Cu3s$_{nano}$ (Si-functionalization) | 0.60 | 3.0 | 3.0 |
| Cu3s$_{nano}$ (P functionalization) | 0.42 | 3.6 | 2.0 |
| Cu3s$_{nano}$ (S functionalization) | 0.40 | 3.6 | 1.8 |

$I_2/I_1$ – the relation of the intensities of the maxima of the multiplet splitting lines;

$\Delta$ – the energy distance between the maxima of the multiplet splitting in the 3s-spectra of the carbon copper-containing nanostructures.

For the studied Si-, P- and S-element-functionalized and unfunctionalized carbon copper-containing nanostructures, the C1s-spectra are similar and they consist of two components C-C with sp²-hybridization and C-C with sp³-hybridization of carbon atoms in the ratio of 1:0.5 (Figure 4.2a).

Cu3s

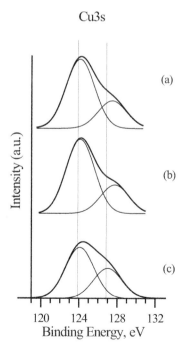

**FIGURE 4.2** C1s-spectra of the unfunctionalized and sp-element-functionalized Cu/C nanostructures. (a) unfunction, and Si-, P-, S-function, Cu/C; (b) Cu/C+N; (c) Cu/C+I.

A small amount of the C-H component in the C1s-spectrum indicates the residual amount of polymer at the nanostructure synthesis.

Thus, on the nanostructure surface, the silicon, phosphorous and sulfur atoms get bound to the atoms of copper and not to the carbon atoms. This is also indicated by the change of the number of noncompensated d-electrons due to the change in the nearest surrounding of the metal atoms. The degree of the covalence of the Me-X bond (where X is Si, P, S) is higher than that of Cu-C in the unfunctionalized nanostructures because the Si, P and S atoms have a larger covalent radius, which value is close to the copper radius value. Consequently, the interaction of the copper atoms with the sp (Si, P, S)-atoms is stronger than that of copper with carbon (Cu-C) and the strength of the bond between the copper atoms and sp-atoms grows at the functionalization.

Next, the investigation of the mechanism of the functionalization of the surface of the carbon copper-containing nanostructures with the N, F and I sp-elements and the influence of the functionalization on the formation of stable complexes of carbon copper-containing nanostructures and the chemical bond between their atoms has been conducted.

In the case when the functional group contains nitrogen, fluorine and iodine, the covalent bond is formed between the sp-element atoms and carbon atoms on the surface of a carbon copper-containing nanostructure, and when the material is modified with the nanostructures, the binding with the material atoms is realized through the atoms of nitrogen, fluorine and iodine.

In the C1s-spectrum (Figure 4.2b) of the nanostructures functionalized with nitrogen-containing groups, the following components are found: C-H (285 eV), C-N(-H) (286.4 eV) and an insignificant peak of the adsorbed component C-OH (289 eV). The N-H bond is relatively weak compared to the stronger bond C-N(H) with the high degree of the covalence of the atoms of carbon and nitrogen having similar covalent radii. The appearance of the C-N(H) bond between the carbon atoms of the carbon copper-containing nanostructures and the nitrogen atoms of the amine groups indicates the formation of stable nanocomposites. The presence of the C-N(H) bond in the N1s-spectrum consisting of two components, N-H (397 eV) and C-N(H) (398.8 eV), also confirms the formation of stable nanocomposites with common covalent bond C-N(H), In the O1s-spectrum, the low-intensity peak corresponds to adsorbed oxygen.

When the functional group contains fluorine, a strong covalent bond between fluorine and carbon atoms is formed in the system, and the bond of the nanostructures with the atoms of the modified material is realized through fluorine atoms. Similar data are obtained for the function-alization of the carbon copper-containing nanostructures with the iodine groups (Figure 4.2c). The interaction of atoms of nitrogen, fluorine and iodine with atoms of copper isn't revealed.

## 4.4   CONCLUSION

Thus, the formation of a strong covalent chemical bond between the atoms of the functional sp-groups and nanostructures is influenced by the similarity of their covalent radii.

Consequently, it can be concluded that the formation of the bond between the atoms of the functional groups and nanostructures is influenced by the possibility of the formation of a strong covalent bond between the atoms on the surface. Therefore, in the present investigation the atoms of silicon, phosphorus and sulfur in the functional groups effectively interact with metal atoms and not carbon atoms.

The study of the mechanism of the functionalization of the surface of the carbon metal-containing nanostructures with the sp-elements atoms and their influence on the variation of the atomic magnetic moment of the carbon metal-containing nanotubes shows the following:

1.   The formation of a stronger covalent bond of the Cu atoms of the nanostructures with the atoms of the sp-elements (Si, P, S) takes place leading to an increase in the activity of the nanotubes surface, which is necessary for the modification of material with them. The activity of the nanostructures at the functionalization with the sp-elements increases at going from S to P and Si. The atomic magnetic moment on the copper atoms also increases at going from S to P and Si during functionalization.
2.   The formation of a strong covalent bond of the carbon atoms of the carbon copper-containing nanotubes with the atoms of nitrogen, fluorine and iodine leads to an increase in the activity of the nano-structure surface.

3. The formation of a strong covalent chemical bond between the atoms of the functional sp-groups and nanostructures is influenced by the similarity of their covalent radii, i.e., by the overlapping of the wave functions of the valence electrons.

4. The method of X-ray photoelectron spectroscopy can be used for studying the mechanism of the functionalization of nanostructures and monitoring their quality.

## KEYWORDS

- atomic magnetic moment
- carbon copper-containing nanostructures
- covalent bond
- functionalization
- satellite structure of C1s- and Cu3s-spectra
- X-ray photoelectron spectroscopy

## REFERENCES

1. Trapeznikov, V. A., Shabanova, I. N., et al. New automated x-ray electron magnetic spectrometers: a spectrometer with technological adapters and manipulators, a spectrometer for melt investigation. Izvestia AN SSSR, Seriya fizicheskaya. 1986, V. 50, №9, pp. 1677–1682.

2. Kodolov, V. I., & Khokhryakov, N. V. Chemical physics of the processes of the formation and transformations of nanostructures and nanosystems. IzhGSHA, Izhevsk. (2009), In two volumes. V. 1, 360 p. v2, 415 p.

3. Shabanova, I. N., & Terebova, N. S. "Application of the X-ray photoelectron spectroscopy method for studying the magnetic moment of 3d metals in carbon-metal nanostructures" Surface and Interface Analysis. 2010, v.42, № 6–7, pp. 846–849.

4. Shabanova, I. N., Kormilets, V. I., & Terebova, N. S. "XPS-studies of the electronic structure of Fe-X (X = Al, Si, P, Ge, Sn) systems" J. Electron Spectroscopy 2001, 114–116, 609–614.

5. Shabanova, I. N., Mitrokhin, Yu. S., & Terebova, N. S. "Experimental and theoretical study of electronic structures of Ni-X(X = Al, Si, P) systems." Electron Spectroscopy and Related Phenomena, 2004, 137–140, 565–568.

# CHAPTER 5

# COMPARATIVE ANALYSIS OF ELEMENTAL AND STRUCTURAL COMPOSITION OF NANOCRYSTALLINE TANTALUM POWDERS

S. YU. TRESCHEV,[1] S. P. STAROSTIN,[2] S. S. MIKHAYLOVA,[1] O. M. KANUNNIKOVA,[1] B. E. PUSHKAREV,[1] F. Z. GILMUTDINOV,[1] M. V. SOBENNIKOVA,[1] V. I. LADYANOV,[1] and V. P. LEBEDEV[2]

[1]*Physic-Technical Institute of the Ural Branch of the Russian Academy of Sciences (PhTI UB RAS), Izhevsk, Russia*

[2]*Elekond Open Joint-Stock Company, Sarapul, Russia*

## CONTENTS

## 5.1  INTRODUCTION

The finely-divided tantalum powders serve as a basis for production of porous tantalum capacitors, that provide high specific charge, low leak currents, high stability and reliable characteristics. The new-generation capacitors made of materials with high specific surface and controllable properties require essentially new tantalum powders with different specific charge and grain-size properties. The powder evaluation is a necessary stage of a research for better technology of capacitor production.

The capacitor tantalum powders are produced through several ways. Among the first commercial methods is the metallothermic way [1]. Under this technology, potassium fluotantalate ($K_2TaF_7$) is used to produce metallurgic tantalum, which is subsequently melted, through 2–3-time electron beam remelting, into ingots, 99.95–99.97%wt. Then these ingots are saturated with hydrogen at a high temperature (hydrogenated), grinded in a mill and dehydrogenized by heating in vacuum at 700–800°C. The result is the powders of the so-called "fragmentation type" with a small amount of impurities, relatively small specific charge up to 5 mC/g, and specific surface up to 0,1 m$^2$/g [2–4]. The tendencies in the capacitor-making technologies demand nanocrystalline powders of up to 99,997 purity, specific surface up to 1–2 m$^2$/g and specific charge up to 100 mC/g. Such powders are produced through metallothermic reduction of the molten salts of the metal. An alkali metal or an alkali-earth metal is put into molten high-purity potassium fluorotantalate under Ar atmosphere. The metal reduction takes place on the phase boundary, where the crystallization nuclei are accumulated. Eventually, nanocrystalline tantalum powders are produced [5–7].

The dimensions, shape and impurity composition of the powders depend on the production method and considerably determine the properties and efficiency of the target device – capacitor. The aim of this paper is to provide a comparative study of the properties of the nanocrystalline tantalum powders produced in Russia/abroad and used to make capacitors.

## 5.2  EXPERIMENTAL RESEARCHES

The subjects of this research were tantalum powders produced through sodiumthermic [1, 8, 9] and fragmentation [2–4] methods (Table 5.1).

**TABLE 5.1** Research Subjects

| Production Method | Name | Producer |
|---|---|---|
| Fragmentation | OcK2.5 | Ulba Metallurgical |
| | TaK2.5 | Plant, Kazakhstan |
| | 2Б | |
| | QR7 | H.C. Starck, USA |
| Sodiumthermic | TaK80 | H.C. Starck, USA |
| (from molten substance) | TaK30 | OTIC, China |
| | K-12 | OOO NPCKM Tantal, Russia |
| | STA 18 | H.C. Starck, USA |

The shape and size of the powder particles (with and without camphora) are studied through scanning electron microscopy (SEM) with a Philips SEM-515 microscope. The impurity content was analyzed through emission spectroscopy with inductively coupled plasma with a Spectroflame spectrometer (Germany). The units used in the tables are ppm – parts-per-million. The content of the surface layers was studied through Auger-electronic microscopy with a JAMP 10-S spectrometer. The X ray diffraction analysis (XRD) was performed with X-ray diffractometers: DRON-3 with CoKα-radiation and Bruker D8 advance – with CuKα-radiation.

The true density of the particles was found through the picnometric method, with heptane as the working fluid. The measurements were taken at 25°C.

## 5.3 RESULTS AND DISCUSSION

### 5.3.1 PARTICLE SHAPE AND SIZE

TaK80 tantalum powder produced from molten salts is spherical porous granules of 10–40 μm with extended surface (Figure 5.1a).

The granules consist of particles smaller than 1 μm (Figure 5.1b). In the powders produced by Tantal Company (K12), there are dendritic structures (Figure 5.2) of hundreds of micrometers; in their turn, these particles consist of smaller particles of hundreds of nanometers.

**FIGURE 5.1**    SEM image of TaK80 powder: a – x150 zoom, b – x5000 zoom.

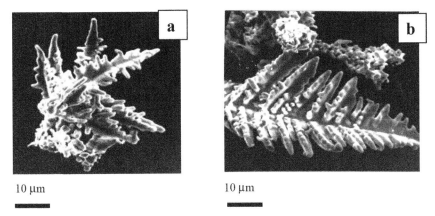

10 μm                                10 μm

**FIGURE 5.2**    SEM image of tantalum powder K12: a – x1500 zoom, b – x1600 zoom (the images have been obtained at the electronic microscopy laboratory at Udmurt State University, Izhevsk, Russia).

TaK30k powder is TaK30 powder with camphora admixture. In this powder, there are individual granules (5–10 μm), as well as agglomerates of 150–200 μm (Figure 5.3a) consisting of granules. The granules consist of smaller particles of 0.2–0.7 μm (Figure 5.3b).

The pressed powders TaK30k, as well as the initial powder (Figure 5.4a,b), are a similar mixture of granules and agglomerates.

The structure of Ock 2.5 tantalum powder (of fragmentation type) is quite different. It consists of big granules of 100–200 μm (Figure 5.5a), which consist from stone-like particles of 11–20 μm (Figure 5.5b). The particle shape (Figure 5.5b) is inherent for the fragmentation-type particles.

TaK2.5K powder represents granules of 50–300 μm (Figure 5.6a), which consist of stone-like particles of 1–40 μm um (Figure 5.6b). Both

**FIGURE 5.3**   SEM image of TaK80k powder: a – x150 zoom, b – x2500 zoom.

**FIGURE 5.4**   SEM-image of pressed powder Tak30k (press): a – x150 zoom, b – x10,000 zoom.

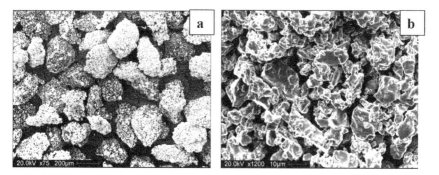

**FIGURE 5.5**   SEM image of Ock80 powder: a – x75 zoom, b – x1200 zoom.

**FIGURE 5.6**    SEM image of TaK2.5k powder: a – x75 zoom, b – x1200 zoom.

stone-like particles and granules have quite different sizes. The pressed TaK2.5k powder (Figure 5.7) has a homogeneous structure consisting of stone-like particles.

## 5.3.2   COMPOSITION OF POWDER PARTICLES

The chemical composition of the examined tantalum powders obtained through two technologies, as well as by Russian and foreign companies is given in Tables 5.2 and 5.3.

According to the quality assessment report, the content of carbon, oxygen, hydrogen, and nitrogen (Table 5.2) is much higher than in the powders obtained through sodium thermic method, as compared to the fragmentation-type powders. It may be due to the porous structure of the sodiumthermic powder particles. Table 5.2 shows the analysis results for the elements which content differs from that in the quality requirements. The content of

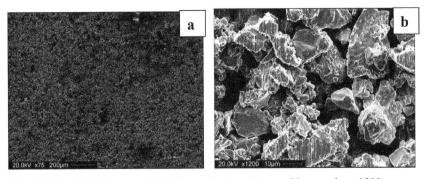

**FIGURE 5.7**    SEM image of TaK2.5k powder (press): a – x75 zoom, b – x1200 zoom.

TABLE 5.2 Chemical Composition of OcK 2.5, TaK2.5, TaK80, and TaK30 Tantalum Powders

| Production method | Fragmentation | | | | Sodiumthermic (from molten substance) | | | |
|---|---|---|---|---|---|---|---|---|
| Powder type | OcK2.5, ppm | | TaK2.5, ppm | | TaK80, ppm | | TaK30, ppm | |
| Chemical element | Quality requirements | Our data | Quality requirements | Our data | Quality requirements | Our data | Quality requirements | Our data |
| C | 10 | – | 10 | – | 57 | – | 13 | – |
| O | 580 | – | 430 | – | 3450 | – | 1920 | – |
| H | 10 | – | 10 | – | 110 | – | 92 | – |
| N | 39 | – | 30 | – | 920 | – | 160 | – |
| Fe | 10 | 0.3 | 6 | 0.3 | 10 | 0.3 | 11 | 0.4 |
| Ni | 3 | 5.4 | 3 | 24.0 | 6 | 35.7 | 11 | 15.5 |
| Cr | 3 | 0 | 3 | 50.6 | 7 | 0 | 9 | 57.9 |
| Si | 10 | 10.8 | 10 | 2.5 | 10 | 15.0 | 10 | 1.6 |
| P | – | – | – | 0 | 100 | – | 76 | 0 |
| K | 4 | 29.0 | 3 | 207.0 | 17 | 0 | 6 | 1.1 |
| Na | 3 | 0 | 3 | 0 | 3 | 0 | 5 | 0 |
| Nb | <30 | – | <30 | 0 | <30 | – | <30 | 0 |
| W | 5 | – | 5 | 0 | 5 | – | 5 | 0 |
| Al | – | 5.7 | – | 29.8 | – | 4.3 | – | 19.7 |
| Co | – | 0 | – | 71.1 | – | 0 | – | 64.3 |
| Pb | – | – | – | 44.0 | – | – | – | 41.0 |

TABLE 5.3   Chemical Composition of Foreign and Russian Tantalum Powders

| Production method | Sodiumthermic (from molten substance) | | | | Fragmentation type | | | |
|---|---|---|---|---|---|---|---|---|
| Powder type | K12 (Russia) ppm | | STA18 (Starck) ppm | | 2Б (UMP) ppm | | QR7 (Starck) ppm | |
| Chemical element | Quality requirements | Our data | Quality requirements | Our data | Quality requirements | Our data | Quality requirements | Our data |
| Cu | <10 | 114.8 | <1 | 51.8 | 20 | 12.7 | <1 | 121.5 |
| Fe | 30 | 168.6 | 7 | 82.5 | 50 | 117.0 | – | 144.0 |
| Mg | <3 | 27.7 | 9 | 14.4 | 3 | 41.6 | <1 | 2.1 |
| Mo | <5 | 21.0 | <2 | 10.6 | – | 24.0 | <10 | 5.6 |
| Nb | 12 | 27.9 | <1 | 33.2 | 2000 | 81.0 | <25 | 42.8 |
| Ni | 8 | 49.3 | 6 | 43.9 | 5 | 62.7 | <5 | 58.2 |
| Si | 10 | 63.6 | <3 | 80.3 | 30 | 200.2 | <8 | 56.0 |
| Sn | <10 | 85.4 | <3 | 31.9 | 5 | 20.9 | <1 | 209.2 |
| Ti | <10 | 0 | <2 | 1.0 | 8 | 19.1 | <5 | 0 |
| W | – | <1 | <5 | 4.2 | – | 115.8 | <40 | 68.1 |
| Zr | <5 | 9.1 | <2 | 8.9 | 5 | 15.8 | – | 12.3 |
| Σ, sum | 103 | 567.4 | 41 | 362.7 | 2126 | 710.8 | 96 | 719.8 |

various impurities in the tantalum powders is not high, except for potassium in TaK2.5 It is noteworthy that cobalt and lead are usually ignored in the quality reports, while they are observed in TaK2.5 and TaK30 in quite remarkable amounts. Besides, there is a plenty of chromium in the powders. A high concentration of cobalt, lead and chromium is observed in the powders manufactured both in Russia and abroad.

The analysis of other powders (Table 5.3) demonstrated that the overall quantity of impurities in the fragmentation-type powders is 1.5–2 times higher than that in the sodiumthermic powders. It may be noted that the content of almost all the impurities exceeds – sometimes considerably (iron, copper, tin, silicon) – the data given in the quality requirements.

The Auger-spectroscopy of the tantalum powder showed that the surface of the particles obtained through various technologies is oxidized.

### 5.3.3 STRUCTURAL AND PHASE COMPOSITION OF POWDERS

The d XRD results demonstrated that all the powders have cubic body-centered lattice (bcc), Im3m space group with a lattice parameter a = 0.3306–0.3309 nm (Table 5.4). It is in accordance with database JCPDS 0.3306 nm [10].

Meanwhile, the lattice microdistortion in the fragmentation-type powders is greater than that in the molten-salt powders. After pressing the powder lattice microdistortion increases. The dimensions of the coherent scattering regions in different powders are slightly different.

**TABLE 5.4** XRPA Analysis Results

| Sample | a, nm | +D, nm | $+\varepsilon^2,{}^{1/2}$, % |
|---|---|---|---|
| OcK2.5 | 0.33061±0.00003 | 63.0±0.8 | 0.166±0.002 |
| TaK2.5K | 0.33058±0.00003 | 56.0±0.8 | 0.100±0.002 |
| TaK2.5K (press) | 0.33059±0.00002 | 53.4±0.6 | 0.134±0.002 |
| TaK80 | 0.33091±0.00002 | 64.5±0.5 | 0.101±0.001 |
| TaK30K | 0.33056±0.00002 | 46.0±0.4 | 0.074±0.004 |
| TaK30K (press) | 0.33061±0.00002 | 46.9±0.6 | 0.175±0.002 |

a – period of lattice, +D – diameter of coherent scattering block, $+\varepsilon^2,{}^{1/2}$ – mean square microdistortion of lattice).

The true density of TaK80 and OcK2.5 powders (17.4 g/cm$^3$ and 17.1 g/cm$^3$, respectively) is higher than the reference density of bulk tantalum [11]. According to Ref. [12], the observed high density is caused by the surface layer compacting of substance. In the observed nanosized powders, the share of the surface layers is high; consequently, the density noticeable grows.

## 5.4 CONCLUSION

The comparative analysis of tantalum powders has demonstrated that the powders obtained through sodiumthermic method are considerably different from the fragmentation-type powders in terms of shape of powder particles. While the fragmentation-type powders represent an irregular shape agglomerate of hundreds of micrometers, which consist of micrometer-scale stone-like particles; the sodiumthermic powders are a spherical agglomerates of dozens of micrometers, which comprised from nanometer-scale particles.

The sodiumthermic powders contain more impurities of C, O, H, N than the fragmentation-type powders. The overall quantity of the remaining impurities in the fragmentation-type powders is higher than in the sodiumthermic powders.

All the observed powders were of α-Ta-structure. In OcK2.5, TaK2.5K, TaK80, TaK30K powders, the parameters of the bcc lattice are slightly different.

The true density of the tantalum powders obtained through fragmentation and sodiumthermic methods is higher than the density of bulk tantalum.

## KEYWORDS

- electronic microscopy
- elemental analysis
- nanocrystalline powder
- tantalum
- X-ray diffraction

## REFERENCES

1. Zelikman, A. N., & Meerson, G. A. The metallurgy of rare metals. M. Metallurgy, 1973, 608 p. (in Russian).
2. Zelikman, A. N. The metallurgy of high-melting rare metals. M. Metallurgy, 1986, 440 p. (p. 259–262) (in Russian).
3. Starostin, C. P., Leontiev, L. I., Kostilev, V. A., Lisin, V. L., Zakharov, P. G., & Petrova, C. A. Microstructure and functional properties of tantalum capacitor anodes of new generation. http://butlerov.com/readings/ date 15.09.2014.
4. Orlov, V. M., Kolosov, V. N., Prokhorova, T. Yu., & Miroshnichenko, M. N. The investigation on technology of highly capacitive tantalum powders. Non-ferrous metals 2011, №11. (in Russian).
5. Pat. 6238456-B1 US, B 22 F 3/12; C 22 C 1/06. Publ. May. 29. 2001.
6. Pat. 4149876-B1 US, B 22 F 9/00. Publ. Apr. 17. 1976.
7. Pat. 5605561-B1 US, B 22 F 9/20. Publ. Feb. 25. 1997.
8. Orlov, B. M., Runtgenen, T. I., & Altukhov, V. G. The influence of heat treatment on propertis of tantalum powders with high specific surface. Physics and Chemistry of Material Treatment. 1999. №2, pp. 73–74 (in Russian).
9. Beliyaev, K. Yu., Orlov, V. M., Prokhorova, T. Yu., & Miroshnichenko, M. N. The influence of melt content on properties of sodiumthermic tantalum powders. Melts. 1998, № 5, pp. 67–72. (in Russian)
10. Hubbard, C. R., McMurdie, H. F., Mighell, A. D., & Wong-Ng, W. JCPDS-ICDD Research Associateship (Cooperative Program with NBS/NIST). J. Res. NIST. 2001, T. 106, C. 1013–1028.
11. Chemycal Encyclopeadia M.: Soviet Encyclopeadia, 1995, V. 4, pp. 494 (in Russian).
12. Semenchenko, V. K. Surface phenomena in metals and alloys, 18. GITTL, M.,, 1957. (in Russian).

## CHAPTER 6

# PHASE TRANSFORMATIONS IN NANOSTRUCTURED STEEL 08H15N5D2T AT HEAT

T. M. MAKHNEVA

*Institute of Mechanics of the Ural Branch of the Russian Academy of Sciences (IM UB RAS), Izhevsk, Russia*

## CONTENTS

### 6.1   INTRODUCTION

During the last two decades, significant successes are observed in the field of intensive action on a material using different methods (laser, irradiation, intensive plastic deformation (IPD), ion implantation, etc.). Among them, a considerable attention is given to the IPD in the chapter. In this connection, the investigation of phase transformations caused both by the change of composition and by external actions is very important. The problem of the IPD influence on phase transformations at heating in closed space is discussed in the present paper in respect to chromium-nickel steel.

High-strength stainless steel VNC-2 (08H15N5D2T) belongs to the class of martensitic low-carbon steels hardened by aging by dispersion copper-hardening in the temperature range of 400°–500°C. After standard thermostrengthening treatment (quenching and aging), in the steel structure the presence of austenite (γ-phase) is observed, the amount of which is determined by the content of austenite-forming elements in the melt [1], the ratio of titanium and carbon (Ti/C) and the temperature of aging [2]. The temperature dependence of $\gamma = f(t_{aging})$ at 3 hour aging has the shape of a curve with a minimum at temperatures of maximal hardening (430°–450°C) and correlates with the precipitation of a strengthening ε-phase [3, 4]. The position of the minimum $\gamma = f(t_{aging})$ depends on the Ti content in the melt and the initial state of steel [3]. It is of interest to investigate the formation of the structure and phase composition of the 08X15N5D2T steel after heat treatment in a CS, which is subjected to IPD with the help of cold rolling by 300%. In the present paper, the amount and stability of retained austenite after quenching in the CS has been determined, the kinetics of the martensite and austenite decomposition at heating in the temperature range of aging, the variation of the lattice spacing of phases and the content of nitrogen in the steel have been studied.

## 6.2   MATERIAL AND METHODS

The material for the investigation was a 25–30 mμ thick foil prepared by cold rolling from forge strips of commercial grade steel with the composition (mass%): 14.46 Cr; 5.34 Ni; 0.02 Ti; 0.09 C; 0.03 N; 1.86 Cu; 0.63 Si; 0.32 Mn. Specimens were cut out from the foil and placed into sealed quartz vessels evacuated to $2.6 \times 10^{-4}$ Pa. The treatment conditions were as follows: quenching from 1000°C (20 min) in the air; aging in vacuum $1.3 \times 10^{-3}$ Pa at the temperatures of 375° , 475°C (0.25–100 h). The methods used were X-ray phase analysis, the magnetic method, Mössbauer effect spectroscopy and the physicochemical method.

## 6.3   RESULTS AND DISCUSSIONS

Intensive plastic deformation (with the use of cold rolling by 300%) forms a fine-grained structure of α-martensite with the grain size 9÷10 nm in the steel under study. After quenching from 1000°C, there are two phases in the steel structure, namely α-martensite and retained austenite ($\gamma_{retained}$). The amount of $\gamma_{retained}$ is ~ 30%; it is three times larger than in massive specimens at similar quenching temperature without a CS. High level of the austenite stabilization can be due to the incompleteness of the process of the return and retention of submicrostructural irregularities in the foil structure, which appear at the IPD stage [5], and the fine grain (9÷10 nm).

At studying phase transitions, it has been established that in the cold-rolled steel VNC2 "reverse" austenite is formed during both the isothermal holding in the temperature range of 375°–475°C and the processes of tempering and aging; the amount of the austenite varies through the maximum depending on the temperature and time of holding (Figure 6.1, *a-curves 1 and 2*): at 1 h holding at 475°C, ~40% of austenite is formed, which does

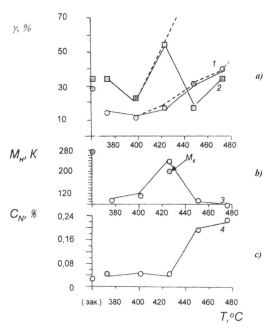

**FIGURE 6.1**   The aging temperature influence on: a – the austenite amount (γ); b – the martensite point ($M_{begin}$) and c – the nitrogen content ($C_N$) in the cold-rolled steel 08H15N5D2T. The points on the ordinate axis are the values after quenching from 1000°C (20 min) in CS. *1* – aging duration is 1 h; *2, 3, 4* – 3 h; *1, 2* are Mössbauer effect spectroscopy data; $M_{begin}$ and $M_{end}$ are the temperatures of the beginning and the end of the (γ→α) transformation.

not undergo martensite transformation ($\gamma\rightarrow\alpha$) at cooling and remains completely stable to room temperature; at 3 h holding, the maximum displaces to the temperature 475°C reaching the values of $\gamma \sim 55\%$ (*curve 2*).

Judging by the martensite point $M_{begin}$ in the temperature range of 375°–400°C (Figure 6.1, *b-curve 3*), the formation of the "reverse" austenite starts at temperatures below 400 °C and is accompanied by its sharp decrease from 280 K at quenching in CS down to 100 and 120 K at 375° and 400°C, respectively. At increasing temperature of aging and the growth of the amount of the "reverse" austenite, the degree of alloying changes, and up to 425°C the stability is decreasing; $M_{begin}$ increases to 240 K, but does not reach the initial values in the quenched condition (Figure 6.1, *b*). At further increase of the aging temperature, the formed austenite (dashed lines) undergoes partly the ($\gamma\rightarrow\alpha$) transformation at cooling. As is seen from the variation of $M_{begin}$, the austenite retained after cooling (solid lines) is more stable when the secondary ($\gamma\rightarrow\alpha$) transformation is more complete and the content of nitrogen is higher, the adsorption capability of which increases several times at the above temperatures [6] (Figure 6.1, *c*).

The investigation of the ($\gamma\rightarrow\alpha$) transformation kinetics at heating of the hardened steel under study allows to find the conditions of the most complete decomposition of $\gamma_{ret}$; at the aging temperatures 400° – 425 °C this process is ended after 1 h holding without reaching the completion (Figure 6.2, *a* – the descending part of *curves 1 and 2*), and at the aging temperature 475 °C, the $\gamma_{ret}$ has almost completely decomposed by the 15th min (the beginning of *curve 3*). The formation of the "reverse" austenite in the aging temperature range obeys the regularities of the ($\gamma\rightarrow\alpha$) transformation at slow heating or in the isothermal conditions as is seen from the kinetic curves (Figure 6.2, *a*) [4, 7].

The analysis of possible reasons promoting the decrease of the critical points $A_{begin}$ and $A_{end}$ of the ($\alpha\rightarrow\gamma$) transformation and, as a consequence, the formation of "reverse" austenite at the aging temperatures shows that the most probable reasons are high rate of heating [8, 9], fine grains [10], the presence non-decomposed austenite of quenching [11] and the chemical inhomogeneity of the $\alpha$-matrix [12, 13].

The probability of the existence of concentrated inhomogeneities in Fe-Cr alloys is confirmed by the electronic structure calculation performed

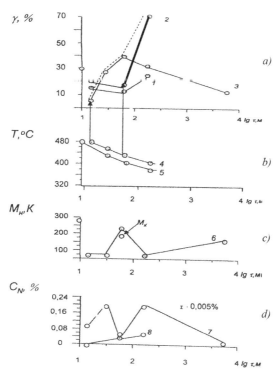

**FIGURE 6.2**   The influence of the duration of aging at the temperatures 400°C (*1*), 425°C (*2*) and 475°C (*3*) on the phase composition (a), the beginning of the formation of "reverse" austenite (*4b*), chromium regions (*5 b*), the martensite point (*6 c*) and the content of nitrogen after aging at 475°C (*7 d*) and 425°C (*8 d*) in the cold-rolled steel 08H15N5D2T. The points on the ordinate axis are the values after quenching in closed space.

by two complementary methods, namely, the method of recursion in straight space and the band TB-LMTO-ASA formalism [14]. A localized state of d-electrons in the chromium concentration range of 20 ÷ 30% has been obtained, which is in good agreement with the experimental data and indicates the probability of the existence of the chromium-enriched regions. It has experimentally been established [12] that the Fe-Cr-Ni matrix layering in the cold-rolled steel takes place in the temperature range of 375°–475°C with the formation of the chromium concentration inhomogeneities (chromium regions). The size of such inhomogeneities after 100 and 1000 h can be estimated as being within 5 nm. The kinetics of the chromium regions formation in the steel under study has the shape of the C-curve (Figure 6.2, *b-curve 5*) and coincides with the temperature range in which the copper precipitation (400°–475°C) and the "reverse" austenite formation (Figure 6.2, *b-curve 4*) take place. As it follows from the comparison of curves 4 and 5, the stage of the precipitation of the chromium-enriched regions precedes the isothermal formation of austenite at aging. In addition, it is known [15] that 1% of Cr decreases the $A_{begin}$ by 17 degrees and the stronger the regions are enriched with chromium, the lower the $A_{begin}$ is. Consequently, the decrease of the $A_{begin}$ critical point from 580°C, which is characteristic of the studied steel, to 400°C can be due to the chromium-inhomogeneity of the solid solution. If we take into account that the enrichment of regions with chromium is possible up to 40% as it was shown by the authors of [16], the formation of the "reverse" austenite at the aging temperatures becomes understandable.

The phase composition and lattice spacing of the α and γ phases of the foils investigated after various aging regimes were determined by the X-ray method. The amount of retained austenite reaches 70% at 3 h holding at the temperature of maximal hardening (425°C) (Figure 6.3, *a*).

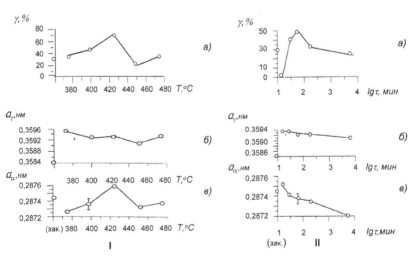

**FIGURE 6.3** The amount of austenite (a) and the variation of the lattice spacing of the γ and α phases (*b* and *c*, respectively) after aging of the cold-rolled steel 08H15N5D2T. I – the duration of the isothermal holding at aging for 3 h. II – aging at the temperature 475 °C. The x-ray data. The points on the ordinate axis are the values after quenching in closed space.

The nickel-enrichment of the "reverse" austenite, which is one of the reasons for its stabilization in the two-phase ($\alpha+\gamma$)-region, is confirmed by the decrease of the crystal lattice spacing of the austenite with the growth of its amount (Figure 6.3.I and II, *a* and *b*) and also by the increase of the austenite resistance to overcooling (Figure 6.1, *b* and Figure 6.2, *c* – after 15 and 30 min of holding at 475°C). The determination of the critical points at the heating rate 100°C/min by the magnetic method confirms the formation of the $\gamma$-phase at the aging temperatures: at 1 h holding, the $\gamma$-phase is formed at 400°C.

The determination of the lattice spacing of the $\alpha$ and $\gamma$ phases has allowed to reveal a number of peculiar features in the character of the crystalline structure of the cold-rolled foils:

- the values of the lattice spacing of both the retained austenite ($a_\gamma$) and the martensite ($a_\alpha$) after quenching of the cold-rolled foils in CS (Figure 6.3.I-II, *b, c* – the values on the ordinate axis) are much higher in comparison with the similar parameters obtained for massive specimens [17];
- the "reverse" austenite lattice spacing in the cold-rolled foils is always higher than the lattice spacing of the austenite of quenching ($a_\gamma = 3.5844$ Å);
- the character of the kinetic curves of the variation of the crystal lattice spacing of the $\gamma$- and $\alpha$-phases shown in Figure 6.3.II, *b* and *c* for the foils corresponds to the character of the similar curves at the temperatures 575°–600°C at which the "reverse" ($\alpha \rightarrow \gamma$) transformation starts and develops in time [17].

The use of closed space in order to decrease the oxidation of the surface during the heat treatment of the foils allows to keep the surface color light; however, it does not preclude the interaction of residual gases ($O_2$, $N_2$) with the foil surface, i.e., the variation of its phase composition, the absorption of nitrogen into the $\gamma$-phase at quenching [18] and into the chromium-alloyed regions at aging temperatures above 400°C [6] (Figure 6.2, *d*). Since the formation of the regions is a primary process (Figure 6.2, *b*, *curve 5*), the interaction of nitrogen with the chromium regions at aging is the most probable process.

Thus, the intensive plastic deformation generates the structure with the blocks of 9–30 nm in size, increases the amount of $\gamma_{retained}$ by a factor of three after quenching and the lattice spacing of the γ- and α-phases, accelerates the process of the α-matrix chromium-decomposition at aging, and hindering the retained austenite decomposition up to the temperature 475°C, promotes the formation of the chemical inhomogeneity at aging temperatures. Due to the above-mentioned, the intensive plastic deformation decreases the critical temperatures of the (α→γ) transformation and leads to the formation of the "reverse" austenite in the structure. Nickel, air nitrogen and mechanical work-hardening from cold plastic deformation provides the desired level of austenite stabilization and overcooling resistance.

## KEYWORDS

- aging
- austenite
- closed space
- cold rolling
- intensive plastic deformation
- lattice spacing
- low-carbon stainless steel
- martensite transformation
- quenching

## REFERENCES

1. Zvigintsev, N. V., Lepekhina, L. I., Mikhailov, S. B., Mikhailova, N. A., & Gapeka, T. M. On the austenite stabilization in steel 08X15N5D2T. Trudy Vuz. RF, Heat treatment and physics of metals. Sbornik: Sverdlovsk. 1978, V. 4, pp. 56–62.
2. Makhneva, T. M., & Makhnev, Ye. S. The influence of titanium on the brittle fracture resistance of maraging steel 08H15N5D2T. MiTOM. 1990, №8, pp. 40–43.
3. Makhneva, T. M. Embrittlement of corrosion-resistant maraging steel 08H15N5D2T at heat treatment. Kand. diss. Izhevsk, 1990, 156 p.
4. Potak, Ya. M. High-strength steels. M.: Metallurgia, 1972, 208 p.
5. Blanter, M. Ye., Garbuzova, N. Ye., et al. Mechanism of disordering of cold-hardened iron at fast heating. MiTOM. 1965. №4, pp. 22–26.

6. Leroy, V., Grass, H., Emond, C., & Habraken, L. Memoirs Scientifiques de la Revue de Metallurgia. 1976, T.73, №10, pp. 599–609.

7. Makhnev, Ye. S., & Makhneva, T. M. Reverse martensite transformation in steel VNC-2YSh. Martensite transformations in metals and alloys. Sb. dokl. Kiev: Naukova Dumka, 1979, pp. 180–184.

8. Zeldovich, V. I., Pinkevich, O. S., & Sadovsky, V. D. Structural and concentration changes at ($\alpha \rightarrow \gamma$)-transformation in steel Fe-23.1%Ni. The influence of the heating rate and retained austenite FMM. 1979, Vol. 47, Vyp. 6, pp. 1201–1212.

9. Makhnev, Ye. S., & Makhneva, T. M. The influence of the heating rate on the temperature of the austenization and phase composition of steel VNC-2USh (08H15N5D2T). M: VILS. Technology of light alloys. 1973, №7, pp. 53–56.

10. Kardonsky, V. M. Austenite stabilization at the reverse ($\alpha \rightarrow \gamma$)-transformation. FMM. 1975, Vol. 40, pp. 1008–1012.

11. Shneiderman, A. Sh. Evaluation of the retained austenite role at the $\alpha \rightarrow \gamma$ transformation in steels. FMM. 1980, Vol. 50. Vyp. 3, pp. 574–582.

12. Makhneva, T. M., Yelsukov, Ye. P., & Voronina, E. V. Kinetics of layering and the phase composition at aging of cold-rolled foils from the H15 alloy and the 08H15N5D2T. FMM. 1991, №5, p. 130.

13. Nizhnik, S. B., Doroshenko, S. P., & Usikova, G. I. The quenching temperature influence on the development of the ($\alpha \rightarrow \gamma$)-transformation and the mechanical properties of the maraging steel. FMM. 1983, Vol. 56, Vyp. 2, pp. 324–333.

14. Mitrokhin, Yu. S., & Makhneva, T. M. The electronic structure and magnetic properties of the Fe-Cr system alloys. "Uspehi fiziki metallov." NAN of the Ukraine. G.V. Kurdumov IMF. 2001, Vol. 2. №2, pp. 109–129.

15. Perkas, M. D. High-strength maraging steels. M.: Metallurgia. 1970. 224 p.

16. Bashkirov, Sh. Sh., Ivanov, N. G., Kurbatov, G. D., Makhnev, Ye. S., et al. The application of Mössbauer effect for the investigation of the inhomogeneity of the chemical and phase composition in the Fe-Cr based alloys. Sb.: Paramagnetic resonance. Kazan. 1980, Vol. 16, pp. 52–71.

17. Roshina, I. N., & Kozlovskaya, V. I. The reverse transformation in steel VNC-2. FMM. 1971, Vol. 31, Vyp. 3, pp. 589–594.

18. Goncharova, N. V. Simulation of phase transformations in Fe-Cr-based alloys at high temperatures. Aftoref. kand. diss. fiz.-math. Nauk. Izhevsk, 2000. 20 p.

# CHAPTER 7

# FORMATION OF THE MATERIAL STRUCTURE FROM THE NANO-SIZED TIO$_2$ POWDER DURING SINTERING PROCESS

A. A. GUROV, S. E. POROZOVA, and A. A. SMETKIN

*Perm National Research Polytechnic University, Perm, Russia*

## CONTENTS

### 7.1 INTRODUCTION

TiO$_2$ powders are widely used for gas sensors, dielectric ceramics, dyes, etc. The increased interest in recent years is the high photocatalytic activity of TiO$_2$, which allows to implement chemical processes using solar energy [1, 2]. Industrial powders are a mixture of various fractions from submicron powders to larger particles. We know a considerable amount of work on the preparation of nano TiO$_2$ [3–5]. However, for various structural solutions are more suitable not a powders but bulk materials, which sintering from nanopowders are not well investigated.

Purpose of work – study the formation of the structure of materials during sintering of the samples obtained by unilateral dry pressing of nano-sized powders of $TiO_2$.

## 7.2   EXPERIMENTAL RESEARCHES

$TiO_2$ powder with a particle size of 25–35 nm (data obtained by means of thermal desorption of nitrogen and scanning electron microscopy) prepared according to the author's technique of aqueous ethanol solutions with polymeric additives [5]. The phase composition of the powder after heat treatment of coagulates represented only by the low-temperature modification anatase.

Previously, compaction $ZrO_2$ nanopowders has been shown that the mechanochemical activation in aqueous medium with additions of surfactants can be substantially improved compactability submicron powder and intensify the sintering process [6]. The powder is activated for 0.5 hours in a planetary mill "SAND" in drums with chalcedony grinding balls at a speed of 160 $min^{-1}$. Activation was carried out in an aqueous medium (at a weight ratio of balls:powder:water = 2:1:1) supplemented with 0.5% (wt.) of agar-agar in the form of a premixed aqueous solution. Samples were formed by cold uniaxial pressing. Compression was performed in a closed mold at a pressure of 200 MPa. The molded samples were annealed and sintered in air at 1000–1400°C with isothermal exposure for 2 hours.

The kinetics of the sintering of the pressed samples were examined by thermomechanical analysis apparatus Sentsys Evolution (France).

Phase composition was determined by Raman spectroscopy in the Fourier spectrometer Senterra (Bruker, Germany) at a wavelength of 532 nm emitting laser.

The microstructure of the sintered samples was studied by the scanning electron microscope Ultra 55 (Carl Zeiss, Germany) at microsections subjected to high-temperature etching and fractures.

## 7.3   RESULTS AND DISCUSSION

Figure 7.1 shows the dependence of the shrinkage of the titanium dioxide sample temperature. Start of shrinkage at low temperatures due to the

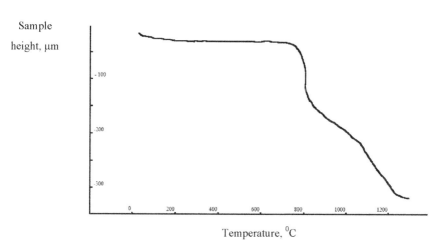

FIGURE 7.1    Dependence on temperature shrinkage of TiO$_2$.

fact that there was a transition of anatase to rutile. Next there is a curve bend, the rate of shrinkage temperature up to 1100°C unchangeable, then observed its increase up to 1250°C. Ending the process of changing the size of the sample observed at 1300°C.

The phase composition of samples by Raman spectroscopy showed that for all material processing temperatures are well crystallized and the main phase is rutile. Figure 7.2 shows the Raman spectrum of the sample after heat treatment at 1000°C.

The material is a mixture of anatase and rutile, anatase content but is very small and is fixed only to the most intense peak Eg at 145 cm$^{-1}$. This peak, in spite of the low intensity, well defined, and is present on all Raman spectra of sintered samples up to 1400°C virtually unchanged. The presence of the anatase found not only on the surface of the sintered samples, but also to fractures, indicating the presence of conditions for maintaining the low temperature phase in the samples subjected to high temperature treatment.

Thus, samples of TiO$_2$ nanopowder are polyphase even after sintering. Figure 7.3 shows SEM images of fracture specimens after sintering at 1100–1400°C.

Samples consist of densely packed agglomerates of particles. The agglomerates formed during synthesis and mechano-chemical activation

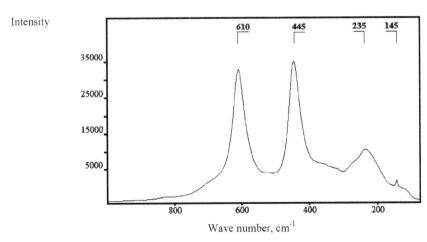

**FIGURE 7.2**   Raman spectrum of sintered $TiO_2$ ($T_{sint}$ = 1000°C).

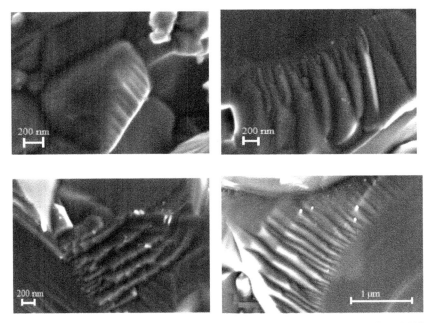

**FIGURE 7.3**   SEM images of fracture specimens: (a) 1100°C, zoom 30,000; (b) 1200°C, zoom 25,000; (c) 1300°C, zoom 25,000; (d) 1400°C, zoom 25,000.

of the powder layers formed nanoparticles arranged in lamellar structures. Formation of agglomerate took place in the presence of a surfactant, and styling particles resembles a layered structure of many natural polymer-containing composite materials such as pearl. The layered structure is well identified on microsections samples (Figure 7.4). The first stages of degradation of such a structure at 1400°C are fixed on microsection.

On the SEM pictures it was possible to determine the dimension and layered structures. Figure 7.5 shows at magnifications 40,000–50,000 SEM image of fracture patterns of samples sintered at 1200–1300°C samples layered structure is best revealed.

The distance between the plates and the plate thickness on average 75–80 nm (Figure 7.5). For demonstration purposes, the figure shows only

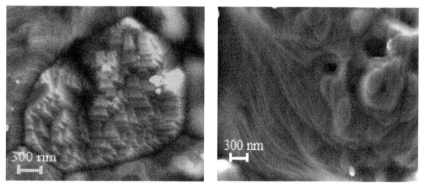

**FIGURE 7.4** SEM image microsections samples: (a) 1200°C, zoom 20,000; (b) 1400°C, zoom 20,000.

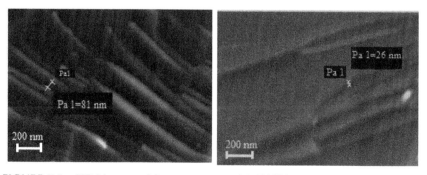

**FIGURE 7.5** SEM images of fracture patterns: (a) 1300°C, zoom 40,000; (b) 1200°C, zoom 50,000.

one dimension. The width of the individual "cores" of which plate consists of 25–40 nm (Figure 7.5b), which corresponds to an average particle size of the precursor powder. This structure prevents the diffusion processes between the layers inside the agglomerates. There are sintering agglomerate surfaces with a gradual "grasping" interior layered areas. The layers are parallel to each other within the agglomerates, but are disordered in the sample as a whole. Areas of the compacted plates are formed in "pseudograin" (Figure 7.4a). In recent boundaries, especially at the joints, marked formation of pores. Similar layered structure formed as a result of the gel casting nanopowder $TiO_2$, when the formation of the structure of the raw billet occurs in a relatively free packing of the nanoparticles. [7].

## 7.4  CONCLUSION

After heat treatment at 1000–1400°C material according to Raman spectroscopy is polyphasic and contains, along with a small amount of rutile anatase.

During sintering of the samples obtained by unilateral dry pressing from a nanopowders $TiO_2$, it forms lamellar structures that are well identified as fractures and microsections material. The distance between the layers and the layer thickness on average 75–80 nm. The width of the individual "cores" that make up the layers of 25–40 nm, which corresponds to the average primary particle size of the powder.

This structure prevents the diffusion processes between the layers, but a prerequisite for obtaining composite laminates. When the sintering temperature is 1400°C marked initial stages of degradation of the layered structure.

## KEYWORDS

- layered structure
- nanosized powders
- sintering
- Titan dioxide

## REFERENCES

1. Sedneva, T. A., & Lokshin, E. P. i dr. Fotokataliticheskaya aktivnost modificirovannogo volframom dioksida titana. DAN. 2012, Vol. 443, no. 2, pp. 195–197 (in Russian).
2. Titania nanotubes go commercial. Potential use in fuel cells, solar panels. URL: http://phys.org/news/2012–11-titania-nanotubes-commercial-potential-fuel.html
3. Masato Kakihana, Makoto Kobayashi, Koji Tomita, Valery Petrykin. Application of Water-Soluble Titanium Complexes as Precursors for Synthesis of Titanium-Containing Oxides via Aqueous Solution Processes. Bull. Chemical Society of Japan. 2010, Vol. 83, No. 11, pp. 1285–1308.
4. Ismagilov, Z. R. Cikoza, L. T. & Shikina, N. V. Sintez i stabilizaciya nanorazmernogo dioksida titana. i dr. Uspexi ximii. 2009, Vol. 78, no. 9, pp. 942–955.
5. Gurov, A. A., Karmanov, V. I., Porozova, S. E. & Shokov, V. O. Sintez i svojstva nanoporoshka dioksida titana dlya polucheniya funkcionalnyx materialov. Vestnik PNIPU. Mashinostroenie, materialovedenie. Vol. 2014, no. 1, pp. 23–29 (in Russian).
6. Ziganshin, I. R., Porozova, S. E., Karmanov, V. I., Torsunov, M. F., & Xafizova, R. M. Izmenenie xarakteristik promyshlennogo poroshka dioksida cirkoniya i materialov na ego osnove mexanoximicheskoj aktivaciej. Izv. vuzov. Poroshkovaya metallurgiya i funkcionalnye pokrytiya. 2009, no. 4, pp. 11–15 (in Russian).
7. Mishhinov B.V., & Porozova S.E. Formirovanie struktury materiala v processe gelevogo litya nanoporoshka dioksida titana. Vestnik PNIPU. Mashinostroenie, materialovedenie. 2014, no. 3, pp. 37–42 (in Russian).

# CHAPTER 8

# THE DEVELOPMENT OF A NEW NANOSCALE KIT ON ALUMINA NANOPOWDER-BASED FOR GENERATOR OF TECHNETIUM-99M AND THE STUDY ON ITS FUNCTIONAL SUITABILITY FOR DIAGNOSIS IN ONCOLOGY

V. SCURIDIN, N. VARLAMOVA, E. STASUK, E. NESTEROV, L. LARIONOVA, R. ZELCHAN, and E. ILINA

*National Research Tomsk Polytechnic University, Tomsk, Russia*

## CONTENTS

## 8.1   INTRODUCTION

Sentinel lymph nodes (SLN) are the first lymph nodes in the path of the outflow of lymph from cancer tumor. It is believed that if SLN are not impressed by metastatic process, all other regional lymph nodes remain intact [1, 2]. The use of radioactive nano-colloids in oncology based on the ability to quickly and effectively identify SLN. The widespread introduction of technology to identify SLN would improve the quality of life of 100,000 of Russians, whom annually not for indications perform mutilation operation in occasion of cancer pathology [1, 3–5]. In world practice, has accumulated considerable experience with nuclear imaging SLN for melanoma and breast cancer. When tumors at other sites the effectiveness of this technique is studied in scientific research [1, 2, 4, 6, 7].

The aim of our studies is the study of the adsorption of technetium-99m alumina nano-sized structures for nano-colloid radiopharmaceutical preparation used in medical diagnosis of cancer.

## 8.2   EXPERIMENTAL RESEARCHES

The best method for identifying SLN is use labeled with technetium-99m colloidal nanomaterials for radiometric or scintigraphic localization of lymph node [6]. Wherein the determining factor in selecting of indicator is the size of radioactive particles. Thus, according to Schauer et al. [5], a colloid particle with size less than 50 nm can accumulate not only in SLN, but also in subsequent nodes. Particles greater than 100 nm slowly migrate from the injection site. Optimal for identifying SLN was recognized colloid particle size from 50 to 80 nm. To date, all known nano-colloid preparations are made on the basis of compounds that form stable hydrosols. At the same time as the starting compounds for their preparation are often chosen organic substances of different structures, the complexity of the synthesis of which largely determines the high cost of such products. Another part of the radiopharmaceutical preparation is inorganic $^{99m}Tc$ complexes with sulfides of rhenium and antimony, which are also obtained by sufficiently sophisticated technology that prevents their widespread use in practical medicine.

Our studies have shown that stable colloidal compounds can be prepared in a simpler way – restored $^{99m}Tc$ adsorption on gamma-alumina [8, 9]. The quantity of adsorption of the radionuclide on the surface of the oxide exceeds 80%. The main prerequisites for use of nanoscale oxide powders of gamma-$Al_2O_3$ as "carrier" tag of $^{99m}Tc$ is its lower toxicity compared with sulfides of rhenium and antimony in combination with good absorption properties, availability, and low cost. A feature of this compound is an organic coating of nanoparticles. During passage through the lymphatic paths nanoparticles lose their organic coating firmly locking in SLN without redistribution in the body.

Thus, the aim of this study was to examine the possibility of an experimental application of new national radiopharmaceutical preparation based on technetium-99m-labeled gamma-alumina for visualization of SLN.

For the preparation of the lyophilizate in sterile 10 mL vials was added 1 mL nano-colloid solution, 0.01 mL of a solution of stannous chloride at a concentration of 7 mg/mL, 0.1 mL of a solution of ascorbic acid in a concentration of 10 mg/mL, and 10 mL of 0.075–0.1% solution of gelatin. The contents of the vials were frozen at liquid nitrogen temperature and was placed in a sublimator. Freeze-drying was performed at preset parameters lyophilizer for 24 hours, followed by subsequent drying for 4.5 hours at 15°C.

The size of nanoparticles was determined by NANOFOX (SYMPATEC GmbH). Also, for express determination of the proportion in the preparation of particles with different sizes were used membrane filters with pore diameter of 25 to 220 nm.

To prepare the radiopharmaceutical preparation under aseptic conditions to the vial of reagent injected with a syringe 4 mL of eluate from the generator of technetium-99m total activity 1120–2000 MBq. The contents of the vial was stirred by shaking to completely dissolve the lyophilisate and incubated at 70–80°C temperature for 30 min. After cooling to room temperature the drug was withdrawn from the vial with a syringe with a sterilizing filter (0.22 micron).

The pharmacokinetics of radiopharmaceuticals studied on white rats, male Wistar. The indicator was injected subcutaneously in the I interdigital interval of right foreleg at a dose of 30 MBq. Before a single subcutaneous injection of radiopharmaceutical animals were anesthetized with ether. The volume of the dose administered was 0.1 mL (volumetric activity 300 MBq/mL). After 1, 3, 5, 10, 15, 30, 45, 60, 75, 90, 105, 120, 150, 180 minutes and 24 hours after injection, the animals were decapitated in groups of 5 animals for each time interval. The recovered bodies were packed into bottles for weighing and direct radiometry. Standard radiopharmaceutical preparation and bioassays have the same volume (after weighing bottles filled to the same level with distilled water) and the geometric shape [10]. At radiometric study of differential discriminator tuned to photopeak 140 keV, with 20% of the window width. According to the results of radiometry determined the level of accumulation of radiotracer in axillary lymph node and the injection site.

The content and participate in the experiment, the animals were performed in accordance with the rules adopted by the "European Convention for the Protection of Vertebrate Animals used for Experimental and other Scientific Purposes."

## 8.3  RESULTS AND DISCUSSION

Radiometry of rats showed that the radiopharmaceuticals based on technetium-99m-labeled gamma-alumina after subcutaneous injection resorbed actively – after 1 hour it was about 2/3 of injected activity. After 24 hours in the subcutaneous depot remains about half the dose. Leaving the injection site, the radiopharmaceutical accumulated in axillary lymph node – 15 minutes after the injection of the indicator medium accumulation therein was 1.19%. The first hour of the study the average radiopharmaceuticals accumulation in the lymph node has reached 8.6% and gradually increased to 12% for 24 hours of observation. Received through the thoracic duct into the blood the radiopharmaceutical was very active captured by the kidneys and excreted in the urine. The kidney accumulation level indicator increased from 2.090% 10 minutes after the injection to 4.182% to almost 30 minutes, and was maintained at this level until 24 hours of study. The liver and spleen accumulation value of the radiopharmaceutical preparation was gradually increased and reached 1.790% and 2.180%, respectively, to 24 hour experiment. It should be noted a low content of radiopharmaceuticals in the heart, lungs and blood, which was registered less than 1% of the injected radioactivity.

In oncology practice the injection of radiopharmaceuticals often located in close proximity to the sentinel lymph node, making it difficult to visualize. Given this fact, it is important to estimate the ratio of accumulation of radiopharmaceuticals based on technetium-99m-labeled gamma-alumina in axillary lymph node and the injection site. In the experiment, the rate is gradually increased from 1.669 to 26.104 15 minutes to one day after the start of the study.

## 8.4   CONCLUSION

Thus, under selected conditions of synthesis is mainly formed of nano-colloid $^{99m}$Tc-Al$_2$O$_3$ with a desired particle size from 50 to 100 nm Experimental study of the pharmacokinetics of radiopharmaceuticals based on technetium-99m-labeled gamma-alumina showed that investigated radiopharmaceutical preparation can be successfully used for lymphoscintigraphy and sentinel node imaging with 15 minutes and 24 hours after subcutaneous administration. At the level of gamma-radiation protection probes, which exceeds 0.1%, the specified ratio (1.5–6.5%) is quite acceptable for efficient intraoperative detection of sentinel lymph nodes using investigational radiopharmaceuticals.

## KEYWORDS

- **gamma-alumina**
- **generator of technetium-99m**
- **nanocolloid**
- **radiopharmaceutical drug**
- **reagent.**
- **sentinel lymph node**

## REFERENCES

1. Chernov, V. I., Sinilkin, I. G., & Shirjaev, S. V. Radionuklidnoe vyjavlenie storozhevyh limfaticheskih uzlov v: Pod red. Ju.B.Lishmanova, V.I.Chernova. Nacional'noe rukovodstvo po radionuklidnoj diagnostike (Radionuclide identification of sentinel lymph nodes in: Ed. Yu.B. Lishmanova, V.I. Chernova. National guidelines for nuclear medicine), Publ.STT, Tomsk, 2010, pp. 336–343 [in Russian].
2. Paredes, P., Vidal-Sicart, S., Zanón, G. et al. Clinical relevance of sentinel lymph node in the internal mammary chain in Breast cancer patients. E.J. Nucl. Med. 2005, 32(11), pp. 1283–1287.
3. Kanaev, S. V., Novikov, S. N., Zhukova, L. A., Zotova, O. V., Semiglazov, V. F., & Krivorot'ko, P. V. Ispol'zovanie dannyh radionuklidnoj vizualizacii individual'nyh putej limfoottoka ot novoobrazovanij molochnoj zhelezy dlja planirovanija

luchevoj terapii [Using data radionuclide imaging of individual pathways of lymph from the breast tumors for radiation therapy planning]. Voprosy onkologii, 2011, 57(5), pp. 616–621 [in Russian].

4. Chernov, V. I., Afanas'ev, S. G., & Sinilkin, A. A. Radionuklidnye metody issledovanija v vyjavlenii "storozhevyh" limfaticheskih uzlov [Radionuclide methods of investigation to identify "sentinel" lymph nodes]. Sibirskij onkologicheskij zhurnal, 2008, 28(4), pp. 5–10 [in Russian].

5. Schauer, A. J. The Sentinel Lymph Node Concept.- Berlin Heidelberg New York: Springer, 2005, p. 565.

6. Maza, S. Peritumoral versus subareolar administration of technetium-99m nanocolloid for sentinel lymph node detection in Breast cancer: preliminary results of a prospective intra-individual comparative study. QJ Nucl Med. 2003, Vol. 30, pp. 651–688.

7. Afanas'ev, S. G., Avgustinovich, A. V., Chernov, V. I., & Sinilkin, I. G. Vozmozhnost' opredelenija storozhevyh uzlov u bol'nyh rakom zheludka [Possible to determine the guard nodes in patients with gastric cancer]. Sibirskij onkologicheskij zhurnal, 2009, 34(4), pp. 27–32 [in Russian].

8. Skuridin, V. S., Stasjuk, E. S., Varlamova, V. N., Rogov, A. S., Sadkin, V. L., & Nesterov, E. A. Poluchenie novogo nanokollodnogo radiofarmpreparata na osnove oksida aljuminija [Getting a new nanokollodnogo radiopharmaceutical based on aluminum oxide]. Izvestija TPU. Himija, 2013, no. 323(3), pp. 33–37 [in Russian].

9. Skuridin, V. S., Stasjuk, E. S., Varlamova, N. V., Postnikov, P. S., Nesterov, E. A., & Sadkin, V. L. Poluchenie i jeksperimental'nye ispytanija mechennyh tehneciem-99m nanokolloidnyh preparatov na osnove gamma-oksida aljuminija i magnitoupravljaemyh chastic Fe@C(IDA) [Preparation and experimental tests labeled with technetium-99m nanokolloidnyh drugs based on gamma-alumina particles and magnetically Fe @ C (IDA)]. Izvestija VUZov, Fizika, 2011, no. 54(11), pp. 332–339 [in Russian].

10. Rukovodstvo po jeksperimental'nomu (doklinicheskomu) izucheniju novyh farmakologicheskih veshhestv (Manual on experimental (preclinical) study of new pharmacological substances.). Moskva. Pod red. chlen-korr.RAMN, professora R.U. Habrieva. 2005, pp. 729–735 [in Russian].

# CHAPTER 9

# NANO STRUCTURES BY SEVERE PLASTIC DEFORMATION (SPD) PROCESSES

EHSAN BORHANI,[1] BAHRAM AZAD,[2] and HASSAN ABDOOS[3]

[1,3]*Department of Nano Technology, Nano Materials Engineering Group, Semnan University, Semnan, Iran, E-mail: Ehsan. borhani@profs.semnan.ac.ir, Hassan.abdoos@gmail.com*

[2]*Department of Material Science and Engineering, Semnan University, Semnan, Iran*

## CONTENTS

## 9.1   INTRODUCTION

Severe plastic deformation (SPD) processes are used for deforming of bulk metals under certain conditions. SPD is a one of the easy technique to develop bulk nanostructured metals. One of the most differences between SPD processes and conventional metal forming processes is the imposed plastic strain. On the other hand, in conventional forming processes, the imposed plastic strain is generally less than 2.0. In addition, the conventional forming processes are carried out up to a plastic strain greater than 2.0, the thickness or diameter of specimens become very thin and are not suitable to be used for structural applications. However, several SPD techniques have been designed for achieving to high strength with minimal changes in the initial sample dimensions and allow us to impose an extremely large strain on the bulk metal without changing the shape.

Grain refinement of metallic materials has been an important subject in research and development of metals and alloys, but the minimum mean grain size that can be achieved in bulky metallic materials has been around 10 µm. On the other hand, development of SPD processes that can apply a quite large plastic strain (above 4–5 in logarithmic equivalent strain) has made it possible to fabricate ultrafine grained (UFG) bulky metallic materials of which mean grain size is much smaller than 1 µm. Based on this technological progress, fundamental studies of UFG materials have been energetically carried out since 1990s, and a number of new and sometimes surprising findings about interesting structures and properties of UFG metals have been acquired.

Two different approaches have been developed to produce UFG metals, which are referred to as "bottom-up" and "top-down" methods. In the bottom-up approach, nano-particles or individual atoms are assembled to a dense solid material. There are different methods of this approach, like inert gas condensation, electro-deposition, isostatic pressing or hot isostatic pressing. These methods are often limited to the production of small samples that are generally not appropriate for structural applications. Additionally, the final product contains some residual porosity and also contamination, which is introduced during the fabrication procedure. In the top-down approach a coarse-grained bulk materials is processed through heavy straining in order to produce UFG materials. Using this approach one can avoid the small product sizes, the residual porosity and the presence of contamination, which are unavoidable using the bottom-up process.

The previous studies on UFG materials fabricated by SPD, however, have mostly used pure metals. Figure 9.1 illustrates the volume fraction of grain boundary as a function of mean grain size. Here the thickness of the grain boundary region is assumed to be 1 nm. The volume fraction of grain boundaries in the coarse-grained materials with mean grain size larger than 10 μm is expected to be almost 0%. This indicates that most of the conventional metallic materials are "rare of grain boundaries." In contrast, the volume fraction of grain boundaries greatly increases when the grain size becomes smaller than 1 μm.

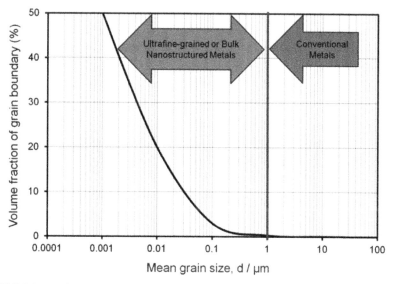

**FIGURE 9.1** The volume fraction of grain boundary as a function of mean grain size.

From another point of view, it is well known that mechanical properties of materials can be improved by grain refinement. So far, however, the minimum grain size of bulky conventional metallic materials that could be achieved is about 10 μm. It is, therefore, expected that UFG materials or nano-structured materials exhibit superior mechanical properties. The yield strength ($\sigma_y$) of polycrystalline metals is related to the grain size by the following Hall-Petch equation:

$$\sigma_y = \sigma_0 + Ad^{-\frac{1}{2}} \tag{1}$$

where $\sigma_0$ is the friction stress, A is a constant and d is the grain diameter.

As the equation shows the yield strength increases with decreasing square root of the grain size. The decrease of the grain size leads to a higher tensile strength without reducing the toughness, which differs from other strengthening methods such as heat treatment.

Valiev et al. pointed out that some UFG or nano-structured materials can manage both high strength and large ductility, though high strength and large ductility are in trade-off relationship in conventional materials. The UFG or nano-structured materials also exhibit peculiar properties that have never been seen in conventional metallic materials having coarse-grained structures. Tsuji et al. reported that yield-drop phenomena with Lüders band deformation appeared during tensile test in pure aluminum when the grain size is smaller than 2 μm. Huang et al. reported a unfamiliar phenomenon of bulk nano-structured Al and Fe, that is, "hardening by annealing and softening by deformation." In general, metallic materials are hardened by deformation and are softened by annealing. The unique properties of the UFG materials prompted materials scientists to study them in more details, in both fundamental and application issues, under the demands for lightweight or high strength materials.

## 9.2  VARIOUS SPD METHODS

Several techniques of SPD have been developed to fabricated UFG materials, such accumulative roll bonding (ARB), cross accumulative roll bonding (CARB), equal channel angular pressing (ECAP), torsion under high pressure (HPT), repetitive extrusion and upsetting (REU) process, tube cyclic extrusion-compression (TCEC) process, repetitive corrugation and straight-ening (RCS) process.

### 9.2.1  ARB PROCESS

ARB process is used to fabricate UFG materials and in comparison to the other methods, the productivity of the ARB process is relatively high. ARB process applies to continuous of bulky sheet materials. ARB process was invented by Saito et al. in 1998. The principle of ARB process is, schematically, shown in Figure 9.2.

ARB process is a SPD process using rolling deformation. Rolling is the most advantageous metal working process for continuous produc-tion of plates, sheets and bars. However, the total reduction applied to the materials is substantially limited in conventional rolling because of the decrease in the thickness of the materials, with increasing the reduction. In ARB process, 50% rolled sheet is cut into two, stacked to the initial dimension after degreasing and wire brushing of the contact surfaces and then rolled again. In order to obtain one-body solid materials, the roll-ing in ARB process is not only a deformation process but also a bonding process (roll bonding). To achieve good bonding, the roll bonding process

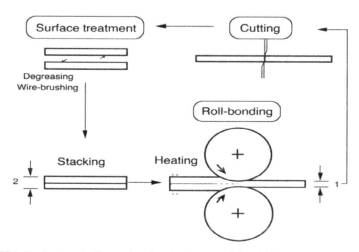

**FIGURE 9.2**   Schematic illustration showing the principle of ARB process.

is sometimes carried out at elevated temperatures below recrystallization temperature of the material. ARB process can apply significant amount of plastic strain into the materials, because the above-mentioned procedures can be repeated limitlessly. The Von-Misses equivalent strain ($\varepsilon_{eq}$) after n cycles of ARB can expressed as:

$$\varepsilon_{eq} = -\frac{2n}{\sqrt{3}} ln\frac{t}{t_0} = -\frac{2n}{\sqrt{3}} ln(1-r) \tag{2}$$

where $t_0$, t and r are the initial thickness of the stacked sheets, the thickness after roll bonding and the reduction in thickness per cycle, respectively.

Sheets of various kinds of metallic materials, such as Al, Cu alloys and steels, can be processed by ARB process to develop UFG structures. The most characteristic feature of the UFGs in the ARB processed materials is the elongated morphology. The microstructure reveals the lamellar boundary structure in heavily deformed materials elongation along the rolling direction (RD). The elongated UFG are the grains surrounded by high angle grain boundaries (HAGBs) having dislocation substructures inside the grains. The formation process of the UFGs during ARB is characterized by ultrafine grain subdivision, recovery inside the UFGs, and short-range grain boundary migration. Table 9.1 summarizes the geometrical changes of the specimen thick sheets are stacked and roll-bonded by a 50% reduction per cycle.

Recently, a novel method based on the ARB process was introduced to fabricate nano-structured metals. This method which was named as the cross-accumulative roll bonding (CARB) process, can potentially affect the morphologies of the microstructure. To process by CARB, the strips are degreased by acetone and scratch brushed at first. Then, two strips

**TABLE 9.1** Summarizes the Geometrical Changes of the Specimen During ARB Process Where Roll-Bonded by 50% Reduction Per Cycle

| Number of cycles, n | 1 | 2 | 3 | 5 | 10 |
|---|---|---|---|---|---|
| Number of layers, m | 2 | 4 | 8 | 32 | 1024 |
| Total reduction, r (%) | 50 | 75 | 87.5 | 96.9 | 99.9 |
| Equivalent strain, $\varepsilon$ | 0.80 | 1.60 | 2.40 | 4.00 | 8.00 |

are stacked over each other and roll-bonded by a 50% reduction. 50% rolled strip is cut into two strips, and after degreasing and brushing, they are stacked over each other and rotated 90° around the normal direction (ND) axis. In fact, the strips are roll-bonded in the transverse direction (TD) of the prior stage, i.e., rolling direction (RD) of the current stage. Schematic illustration of CARB process is shown in Figure 9.3.

## 9.2.2 ECAP PROCESS

ECAP, also known as equal channel angular extrusion (ECAE), is an especially attractive processing technique because it is a relatively simple procedure that allows one to obtain homogeneous UFGs in large billets of materials with different crystal structures and has the potential for developing in commercial metal-processing procedures. During ECAP a sample in the form of a rod or bar is pressed through a die having two interesting channels, which are equal in cross-sectional area. When the work-piece is side extruded through the channel, the total strain is:

$$\varepsilon = \frac{1}{\sqrt{3}}\left\{2\cot\left(\frac{\varnothing}{2}+\frac{\Psi}{2}\right)+\Psi cosec\left(\frac{\varnothing}{2}+\frac{\Psi}{2}\right)\right\} \qquad (3)$$

where $\varnothing$ is the angle of intersection of two channels and $\Psi$ is the angle subtended by the arc of curvature at the point of intersection. When $\varnothing = 90°$ and $\Psi = 0°$, the total strain from the above equation is $\varepsilon = 1.15$ and after n passes, it becomes $n*\varepsilon$.

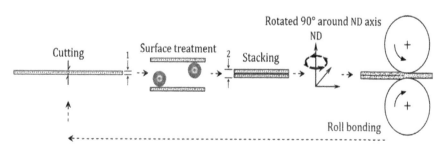

**FIGURE 9.3** Schematic illustration showing the principle of CARB process.

The schematic representation of ECAP process is shown in Figure 9.4. As can be seen in this figure, the specimen is side extruded through the shear deformation zone with the dead zone in outer corner of the channel. The main advantages of this procedure are that it can be applied to fairly large samples and the product has roughly the same dimensions as the original billet. Therefore, the process can be repeated and the homogeneity of the sample can be assured by repetition of the procedure continued until sufficiently high strains are reached.

The retention of the same cross-sectional area when processing by ECAP, despite the introduction of very large strains, is the important characteristic of SPD processing and it is a characteristic that distinguishes this type of processing from conventional metalworking operations such as rolling, extrusion and drawing.

Since the cross-sectional area remains unchanged, the same sample may be pressed repetitively to attain exceptionally high strains. Figure 9.5 shows the fundamental process of metal flow during ECAP. The channel is bent through an angle equal to 90° and the specimen is inserted within the channel and it can be pressed through the die using a punch. Four fundamental processing routes have been identified in ECAP: route A, route $B_A$, route $B_C$ and route C. In route A the specimen is pressed without rotation, in route $B_A$ the specimen is rotated by 90° in an alternate direction between

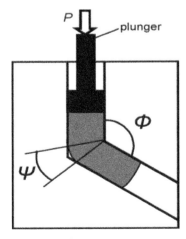

**FIGURE 9.4**   Schematic illustration of ECAP process.

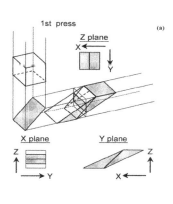

(a)

| Route | Plane | Number of pressings | | | | | | | | | (b) |
|---|---|---|---|---|---|---|---|---|---|---|---|
| | | 0 | 1 | 2 | 3 | 4 | 5 | 6 | 7 | 8 | |
| A | X | | | | | | | | | | |
| | Y | | | | | | | | | | |
| | Z | | | | | | | | | | |
| $B_A$ | X | | | | | | | | | | |
| | Y | | | | | | | | | | |
| | Z | | | | | | | | | | |
| $B_C$ | X | | | | | | | | | | |
| | Y | | | | | | | | | | |
| | Z | | | | | | | | | | |
| C | X | | | | | | | | | | |
| | Y | | | | | | | | | | |
| | Z | | | | | | | | | | |

**FIGURE 9.5**  Fundamental process of metal flow during ECAP. (a) The deformation of a cubic element on a single pass, (b) Shearing characteristic for four different processing routes.

consecutive passes, in route $B_C$ the specimen is rotated 90° counterclockwise between each pass and in route C the specimen is rotated by 180° between passes. Detailed investigations of the dislocation structure, grain sizes and shapes showed that the optimal route to produce equiaxed very fine grain structure is route $B_C$ at least for Al and Ti alloys.

There are several different factors affect on the workability and microstructure evolution of the materials processed using ECAP. First, factors associated directly with the experimental ECAP facilities, such as the values of the angles within the die between the two parts of the channel and at the outer arc of curvature, where the channels intersect. Second, there are experimental parameters related to the processing regimes where some control may be exercised by the experimentalist, for example including: the speed of pressing, the temperature of the pressing operation and the presence or absence of any back-pressure. Third, there are also other processing factors that may play a role in influencing the extent of the grain refinement and the homogeneity of the as-pressed microstructure, including the nature of the crystallographic texture and the distribution of grain misorientations in the unpressed material. These microstructural characteristics and other processing parameters determined by the operator, finally designate the structure and properties of as-pressed materials. Currently, ECAP is the most well developed of all potential SPD techniques.

## 9.2.3 HPT PROCESS

HPT (High Pressure Torsion) is another commonly used SPD technique. In this method, a disk-shape sample is placed between two anvils, a high pressure (typically several GPa) is applied, then one of two anvils is rotated. Figure 9.6, shows the HPT procedure, schematically.

The shear strain ($\gamma$) imposed by HPT can be written in the form of:

$$\gamma = \frac{2\pi nr}{h} \qquad (4)$$

Where n is the number of rotations, r is the radius and h is the height of the disc. The equivalent strain ($\varepsilon_{eq}$) can be calculated as:

$$\varepsilon_{eq} = \frac{\gamma}{\sqrt{3}} \qquad (5)$$

This expression, however, does not deal with the reduction of the height of the sample caused by the high pressure during the deformation. Hence, for large strains ($\gamma \geq 0.8$), where the height reduction is obvious usually another formula is used:

$$\varepsilon = \ln\left(\frac{2\pi nrh_0}{h_f^2}\right) \qquad (6)$$

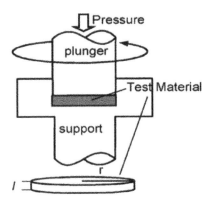

**FIGURE 9.6**   Schematic illustration of the thin disc-HPT process.

where $h_0$ and $h_f$ denote the initial and the final thickness of the sample, respectively. According to this equation, the imposed equivalent strain depends logarithmically on the radius, which is important for smaller number of revolutions, but has less effect on the microstructure after a large number of turns.

The typical applied pressure during HPT (GPa range) is high enough to induce plastic deformation without any torsional straining, therefore the pressure itself can cause significant microstructural changes.

## 9.2.4 REU PROCESS

One of the a new SPD methods for grain refinement of bulk materials is REU (Repetitive Extrusion and Upsetting) process, containing of successive cycles of conventional direct extrusion followed by upsetting. This process is started by extruding a bar in order to elongate the initial grains. As the deformation zone remains non-extruded, a second bar is inserted into the container to remove it. The first bar is upset until the initial diameter is reached again, so the extrusion process can be repeated. Schematic illustration of showing the principle of REU process is seen in Figure 9.7. REU consists in the combination of two well-known conventional plastic deformation processes without using any additional tools and/or devices.

Some of REU advantages compared to other SPD process are a higher strain per cycle, more shear planes with different orientation and consequently a more effective grain refinement and no additional machining of the specimens is required. The extrusion combined with upsetting achieves the highest effective strain per cycle between all SPD processes. The effective strain/cycle is given as a sum of strain in extrusion and upsetting:

$$\varepsilon = 4 \ln \frac{D}{d} \qquad (7)$$

For n cycles the accumulated strain is:

$$\varepsilon = 4n \ln \frac{D}{d} \qquad (8)$$

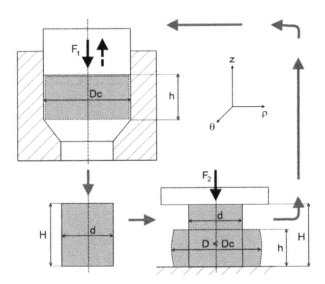

**FIGURE 9.7**   Schematic illustration of showing the principle of REU process.

where D and d are indicated in Figure 9.7.

The velocity discontinuities and the slip line field in upsetting process are explained the grain refinement mechanism during REU process. The velocity discontinuities increases the tangential stress ($\tau$) that can create the bending or shearing of grains or sub-grains. Along the slip lines the tangential stress reaches its maximum values, i.e., yield strength in pure shear (k). The grain refinement during REU process is explained by three mechanism. First, the elongation of grains during extrusion. Second, the development of new boundaries inside a grain when crossing the velocity discontinuities surfaces. Third, the fragmentation of grains by the maximum tangential stress which produce the shearing on the slip lines during upsetting. By repeating REU several times, an increase in the density of dislocations is expected. More and more grains will be subdivided, the textured structure will be broken and a new structure with small grains will arise.

### 9.2.5  TCEC PROCESS

TCEC are developed as a novel SPD process for tubular materials. TCEC is a one of severe plastic deformation processes that cylindrical tubes are deformed through TCEC to large strains, without dimensional and geometrical changes. Schematic illustrations of the TCEC principles are shown in Figure 9.8.

The initial tube is placed between the mandrel and chamber with a small neck zone. Then, two end caps were fastened to the top and bottom sides of the mandrel. These caps are fitted to the chamber with a small clearance and can easily slide and move inside it. The developed space among the mandrel, chamber and two end caps always has a constant volume and fully encloses the tubular sample, during the deformation. The enclosed tubular sample can be processed by pushing the mandrel toward the neck zone of the chamber. The TCEC method is a cyclic process in which the cross section of the processed tube is first reduced while extruding from the neck zone (narrow section) of the chamber and subsequently reached to initial thickness after passing the neck zone. Some advantages of TCEC over other SPD methods of processing tubular components are no need for back pressure, low amount of friction, no restriction on the cross-section and length of tubes, simplicity of tools and accordingly process operation and low cost.

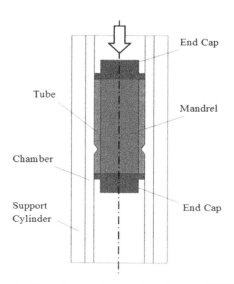

**FIGURE 9.8**  Schematic illustration of showing the principle of TCEC process.

The TCEC method allows large strain deformation of a tubular sample with preservation of the original shape of sample after N cycles. Thus, the total equivalent strain after N cycles of TCEC can be expressed in a general form by the following equation:

$$\varepsilon = 2N \left[ \ln\left( \frac{R_{0-r_0^2}^2}{R^2 - r^2} \right) + \frac{4}{\sqrt{3}} \cot\left( \frac{\varnothing}{2)} \right) \right] \tag{9}$$

## 9.2.6 RCS PROCESS

In this process a work piece is repeatedly bent and straightened without significantly changing its cross-section. Large plastic strains are imparted to the material, which lead to the refinement of the microstructure. The RCS process can be easily adapted to large-scale industrial production. The schematic illustration of RCS process is shown in Figure 9.9.

The equivalent strain per one operation is given by:

$$\varepsilon = 4 ln \frac{[(r+t)/(r+0.5t)]}{\sqrt{3}} \tag{10}$$

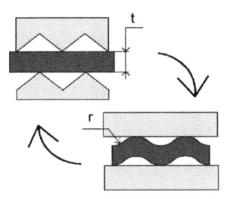

**FIGURE 9.9**   The schematic illustration of showing the principle of RCS process.

where t is the thickness of sample and r is the curvature of bent zone (Figure 9.9). By repeating these processes in a cyclic manner, high strains can be introduced in the work piece.

## 9.3  MICROSTRUCTURE EVOLUTION AND MECHANICAL PROPERTIES DURING SPD

### 9.3.1  DISLOCATION DENSITY AND GRAIN SIZE

As the grain size decreases during SPD, other important microstructural parameters change. Dislocation density increases during SPD process and it leads to the increase of the formation rate of high angle grain boundaries (HAGBs) in grain subdivision process.

At high-imposed strain during SPD process, the dislocation density as well as the grain size reaches a saturation level. SPD method and processed material effect on the saturation level. The saturation dislocation density during SPD process is determined as the equilibrium between dynamic recovery and formation of dislocations. The higher saturation dislocation density corresponds to a more difficult dynamic recovery due to: (i) the low homologous temperature of SPD, (ii) the solid solution alloying, (iii) the second phase particles, and (iv) the high degree of dislocation dissociation due to the low stacking fault energy. As a product of high strain deformation, the deformation-produced nanostructures always contain a certain amount of interior dislocations. The dislocation density varies depending on the metal type and the processing conditions. The existence of interior dislocations may play a key role in determining the mechanical behavior of nanostructured metals.

The increase in yield strength due to grain refinement is usually examined with the well-known Hall-Petch equation, as mentioned before in Eq. (1). It has been found that the Hall-Petch relation is not valid for very small grain sizes; a deviation starts below grain sizes less than $\sim 20$ nm. This is illustrated in Figure 9.10 in a double-logarithmic schematic representation.

Below a critical grain size –which is an optimum grain size where the strength is maximum– the Hall-Petch exponent changes from $-1/2$ to a positive value of about 3; the latter is called the inverse Hall-Petch relation or grain-boundary weakening. The reason for the decrease in flow stress

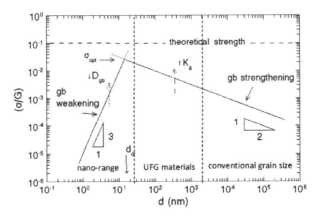

**FIGURE 9.10**   Schematic figure to illustrate the transition from hardening to softening as a function of grain size.

is the absence of dislocation structures and the enhancement of diffusion mechanisms related to the relatively large grain boundary surface (principally Coble-creep). The inverse Hall-Petch relation cannot be observed in UFG materials obtained by SPD, simply because the grain sizes are larger than the optimum grain size (*see* Figure 9.10). However, such nano-materials that have smaller grain sizes than the critical one and produced by non-SPD techniques, can be subjected to SPD deformation where their mechanical behavior are controlled by the inverse Hall-Petch relation.

### 9.3.2   EFFECT OF PRE-EXISTING PRECIPITATES ON MICROSTRUCTURAL EVOLUTION AND MECHANICAL PROPERTIES DURING SPD PROCESS

The previous studies on UFG materials fabricated by SPD, however, have mostly used pure metals. The interaction between UFG structures and phase transformation including precipitation would be an interesting subject in this field. When the materials have pre-existing precipitates before the SPD process, it would affect the evolution of UFG structures and the change in mechanical properties during the process. They can be compared with the microstructures and mechanical properties of the materials without precipitates.

It is well-accepted now that the UFG structures develop in single-phase alloys deformed to high strain levels by SPD processes. However, the presence of second-phase particles could potentially have a significant effect on formation of UFG structures during SPD process. For example, it is expected that precipitates can increase the rate of dislocation generation and large local misorientation gradients. This can lead to an increased rate in the formation of high angle grain boundaries (HAGBs), so that ultrafine grains may be obtained at considerably lower strain than in precipitates-free alloys. At the same time, the second-phase particles can be effective in dislocation pinning. Therefore, the finer structure is expected in materials containing pre-existing precipitates compared to those in precipitates-free materials. The effect of the particles or precipitates on grain boundary is schematically shown in Figure 9.11.

Also, the grain refinement is accelerated by pre-existing precipitates during SPD process. This acceleration is presumably attributed to the inhibition of dislocation motion by the precipitates. In addition, the smaller precipitates seem more effective for grain-refinement. Some of the experiments show that the fine pre-existing precipitates have a significant effect on microstructure refinement at relatively low strains. Also, the rate of HAGBs formation is also affected by the pre-existing precipitates, and that the finer precipitates are more effective for the formation of HAGBs during the ARB process.

It is well known that mechanical properties of materials can be improved by grain refinement. The mechanical properties obtained experimentally

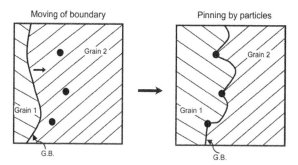

**FIGURE 9.11**   Schematic illustration showing pinning effect of pre-existing precipitates on grain boundary migration.

are correlated to the microstructural parameters, such as grain size and dislocation density, based on strengthening mechanisms, i.e., dislocation strengthening, grain boundary strengthening and precipitation strengthening. The increase in strength and decrease in elongation are typical changes in mechanical properties of metals during SPD, which seems to occur regardless of the kinds of SPD processes and metals. This characteristic change the mechanical properties depends on the starting microstructures. For example, mechanical behavior of Al-0.2wt%Zr alloy and Al-0.2wt%Sc alloy during ARB process was investigated and the results showed that the fine precipitates were formed after aging process and improved mechanical properties in the Aged-specimen compared to the solution treated (ST) specimen. Also, the results indicated that finer pre-existing precipitates were more effective on mechanical properties of the specimens. Although by continuing process, the precipitates probably dissolved due to heavy deformation. Some of researchers have been reported that the strength of specimens during ARB process is attributed to work hardening by increase in dislocation density and formation of subgrains. However, the occurrence of dynamic recovery during the following processing decreases the strength rise rate.

### 9.3.3   FATIGUE STRENGTH

Some characteristics of SPD-processed alloys such as fatigue study still need an in-depth research. High cycle fatigue (HCF) life of most SPD materials is enhanced compared to their coarse grained counterparts. On the other hand, in low cycle fatigue (LCF), the fatigue resistance of conventional coarse-grained materials is superior to that of SPD materials. This phenomenon is schematically shown in Figure 9.12.

The recent studies on the fatigue of UFG metals prepared by ECAP have shown in several cases that the HCF strength is enhanced, compared to coarse grained material, while the LCF performance is reduced because of the lower ductility of the UFG material. In HCF the elastic component of the strain amplitude is dominant and the fatigue life is dictated by the fracture strength of the material, so that in general the fatigue limit of the material increases with increasing strength, which is in favor of SPD materials.

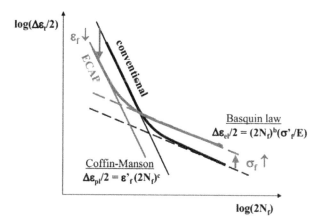

**FIGURE 9.12**   Schematic illustration of the influence of ECAP deformation on the fatigue life of a material.

### 9.3.4  SUPERPLASTICITY

Superplasticity refers to the ability of a crystalline material to exhibit very high tensile elongation prior to failure (hundreds and thousands elongation during tension). This phenomenon has considerable industrial potential for the manufacture of complex sheet structures. Superplasticity is observed in microcrystalline materials with a grain size being usually less than 10 μm and deformed in a definite temperature, strain rate interval, as a rule, $T=0.5–0.6T_m$ at strain rates $10^{-3}–10^{-4}$ s$^{-1}$. It has been widely accepted that the constitutive relationship for superplasticity can be expressed in the following generalized form:

$$\dot{\varepsilon} = A \frac{D_0 Eb}{KT} \exp\left(-\frac{Q}{RT}\right)\left(\frac{b}{d}\right)^p \left(\frac{\sigma - \sigma_0}{E}\right)^n \tag{11}$$

where $\dot{\varepsilon}$ is the strain rate, A is a dimensionless value, $D_0$ is the pre-exponential constant for diffusivity, E is Young's modulus, b is Burger's vector, k is Boltzmann's constant, T is the absolute temperature, Q is the activation energy dependent on the rate-controlling process, R is the gas constant, d is the grain size, p is the grain size exponent, σ is the applied stress, $\sigma_0$ is the threshold stress, and n is the stress exponent.

Based on mentioned equation, the grain size is the sole microstructural determining parameter of the superplastic properties and therefore grain refinement is considered to be the sole method for increasing the superplastic deformation rate or reducing deformation temperature. The grain size less than 1 μm generally shows excellent high strain rate superplasticity. SPD methods make a good opportunity to obtain enhanced superplastic properties with low temperature and high strain rate superplasticity. The presence of mainly HAGBs is one of the most important requirements to obtain high strain rate superplasticity in a number of alloys.

## 9.4  SUMMARY

SPD is a major techniques for manufacturing of ultra fine grained or nano structured bulk metals. These materials exhibit superior mechanical properties due to reduced grain size. Several methods used for creation of persistence plastic deformation depend on material shape and type such as: ARB, CARB, ECAP, HPT, REU, TCEC and RCS. The microstructure and mechanical properties of deformed metals can be effected by final grain size, dislocation density and pre-existing precipitates with different routes. Furthermore, SPD process can enhance the fatigue strength and superplasticity of deformed metals.

## KEYWORDS

- **SPD Processes**
- **Nanostructures**
- **Microstructural Evolution**
- **Fatigue Strength**
- **Superplasticity**

# REFERENCES

1. Xiaoxu Huang, *J. Mater. Sci.* 2007, 42, 1577–1583.
2. Mohammad Reza Kamali Ardakani, Sajjad Amirkhanlou, Shohreh Khorsand, *Materials Science & Engineering A*, 2014, 591, 144–149.
3. Zaharia, L., Comaneci, R., Chelariu, R., & Luca, D. *Materials Science & Engineering A*, 2014, 595, 135–142.
4. Borhani, E., Jafarian, H., Shibata, A., & Tsuji, N. *Mater. Trans.*, 2012, 53, pp. 1863–1869.
5. Morteza Alizadeha, Erfan Salahinejad, *Materials Science & Engineering A*, 2014, 595, 131–134.
6. Azushima, A., Kopp, R., Korhonen, A., Yang, D. Y., Micari, F., Lahoti, G. D., Groche, P., Yanagimoto, J., Tsuji, N., Rosochowski, A., & Yanagida, A. *CIRP Annals – Manufacturing Technology* 2008, 57, 716–735.
7. Borhani, E., Jafarian, H., Sato, T., Terada, D., Miyajima, N., & Tsuji, N. *Proc. 12th Int. Conf. on Aluminum Alloys, 2168 Japan* (2010).
8. Borhani, E., Jafarian, H., Terada, D., Adachi, H., & Tsuji, N. *Mater. Trans.*, 2011, 53, pp. 72–80.
9. Mughrabi, M., Höppel, H. W., & Kautz, M. *Scripta Mater.* 2004, 51, 807–812.
10. Vinogradov, A., & Hashimoto, S. (editors.). M.J. Zehetbauer and R.Z. Valiev, Wiley-VCH, 2004, 663–676.
11. Borhani, E., Jafarian, H., Adachi, H., Terada, D., & Tsuji, N. *Mater. Sci. Forum*, 2010, 667, 211–216.
12. Nieh, T. G., Wadsworth, J., & Sherby, O. D. In *Superplasticity in Metals and Ceramics*. Cambridge: Cambridge University Press, 1997. p. 290.
13. Kaibyshev, O. A. In *Superplasticity in Metals, Intermetallics and Ceramics*. Frankfurt: Springer, 1992.
14. Ma, Z. Y., Liu, F. C., & Mishra, R. S. *Acta Mater.*, 2010, vol.58, pp. 4693–4704.
15. Mabuchi, M., & Higashi, K. *Int. J. Plast.*, 2001, vol. 17, pp. 399–407.
16. Saito, Y., Utsunomiya, H., Tsuji, N., & Sakai, T. *Acta Mater.* Vol. 47, No. 2, pp. 579–583, 1999.
17. Azad, B., Borhani, E., & Mohammadian H. R. Semnani, Kovove Mater. December 2015, Volume 48, No.2, pp. 125–132.
18. Valiev, R. Z., Islamgaliev, R. K., & Alexandrov, I. V. *Progress in Materials Science*, 2000, 45, 103–189.
19. Kamali Ardakani, M. R., Khorsand, S., Amirkhanlou, S., & Nayyeri, M. J. *Materials Science & Engineering A*, 2014, 592, 121–127.
20. Alizadeh, M., & Salahinejad, E. *Materials Science & Engineering A*, 2014, 595, 131–134.
21. Zhilyaev, A. P., & Langdon, T. G. *Progress in Materials Science*, 2008, 53, 893–979.
22. Jonas, J. J., Ghosh, C., & Toth, L. S. *Materials Science & Engineering A*, 2014, 607, 530–535.
23. Zaharia, L., Comaneci, R., Chelariu, R., & Luca, D. *Materials Science & Engineering A*, 2014, 595, 135–142.

24. Babaei, A., Mashhadi, M. M., & Jafarzadeh, H. *Materials Science & Engineering A*, 598 (2014) 1–6.
25. Huang, J., Zhu, Y. T., Alexander, D. J., Liao, X., Lowe, T. C., & Asaro, R. J. *Materials Science and Engineering A*, 2004, 371, 35–39.

# CHAPTER 10

# CONTROLLING ELECTROSPUN POLYMER NANOFIBERS STRUCTURE USING MICROSCOPIC MODEL FOR IDEAL TISSUE ENGINEERING SCAFFOLDS

S. PORESKANDAR and SH. MAGHSOODLOU

*University of Guilan, Rasht, Iran*

## CONTENTS

## ABSTRACT

Electrospinning has emerged recently as a very widespread technology to produce synthetic nanofibrous structures and best candidates for many important applications like scaffolds in tissue engineering. Nanofibrous scaffolds are intended to provide improved environments for cell attachment, migration, proliferation, and differentiation when compared with traditional scaffolds. Producing and creating porosity is the primary challenge of tissue engineering scaffolds. Also, the most significant challenge in this process is to attain uniform nanofibers consistently. For these causes, controlling, producing of electrospun nanofiber becomes important. However, Analysis dynamics of jet formation and its instability is difficult during the process. This instability made unsuitable aligned nanofibers, which are vital for producing scaffolds. For making more suitable and applicative scaffolds, we should control this instability. More suitable and simplest and fastest way for controlling instability is using modeling and computer simulations. The main aim of this article is simulating behavior of the electrospinning process using bead-spring model for better observation.

## 10.1  INTRODUCTION

Tissue engineering is an interdisciplinary field that utilizes the principles of chemistry, physics, materials science, engineering, cell biology and medicine to the development of biological substitutes that restore, maintain or improve tissue/organ functions [1]. It involves the design and manufacture of three-dimensional substitutes to mimic and restore the structural and functional properties of the original tissue. First, cells are harvested from the patient and are expanded in cell culture medium. After sufficient expansion, the cells are seeded into a porous scaffold along with signaling molecules and growth factors, which can promote cell growth and proliferation. The cell-seeded scaffold will be then placed into a bioreactor before being implanted into the patient's body. Nevertheless, in tissue engineering, the generated tissue should have similar properties to the native tissue in terms of biochemical activity, mechanical integrity and function [2]. The Principle of tissue engineering is shown in Figure 10.1.

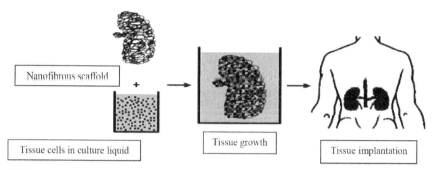

**FIGURE 10.1**   Principle of tissue engineering.

The purpose of tissue engineering is to repair, replace, maintain, or increase the use of a particular tissue or organ. The core technologies intrinsic to this effort can be devised into three fields [3]:

- Cell technology;
- Scaffold frame technology; and
- Technologies for *in vivo* integration.

Making and creating porosity is the main challenge of tissue engineering. In this relation, the percentage of pores, the diameter of the hollows and the value of porosity are important [4, 5]. Today, scientists in tissue engineering have turned to nanotechnology, specifically nanofibers, as the solution to the development of tissue engineering scaffolds. Characterization of these nanofibrous structures for tissue engineering applications is essential [6]. In addition, Successful tissue engineering needs synthetic scaffolds to bear similar chemical compositions, morphological, and surface functional groups to their natural counterparts [7]. In this case, Different methods are functional for creating scaffolds such as electrospinning, self-assembly, and phase separation [8]. The applied science of producing nanofibers especially electrospinning, has drawn more attention from scientists and researcher because of three-dimensional scaffolds is similar to the extra-cellular matrix of natural tissue [5]. The most important challenge in the electrospinning process is to attain uniform nanofibers consistently and reproducibly [7, 9–11]. The mechanics of this process deserving a specific attention and necessary to predictive tools or direction for better under-standing and optimization and controlling process [12]. The principal focus in this research assignment will be to introduce mathematical modeling and simulation of the electrospun nanofibers process for controlling diameter and morphology of electrospun nanofibers used in tissue engineering scaffolds. In this article, features of an ideal scaffold are described at first. Then, electrospinning process and the reasons of modeling are mentioned. At last, modeling and simulation of the process with Matlab script are reviewed.

## 10.2  THE IMPORTANCE OF ELECTROSPUN NANOFIBERS IN TISSUE ENGINEERING SCAFFOLDS

A significant attention of scientists in tissue engineering has turned to nanotechnology, specifically nanofibers, as the answer to the development of tissue engineering scaffolds. Characterization of these nanofibrous structures for tissue engineering applications is essential in understanding its structure [6]. The scaffold frame technology focuses on these aims [3]:

- Designing;
- Manufacturing;
- Characterizing three-dimensional scaffolds for cell seeding;
- In vitro or in vivo culturing.

Many natural polymers like collagen, chitin and chitosan and synthetic biodegradable polymers or a composite mixture of ECM have been widely investigated for its potential applications [13–15]. Examples of applicable varied polymers in tissue engineering are bringing in Table 10.1.

Ideal scaffolds probably approximate the structural morphology of the natural collagen base in the target organ. The ideal scaffold must satisfy a number of often conflicting demands [17]:

1. Appropriate levels and sizes of porosity allowing for cell migration;
2. Sufficient surface area and a variety of surface chemistries that encourage cell adhesion, growth, migration, and differentiation;
3. A degradation rate that closely matches the regeneration rate of the desired natural tissue.

Different methods available for creating scaffolds like molecular self-assembly, hydrogels, the solvent casting–particulate leaching

**TABLE 10.1**   Utilized Polymers in Electrospinning Process for Different Applications in Tissue Engineering [16]

| Fiber diameter | Solvent | Polymer | Application |
|---|---|---|---|
| 200–350 nm | 2,2,2-trifluoroethanol<br>Water | Poly($\varepsilon$-caprolactone) (shell)+<br>Poly(ethylene glycol) (core) | |
| 1–5 µm | Chloroform<br>DMF<br>Water | Poly($\varepsilon$-caprolactone) and<br>poly(ethylene glycol) (shell)<br>Dextran (core) | |
| 500–700 nm | Chloroform<br>DMF<br>Water | Poly($\varepsilon$-caprolactone) (shell)<br>Poly(ethylene glycol) (core | Drug Delivery System |
| ~4 µm | DCM<br>PBS | Poly($\varepsilon$-caprolactone-co-ethyl ethylene phosphate) | |
| 260–350 nm | DMF | Poly(D,L-lactic-co-glycolic acid), PEG-b-PLA‹ PLA | |
| 1–10 µm | DCM | Poly(D,L-lactic-co-glycolic acid) | |
| 690–1350 nm | Chloroform | Poly(L-lactide-co-glycolide) and PEG-PLLA | |
| 2–10 nm | Chloroform<br>methanol | Poly($\varepsilon$-caprolactone) | |
| 500–900 nm | Chloroform<br>DMF | Poly($\varepsilon$-caprolactone) (core)+<br>Zein (shell) | |
| 500 nm | 2,2,2-trifluoroethanol | Poly($\varepsilon$-caprolactone) (core) +<br>Collagen (shell) | |
| 500–800 nm | DMF<br>THF | Poly(D,L-lactic-co-glycolic acid) and PLGA-b-PEG-NH2 | General Tissue Engineering |
| 1–4 mm | DMF<br>acetone | Poly(ethylene glycol-co-lactide) | |
| 0.2–8.0 mm | 2-propanol and water<br>Water | Poly(ethylene-co-vinyl alcohol) | |
| 180–250 nm | HFP | Collagen | |
| 0.29–9.10 mm | 2,2,2-trifluoroethanol | Gelatin | |
| 120–610 µm | HFP | Fibrinogen | |
| 130–380 nm | HFP | Poly(glycolic acid) and chitin | |

**TABLE 10.1** Continued

| Fiber diameter | Solvent | Polymer | Application |
|---|---|---|---|
| 0.2–1 nm | Chloroform DMF | Poly(ε-caprolactone) | |
| 200–800 nm | Acetone | Poly(L-lactide-co-ε-caprolactone) | Vascular |
| 5 μm | Chloroform | Poly(propylene carbonate) | Tissue |
| 300 nm | 1,4-dioxane DCM | Poly(L-lactic acid) and Hydroxylapatite | Engineering |
| 8.77–0.163 m | HFP | Chitin | |

technique, thermally induced phase separation and the electrospinning. Electrospinning has evolved to allow the fabrication of nanofiber scaffolds in this size range and beyond [18]. On the other hand, it is a suitable technique in which the mechanical and biological properties of fibers can easily be tuned by varying the composition of a mixture, which is not easily possible in other scaffold fabrication methods. Nowadays, the interest in the electrospinning technique for the fabrication of tissue engineering scaffolds has increased exponentially. The electrospinning technique also offers precise control over the composition, dimension and alignment of fibers that have an impact on the porosity, pore size distribution and architecture of scaffolds [19]. Also, electrospinning have attracted more attention from scientists and researcher because of three-dimensional scaffolds is similar to the extracellular matrix of natural tissue [5, 8]. Among all biomedical materials under evaluation, electrospun nanofibrous scaffolds have presented great performances in cell attachment, increased and penetration [20]. The electrospun scaffolds mean for biomedical applications are seeing in 1978 for the first time. It extended for cell growing on three-dimensional polymer scaffolds in 1980' researches [17, 21]. The electrospinning technique has been widely used in the fabrication of scaffolds; however, with a few exceptions, the majority of the investigations have been limited to in vitro characterizations [22], and elastin, type I collagen and poly (D,L-lactide-co-glycolide) PLGA polymers [23]. Likewise, the scaffold was helpful in biocompatibility, mechanical property, porosity, degradability in the human physical structure [7]. The ability of the electrospinning technique to combine the advantages of synthetic and natural materials makes it especially attractive, where a high mechanical durability, in terms of high burst strength and compliance (strain per unit load), is required [24, 25]. Also, electrospinning offers the ability to fine-tune mechanical properties during the manufacturing process while also controlling the necessary biocompatibility and structure of the tissue engineered grafts [26]. The mechanical properties of scaffolds are controlled by changing several parameters such as fiber diameter porosity and alignment have been used to regulate cell survival, migration and proliferation during tissue formation [19]. Thus the microstructure and mechanical properties of the resultant scaffolds were controlled by the fiber diameter. Making and creating porosity is the main challenge of tissue engineering. In this relation, the percentage of the pore, the diameter of the hollows and the value of porosity are important [4, 5]. The dynamic mechanical properties of scaffolds are equally important. Therefore, the dynamic mechanical properties and dynamic compliance like burst strength, fatigue and fracture behavior, relaxation and creep of electrospun scaffolds and tissue engineered grafts require more attention. In addition, it is important to conduct further research to overcome many of these limitations, particularly the issue of increasing porosity with minimal impact on the fiber diameter of scaffolds [19]. Nonetheless, there is one big drawback. An unstable dynamical behavior of the liquid jet is made during the electrospinning process. This instability inhibits the fibers to be aligned in a regular way, which is important to satisfy the scaffold specific attentions. For these reasons, the most significant challenge in this procedure is to attain uniform nanofibers consistently and reproducibly [7, 9–11]. Also, electrospun nanofibers are most commonly collected as randomly oriented or parallel-aligned mats [27, 28]. Comparing between the randomly-oriented nanofibrous scaffolds and the aligned ones exhibit that aligned ones have a higher tensile strength and a lower elongation at break [28]. For this reason, the mechanics of this process deserving a specific attention and necessary to predictive tools or way for better understanding and optimization and controlling process [12]. When the instability of the jet can be controlled, electrospinning cannot only be adjusted to produce high-quality scaffolds for tissue engineering, but also for many other applications. Nevertheless, for creating necessary features for scaffolds, it needed to control the diameter and morphology of production nanofibers. In this instance, we can create suitable scaffolds by using modeling and simulation. This way can be used to analyze the dynamical behavior and, in particular, the instability of the jet and how this is affected by changes in the system's parameters without doing expensive experiments time after time. In the next part of the article, we briefly consider the electrospinning process of nanofibers.

## 10.2.1  ELECTROSPINNING OF NANOFIBERS MATS

Nowadays nanofibers have varied characteristics such as a high area to surface and high porosity, are widely attracted the attention of used in varied applications like making scaffolds in tissue engineering [3, 29, 30]. Also, nanofibers have a desirable path for sending and receiving biochemical symbols. In addition of this feature, a nanofibrous structure is more suitable for adhesion and cell increasing [8]. Electrospinning is a simple, inexpensive and straightforward method compare to other methods for producing fibers in nano-scale [31]. As depicted in Figure 10.2, in this procedure, the polymer solution receives electrical charges from a high voltage supply.

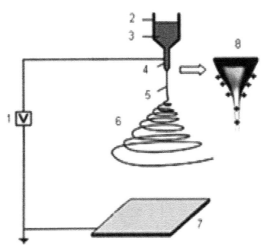

**FIGURE 10.2** Sections of electrospinning process; 1) High voltage 2) Polymer 3) Syringe 4) Needle 5) Straight jet 6) Whipping instability 7) Collector 8) Taylor cone.

These charges are carried by ions through the fluid. If the repulsive force between the charged ions overcomes the fluid surface tension, an electrified liquid jet could be formed and elongate toward the collector [32–35]. When this jet moves straight toward the collector, then, bending instability develops into a series of loops expanding with time. Then, the solvent evaporates during the jet moves. At last nanofibers are collected on a plate or collector [34–36]. As mention above, when jet spirals toward the collector, higher order instabilities reveal themselves result in spinning distance. This instability makes the jet looping in spirals with increasing radius [37].

Depending on several solution parameters, different results can be held using the same polymer and electrospinning setup [38]. Factors that are analyzed to have a primary effect on the formation of uniform fibers are the process parameters, environmental parameters and solution parameters [7, 15, 39–41]. These dates bring in the Figure 10.3. In addition, many researcher studies effects of these parameters on final fibers. A summary of most important parameters is bringing in Table 10.1.

Analysis dynamics instability of jet formation is difficult during the process. This instability made unsuitable aligned nanofibers, which are vital for producing scaffolds. For making more suitable and applicative scaffolds, we should control this instability [42]. The main objective of this project is simulation behavior of the electrospinning process using mathematical models of bead-spring using Matlab script for better observation. By using this method made easy and simple and cost-effective way of analyzing and controlling instability and process. In the next section of the article, we consider about modeling and simulation of the electrospinning process.

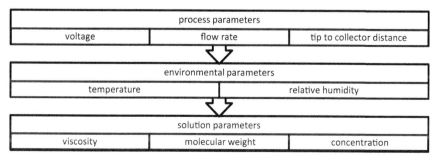

**FIGURE 10.3**    Parameters affect the morphology and size of electrospun nanofibers.

## 10.2.2   MICROSCOPIC MODEL

One of the aims of computer simulation is to reproduce the experiment to shed light on the invisible microscopic details and further explain the experiments. The most widely used simulation methods for molecular systems are Monte Carlo, Brownian dynamics and molecular dynamics (MD). However, Molecular dynamic are the most detailed molecular simulation method, which computes the movements of individual molecules. Molecular dynamics efficiently evaluates different configurationally properties and dynamic quantities which cannot generally be obtained by Monte Carlo [43, 44]. There are five key elements to an MD simulation [45, 46]:

a)   boundary condition;
b)   initial condition;
c)   force calculation;
d)   integrator/ensemble;
e)   property calculation.

The main aim of the mathematical and theoretical analysis of engineering process is access to an algorithm or equation for creating a relation between input and output variables. In addition, for observation of the physical behavior of a sample we should respect this behavior in many samples [47, 48]. For understanding and investigating computer modeling and simulations of complex fluid, it must be used molecular dynamic. This method is the careful and sensitive method of simulations and can be calculated the apparent movement of molecules [44, 49]. In summation, for better controlling the electrospun nanofibers process, we use modeling. In this instance, for simplified work and avoiding of making problems in calculations, we should know about dimensionless analysis.

## 10.2.3   THE IMPORTANCE OF DIMENSIONLESS ANALYSIS

One of the simplest and most powerful tools in physics is dimensional analysis. Also, a good approach systematically getting to grip with such problems is tools of dimensional analysis. Dimensionless quantities, which have no associated physical dimensions, are widely used in mathematics, physics, engineering, economics, and in everyday life (such as in counting). Numerous well-known quantities, such as $\pi$ and $e$ are dimensionless. They are pure numbers and always have a dimension of 1. In non-dimensional scaling, there are two key steps: (a) Identifying a set of physically relevant dimensionless groups, (b) Determining the scaling exponent for each one. The dominant balance of forces controlling the dynamics of any process depends on the relative magnitudes of each underlying physical effect entering the set of governing equations. For achieving a simplified form of the equations and reducing the number of unknown variables, the parameters should be subdivided into characteristic scales in order to become dimensionless. The governing and constitutive equations can be transformed into a dimensionless form, using the dimensionless parameters and groups [31]. The most general characteristic parameters used in the dimensionless analysis in electrospinning are introduced in Table 10.2.

**TABLE 10.2** General Definitions of Dimensionless Parameters in the Equations of Modeling Process

| Parameters | Dimensionless Equations | Parameters | Dimensionless Equations |
|---|---|---|---|
| Elastic modulus | $G \approx \dfrac{\mu^2.\pi.a_0^2}{m.G.L_{el}}$ | Velocity | $\bar{v} \approx \dfrac{v.\mu}{G.L_{el}}$ $\bar{w} \approx \dfrac{w.\mu}{G.L_{el}}$ |
| Stress | $\bar{\sigma} \approx \dfrac{\sigma}{G}$ | Time | $\bar{t} \approx \dfrac{t.G}{\mu}$ |
| Place (Length, Distance) | $H, \bar{X}, \bar{Y}, \bar{Z}, \bar{L}, \bar{r} \approx \dfrac{h, x, y, z, r, l}{L_{el}}$ | Electrical force | $Q \approx \dfrac{e^2.\mu^2}{m.L_{el}^3.G^2}$ |
| Surface tension | $A \approx \dfrac{\alpha.\pi_0^2\mu^2}{m.L_{el}^2.G^2}$ | Voltage | $V \approx \dfrac{e.\mu^2 V_0}{m.G.h.L_{el}}$ |
| Frequency | $K_s \approx \dfrac{\omega.\mu}{G}$ | Area | $\bar{a} \approx \dfrac{a}{a_0}$ |

## 10.2.4 GOVERNING RELATIONS IN BEAD-SPRING MODEL

For investigating the fiber motion and investigating the whipping instability of jet, we proposed a bead-spring model. As mention in Figure 10.4, we consider the jet as a chain that consisted of beads joined by springs with a viscoelastic model in spinning distance (between nozzle and collector). In this written report, we consider microscopic Reneker's model for investigating the behavior of whipping instability in electrospinning. We neglect mass losses due to evaporation in this model [50–52]. The total number of beads, N, increases over time as new electrically charged beads are introduced to represent the flow of solution into the jet. We consider the first formation drop in nozzle part as i=N and the last formation drop as i=1. For investigation all of the forces act along the jet, we consider three beads as i, i +1, i−1.

The distance between i, i+1 with u index and i−1, I with the d index are shown. Because all beads are motions in different directions and axis, then in generally the momentum equation are included many forces (Eq. 1) like coulomb force, electric force, viscoelastic force and surface tension force according to second Newton law (Eq. 2) in three dimensions [50, 52].

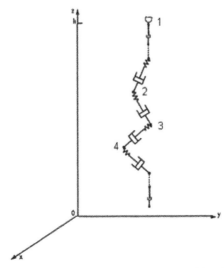

**FIGURE 10.4** Path of the Jet in 3 dimensions; 1) N bead 2) i+1[th] bead 3) i[th] bead 4) i-1[th] bead.

$$\Sigma F_i = \Sigma F_{total} \tag{1}$$

$$m\frac{d^2 R_i}{dt^2} = \sum_{\substack{j=1 \\ j\neq i}}^{N} \frac{e^2}{R_{ij}^3}\left[R_i - R_j\right] - e\frac{V_0}{h}K + \frac{\pi a_{ui}^2 \sigma_{ui}}{L_{ui}}$$

$$\cdot\left[R_{i+1} - R_i\right] - \frac{\pi a_{di}^2 \sigma_{di}}{L_{di}}\left[R_i - R_{i-1}\right] - \frac{\alpha\pi(a)_{av}^2 K_i}{\left(X_i^2 + Y_i^2\right)^{\frac{1}{2}}}$$

$$\cdot\left[i|X_i|sign(X_i) + j|Y_i|sign(Y_i)\right] \tag{2}$$

$R_{ij}$ is the distance between bead $i$ and bead $j$ and is expressed as [50, 52].

$$R_i = iX_i + jY_i + kZ_i \tag{3}$$

The dimensionless momentum equations in 3 directions ($x$, $y$, $z$) of the $i^{th}$ bead define as [50, 52].

$$\frac{d^2\overline{R}_i}{dt^{-2}} = Q\sum_{\substack{j=1 \\ j\neq i}}^{N}\frac{\overline{R}_i - \overline{R}_j}{\overline{R}_{ij}^{-3}} - VK + F_{Ve}\left[\frac{\overline{a}_{ui}^{-2}\overline{\sigma}_{ui}}{\overline{l}_{ui}}\left(\overline{F}_{i+1} - \overline{F}_i\right) - \frac{\overline{a}_{di}^{-2}\overline{\sigma}_{di}}{\overline{l}_{di}}\left(\overline{F}_i - \overline{F}_{i-1}\right)\right] -$$

$$A\frac{\overline{K}\left(\overline{a}_{ui} + \overline{a}_{di}\right)}{4\left(\overline{X}_i^2 + \overline{Y}_i^2\right)^{\frac{1}{2}}}\left[|\overline{X}_i|\left(sign\overline{X}_i L_{el}\right) + |\overline{Y}_i|\left(sign\overline{Y}_i L_{el}\right)\right] \tag{4}$$

## 10.2.5 SIMULATION OF BEAD-SPRING MODEL

Since all the differential equations are defined, the Matlab script can be planned. There are several different ways to do this. In each case, the main feature in the book is its ability to cope with the introduction of new beads into the system and the removal of beads that have reached the collector plate. Steps and difficulties of this program can be discussed as below:

a) New beads are introduced into the system and removed if they progress to the collector: This is the most complex situation since the state vector has a varying size. Moreover, another diffculty is that bead which has reached the collector must be removed from the state vector.

b) New beads are introduced into the system and are reintroduced into the system after reaching the collector: In this way the a number of beads and thus a number of differential equations stays bounded within a specific amount of beads. The integrations start with one bead and the system grow after the first appearance of new beads each time. When the first bead reaches the collector, it will be reintroduced into the system, which leaves the size of the state vector unchanged. This is the most uncomplicated way to describe the system, as the size of the state vector is specific before the integration starts and does not change during the simulation. We simulation, the fluid jet with 20 cm spinning distance as shown in Figures 10.5 and 10.6 with Matlab software.

## 10.3  CONCLUSION

Nanotechnology and especially electrospinning, as nanofibers production technology is attracting much attention of researchers because of producing three dimension pores scaffolds in nano-scale. A very simple technique for producing nanofibers by applying electrostatic forces overcomes the fluid

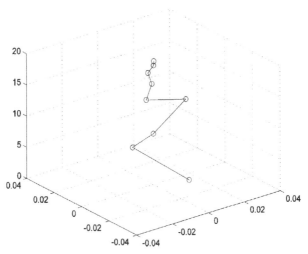

**FIGURE 10.5**  Simulation of fluid jet for N=10 beads from 20 cm distance between spinneret and collector.

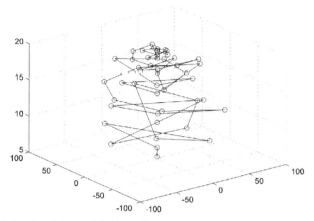

**FIGURE 10.6** Simulation of fluid jet for N=50 beads from 20 cm distance between spinneret and collector.

surface tension. For creating an ideal scaffold, it is necessary to control the morphology of these. For achieving this goal, using simple models would be useful. In this article, a microscopic model of bead-spring was used for simulation and describes the instability behavior of electrospun jet.

## KEYWORDS

- **electrospinning**
- **modeling**
- **nanofibers**
- **scaffolds**
- **simulation**
- **tissue engineering**

## REFERENCES

1. Cui, W., Zhou, Y., & Chang, J. *Electrospun Nanofibrous Materials for Tissue Engineering and Drug Delivery.* Science and Technology of Advanced Materials, 2010, 11(1), 14108–14119.

2. Temenoff, J. S., & Mikos, A. G. *Review: Tissue Engineering for Regeneration of Articular Cartilage.* Biomaterials, 2000, 21(5), 431–440.
3. Fang, J., et al., *Applications of Electrospun Nanofibers.* Chinese Science Bulletin, 2008, 53(15), 2265–2286.
4. Mouthuy, P. A., & Ye, H. *Biomaterials: Electrospinning* in *Comprehensive Biotechnology.* 2011, Elsevier: Oxford. 23–36.
5. Weber, N., et al., *Characterization and in vitro Cytocompatibility of Piezoelectric Electrospun Scaffolds.* Acta Biomaterialia, 2010, 6(9), 3550–3556.
6. Barnes, C. P., et al., *Nanofiber Technology: Designing the next Generation of Tissue Engineering Scaffolds.* Advanced Drug Delivery Reviews, 2007, 59(14), 1413–1433.
7. Lu, P., & Ding, B. *Applications of Electrospun Fibers.* Recent patents on nanotechnology, 2008, 2(3), 169–182.
8. Smith, L. A., & Ma, P. X. *Nano-Fibrous Scaffolds for Tissue Engineering.* Colloids and Surfaces B: Biointerfaces, 2004, 39(3), 125–131.
9. Li, Z., & Wang, C. *Effects of Working Parameters on Electrospinning,* in *One-Dimensional nanostructures.* 2013, Springer. 15–28.
10. Bognitzki, M., et al., *Nanostructured Fibers via Electrospinning.* Advanced Materials, 2001, 13(1), 70–72.
11. De. V. S., et al., *The Effect of Temperature and Humidity on Electrospinning.* Journal of materials science, 2009, 44(5), 1357–1362.
12. Yarin, A. L., Koombhongse, S., & Reneker, D. H. *Bending Instability in Electrospinning of Nanofibers.* Journal of Applied Physics, 2001, 89(5), 3018–3026.
13. Baji, A., et al., *Electrospinning of Polymer Nanofibers: Effects on Oriented Morphology, Structures and Tensile Properties.* Composites Science and Technology, 2010, 70(5), 703–718.
14. Hagvall, H. S., et al., *Three-Dimensional Electrospun ECM-based Hybrid Scaffolds for Cardiovascular Tissue Engineering.* Biomaterials, 2008, 29(19), 2907–2914.
15. Bhardwaj, N. & Kundu, S. C. *Electrospinning: A Fascinating Fiber Fabrication Technique.* Biotechnology Advances, 2010, 28(3), 325–347.
16. Khan, N., *Applications of Electrospun Nanofibers in The Biomedical Field.* Studies by Undergraduate Researchers at Guelph, 2012, 5(2), 63–73.
17. Lannutti, J., et al., *Electrospinning for Tissue Engineering Scaffolds.* Materials Science and Engineering: C, 2007, 27(3), 504–509.
18. Greenwald, S. E. & Berry, C. L. *Improving Vascular Grafts: The Importance of Mechanical and Haemodynamic Properties.* The Journal of pathology, 2000, 190(3), 292–299.
19. Hasan, M. D., et al., *Electrospun Scaffolds for Tissue Engineering of Vascular Grafts.* Acta Biomaterialia, 2013, 1–15.
20. Fang, J., Wang, X., & Lin, T. *Functional Applications of Electrospun Nanofibers.* Nanofibers-production, properties and functional applications, 287–326.
21. Kramschuster, A. & Turng, L. S. *Fabrication of Tissue Engineering Scaffolds.* Handbook of Biopolymers and Biodegradable Plastics: Properties, Processing and Applications. Vol. 17. 2013: Elsevier.
22. Kidoaki, S., Kwon, I. K., & Matsuda, T. *Mesoscopic Spatial Designs of Nano-and Microfiber Meshes for Tissue-Engineering Matrix and Scaffold based on Newly*

*Devised Multilayering and Mixing Electrospinning Techniques.* Biomaterials, 2005, 26(1), 37–46.

23. Stitzel, J., et al., *Controlled Fabrication of a Biological Vascular Substitute.* Biomaterials, 2006, 27(7), 1088–1094.

24. Hasan, M. D., et al., *Electrospun Scaffolds for Tissue Engineering of Vascular Grafts.* Acta Biomaterialia, 2013.

25. Metter, R. B., et al., *Biodegradable Fibrous Scaffolds with Diverse Properties by Electrospinning Candidates from a Combinatorial Macromer Library.* Acta Biomaterialia, 2010, 6(4), 1219–1226.

26. Liang, D., Hsiao, B. S., & Chu, B. *Functional Electrospun Nanofibrous Scaffolds for Biomedical Applications.* Advanced Drug Delivery Reviews, 2007, 59(14), 1392–1412.

27. Beachley, V. & Wen, X. *Effect of Electrospinning Parameters on the Nanofiber Diameter and Length.* Materials Science and Engineering: C, 2009, 29(3), 663–668.

28. Meng, Z. X., et al., *Electrospinning of PLGA/gelatin Randomly-Oriented and Aligned Nanofibers as Potential Scaffold in Tissue Engineering.* Materials Science and Engineering: C, 2010, 30(8), 1204–1210.

29. Huang, Z. M., et al., *A Review on Polymer Nanofibers by Electrospinning and their Applications in Nanocomposites.* Composites science and technology, 2003, 63(15), 2223–2253.

30. Ramakrishna, S., et al., *Electrospun Nanofibers: Solving Global Issues.* Materials Today, 2006, 9(3), 40–50.

31. Rafiei, S., et al., *Mathematical Modeling in Electrospinning Process of Nanofibers: A Detailed review.* Cellulose Chemistry and Technology, 2013, 47(5–6), 323–338.

32. Ghochaghi, N., *Experimental Development of Advanced Air Filtration Media based on Electrospun Polymer Fibers,* in *Mechanical and Nuclear Engineering.* 2014, Virginia Commonwealth. 1–165.

33. Ziabari, M., Mottaghitalab, V., & Haghi, A. K. *Evaluation of Electrospun Nanofiber Pore Structure Parameters.* Korean journal of chemical engineering, 2008, 25(4), 923–932.

34. Sawicka, K. M., & Gouma, P. *Electrospun Composite Nanofibers for Functional Applications.* Journal of Nanoparticle Research, 2006, 8(6), 769–781.

35. Li, W. J., et al., *Electrospun Nanofibrous Structure: A novel Scaffold for Tissue Engineering.* Journal of Biomedical Materials Research, 2002, 60(4), 613–621.

36. Yousefzadeh, M., et al., *A Note on The 3D Structural Design of Electrospun Nanofibers.* Journal of Engineered Fabrics & Fibers (JEFF), 2012, 7(2), 17–23.

37. He, J. H., Wu, Y., & Zuo, W. W. *Critical Length of Straight Jet in Electrospinning.* Polymer, 2005, 46(26), 12637–12640.

38. Sill, T. J., & von H. A. R., *Electrospinning: Applications in Drug Delivery and Tissue Engineering.* Biomaterials, 2008, 29(13), 1989–2006.

39. Angammana, C. J., *A Study of the Effects of Solution and Process Parameters on the Electrospinning Process and Nanofiber Morphology.* 2011, University of Waterloo.

40. Rafiei, S., et al., *New Horizons in Modeling and Simulation of Electrospun Nanofibers: A Detailed Review.* Cellulose Chemistry and Technology, 2014, 48(5–6), 401–424.

41. Tan, S. H., et al., *Systematic Parameter Study for Ultra-Fine Fiber Fabrication via Electrospinning Process.* Polymer, 2005, 46(16), 6128–6134.

42. Zhou, H., *Electrospun Fibers from Both Solution and Melt: Processing, Structure and Property.* 2007, Cornell University.

43. Rossky, P. J., Doll, J. D., & Friedman, H. L. *Brownian Dynamics as Smart Monte Carlo Simulation.* The Journal of Chemical Physics, 1978, 69, 4628–4633.

44. Chen, J. C., & Kim, A. S. *Brownian Dynamics, Molecular Dynamics, and Monte Carlo Modeling of Colloidal Systems.* Advances in Colloid and Interface Science, 2004, 112(1), 159–173.

45. Li, J., *Basic Molecular Dynamics,* in *Handbook of Materials Modeling.* Springer. 2005, 565–588.

46. Rapaport, D. C., *The Art of Molecular Dynamics Simulation.* 2 ed.: Cambridge University Press, 2004, p. 564.

47. Denn, M. M., *Issues in Viscoelastic Fluid Mechanics.* Annual Review of Fluid Mechanics, 1990, 22(1), 13–32.

48. Kröger, M., *Simple Models for Complex Nonequilibrium Fluids.* Physics reports, 2004, 390(6), 453–551.

49. Rossky, P. J., Doll, J. D., & Friedman, H. L. *Brownian Dynamics as Smart Monte Carlo Simulation.* The Journal of Chemical Physics, 1978, 69(10), 4628–4633.

50. Zeng, Y., et al. *Numerical Simulation of Whipping Process in Electrospinning.* in *WSEAS International Conference. Proceedings. Mathematics and Computers in Science and Engineering.* 2009: World Scientific and Engineering Academy and Society.

51. Karra, S., *Modeling Electrospinning Process and a Numerical Scheme using Lattice Boltzmann Method to Simulate Viscoelastic Fluid Flows.* 2007, Texas A&M University.

52. Dasri, T., *Mathematical Models of Bead-Spring Jets during Electrospinning for Fabrication of Nanofibers.* Walailak Journal of Science & Technology, 2012, 9(4), 287–296.

# PART II

# POLYMERS

# CHAPTER 11

# ASSESSMENT OF STRUCTURAL CONDITION OF AMORPHOUS ALLOYS THIN SURFACE LAYERS

O. M. KANUNNIKOVA, O. YU. GONCHAROV, and V. I. LADYANOV

*Institute of Mechanics of the Ural Branch of the Russian Academy of Sciences (IM UB RAS), Izhevsk, Russia*

## CONTENTS

## 11.1   INTRODUCTION

The examination of the concentration profiles of the elements distribution in the depth of the surface layer is among the most common tasks for X-ray photoelectron spectroscopy. The surface layers are successively removed through bombardment with ions of noble gasses (argon) in the ultrahigh vacuum chamber of a spectrometer. The associated effects are examined thoroughly enough [1–4]. The works [4, 5] observed how these effects influence the concentration-profile shape. Most attention was paid to the factors distorting the profiles true shape and to the change of the components chemical condition. Among the reason of the distortion of the concentration-profile shape is the difference between the sputtering yields of pure elements and their compounds. The sputtering yields of the elements within compounds are determined by the nature of the interatomic bond. For example, the sputtering yields of nitrides are higher than of pure metals and corresponding borides, even though their sublattices are equal [1]. Such dependence of the sputtering yield on the elements chemical condition supposes that ion bombardment of a multialloy will give nonuniform etching and phases with lower etching coefficients will be represented as prominent unevenness; so they will create or aggravate surface roughness. The roughness changes the surface layers thickness that contributes to the X-ray photoelectron spectrum and influences the proportion of the signals from the substrate and walls of the unevennesses [4]. The work [6] shows that the spectral lines are mostly influenced by the structural inhomogeneities of dozens and hundreds of nanometers. The structural inhomogeneities smaller than 10 nm and bigger than 1000 nm produce far less or no effect on the signal intensity; therefore, they should not deform concentration profiles

This paper observes how the structural condition of the amorphous alloys surface influences the shape of the concentration profiles reflecting the distribution of the components in the depth of the alloy.

## 11.2   EXPERIMENTAL RESEARCHES

The study considered amorphous alloys $Fe_{70}Cr_{15}B_{15}$ and $Fe_{76.1}Si_{13.8}B_{6.1}Nb_3Cu$, obtained through melt spinning.

The samples of alloys $Fe_{76.1}Si_{13.8}B_{6.1}Nb_3Cu$ (F1, F2, F3, F4) were obtained through spinning at the Central Research Institute of Iron and Steel Industry. F1 strip was cast under standard technology; F2 strip was cast after holding at 1673 K (for 1 min); F3 and F4 strips were cast after gradual heat treatment as follows: heating over 1773 K (5-min holding), cooling down to 1693K and 1573K, respectively, holding at these temperatures and hardening. $Fe_{70}Cr_{15}B_{15}$ alloy was produced from melt after holding at 1673K.

The X-ray photoelectron spectra (XPS) were excited with nonmono-chromatic MgKα-radiation in an upgraded spectrometer ES-2401 [7]. The surface was bombarded with argon ions with an energy of 0.9 keV. The quantitative analysis of the surface layers followed the standard method [8]. The concentration accuracy was 5% of the measured value. The mathematical treatment of the spectra relied on the method based on Fourier transformation with improved convergence procedure [9]. The nucleus was an XPS C1s-spectrum of n-docasanol, obtained with the same spectrometer.

The thermodynamic calculation of the components equilibrium concentrations in the multicomponent heterogeneous systems relied on the known algorithms [10]; such algorithms are based on the simultaneous equations derived from maximum entropy condition and implemented in computer programs. We used ASTRA software complex (by Trusov B., Bauman Moscow State Technical University) and the thermodynamic model of multilayer structures [11] where the scale is represented with a set of layers, parallel to the surface and different in composition. The layers were considered as locally equilibrium subsystems with individual equilibrium parameters. Correspondingly, each such subsystem may have its independent equilibrium-thermodynamic calculation.

The calculations considered high-temperature interaction of $Fe_{76}Si_{14}B_6Nb_3Cu$ alloys and air under $P = 1.01$ Pa. The set parameters:

- temperature $T$ – ranged between 1073 and 1873 K;
- content of air components (oxygen, nitrogen) linearly decreased from the outer surface to the depth of the scale: the weight fraction of the air components – $m_{gas}/M$ ($M = m_{solid} + m_{gas}$) ranged from 0,8 in the outer layer to 0,01 in the layer adjacent to the alloy.

We regarded two perfect solid solutions: one of the alloy components and one of the product of the interaction between the alloy and gas medium.

## 11.3   RESULTS AND DISCUSSION

### 11.3.1   $FE_{76.1}SI_{13.8}B_{6.1}NB_3CU$ ALLOYS

The results of the thermodynamic modeling of the content of the system "$Fe_{76}Si_{14}B_6Nb_3Cu$ – air" under $P = 1.01$ Pa are given in Figures 11.1 and 11.2.

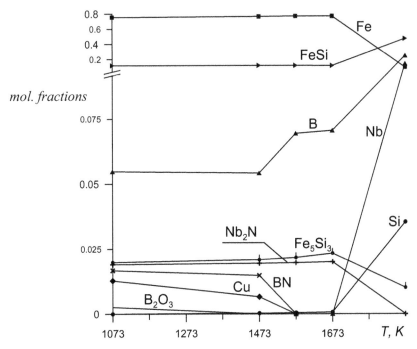

**FIGURE 11.1** Temperature dependence of the components content in the system "$Fe_{76}Si_{14}B_6Nb_3Cu$ – air" under $P = 1.01$ Pa (air content is 0.01 of weight fraction).

The temperature dependence of the systems composition (Figure 11.1) was analyzed to show that, upon heating and alloy-gas equilibrium (such equilibrium is quite possible in near-surface layers), the alloy composition changes due to its interaction with the gas medium and evaporation of the components. As a result of the stated processes, at a temperature below ~1473 K in the scale on the alloy surface, niobium nitrides, boron nitrides, and boron oxide are expected to appear to disappear with further heating. At temperatures higher than ~1673 K, the alloy surface will be enriched with boron, niobium and silicide FeSi; meanwhile, the copper compounds and oxides will disappear completely. It may be supposed that the near-surface composition of the alloy produced through quenching, in the considered conditions, will be close to the composition as between 1473 and 1873 K.

The temperature-dependence modeling for the components content (Figure 11.1) enabled us to demonstrate the possible changes of the

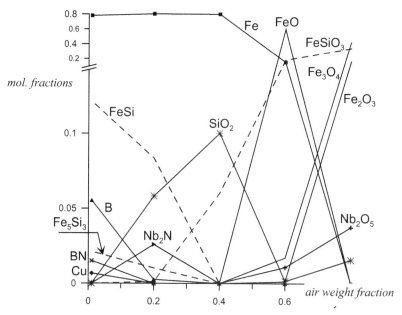

**FIGURE 11.2**   Dependence of the components content according to the depth of the alloys near-surface layers as related to air weight fraction at T = 1473 K under P = 1.01 Pa, for Fe$_{76}$Si$_{14}$B$_6$Nb$_3$Cu alloy.

composition of the alloys surface layers when melted or nearly melted. However, the composition of the near-surface layers resulting from hardening is formed at lower temperatures and, moreover, subject to air environment. To interaction between the alloy and the gas phase was modeled at 1473 K, the temperature being far lower than the alloys minimal pouring temperature and, probably, close to the hardening temperature.

The results of the thermodynamic modeling of the gas environments effect are given in (Figure 11.2). It shows how the components content change according to the depth of the alloys near-surface layers as related to air weight fraction at T = 1473 K and under P = 1,01 Pa for Fe$_{76}$Si$_{14}$B$_6$Nb$_3$Cu composition.

It is obvious (Figure 11.2) that even with relatively low content of the air components, i.e., near the "scale-alloy" border, iron silicides may disintegrate, silicon oxides may emerge and an inner-nitrogenization layer consisting of boron nitrides and niobium nitrides may form. The increase of the air-components share in the alloys near-surface layers results in big amounts of FeO oxide and $FeSiO_3$ silicate. Finally, on the scale surface, $FeSiO_3$, $Fe_3O_4$ and higher oxides of iron, niobium and silicon may form. Of course, the stated components are relative and they may compose quite complex solid solutions.

Copper in $Fe_{76}Si_{14}B_6Nb_3Cu$ alloy does not produce any compounds with other components; it is included in the alloy to provide crystallization centers during the formation of nanocrystalline condition.

The X-ray diffraction (XRD) examination [12] of the structure of F1, F2, F3, F4 alloys confirms the calculations for the compositions with nearly zero air weight factor, i.e., in the alloy bulk (Figure 11.2). It is found out that the melt structure at 1673 K corresponds with the certain correlation of the cluster with short-distance coordination, similar to the coordination of α-Fe, $Fe_3Si$, $Fe_2B$, $Nb_2FeB_2$ crystals. A higher (over critical) temperature dissolves the clusters where boron fixes iron atoms; when cooled, their restoration rate is far lower than for clusters of other types. Consequently, there are more iron atoms not fixed with boron. When the melt is being cooled, they may compose microgroups like $Fe_3Si$ or body-centered cubic phase. This melt structure adjustment will promote sooner formation of body-centered cubic phase crystals due to faster formation of the crystallization centers based on $Fe_3Si$ or BCC-Fe. The adjustment will also speed up crystal growth due to higher diffusion mobility of the iron atoms. Consequently, it thickens the crystallized layer on the surface of the films produced from the superheated melt.

The XRD results demonstrated that the crystallization of F1 film on both sides is lower than of F2, F3, F4-films; meanwhile, the crystalline lines intensity for F1 is lower on the free side than on the contact side; and vice versa for F2, F4, F5.

Upon comparison of the diffractograms with $CoK_\alpha$ and $CuK_\alpha$-radiation, it may be stated that crystals are accumulated in the surface layer, about 3 um thick. When this layer is etched out in a solution of hydrofluoric acid and hydrogen peroxide, the diffractograms look like amorphous ones.

The surface morphology have been surveyed with atomic force microscopy [13] to show the following: the outer surface of F1 alloys contact side is completely crystallized; the average grain size is 80 nm; the free side is amorphous with small crystallized areas; the grain size on the crystallized areas is 20 nm; both sides of F2, F3, F4 alloys are crystallized (Figure 11.3; Table 11.1).

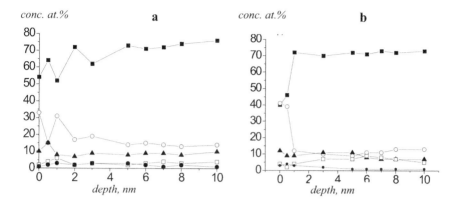

**FIGURE 11.3**   Concentration profiles of element distribution in the outer surface layers of F1 alloys contact (a) and free sides (b).

**TABLE 11.1**   Characteristics of Alloy Surface Topography

| Sample | Observed side | $S_q$ [13] | $d_{avg}$, nm [13] | Con. profile shape |
|---|---|---|---|---|
| $Fe_{76.1}Si_{13.8}B_{6.1}Nb_3Cu$ | Contact side | 9.3 | 50 | Oscillatory |
| | Free side | 1.5 | 24 | Monotonous |
| $Fe_{76.1}Si_{13.8}B_{6.1}Nb_3Cu$ | Contact side | 7.3 | 80 | Oscillatory |
| | Free side | 0.6 | 52 | Oscillatory |
| $Fe_{76.1}Si_{13.8}B_{6.1}Nb_3Cu$ | Contact side | | | Oscillatory |
| | Free side | | | Oscillatory |
| $Fe_{76.1}Si_{13.8}B_{6.1}Nb_3Cu$ | Contact side | 6.9 | 52 | Oscillatory |
| | Free side | 2.8 | 57 | Oscillatory |
| $Fe_{70}Cr_{15}B_{15}$ | Contact side | 1.9 | 40 | Monotonous |
| | Free side | 184.0 | 55 | Oscillatory |

The XPS has found the following. The surface layers composition and volume differ because of the thermostimulated processes that redistribute the surface layers components. According to [14, 15], heating changes the surface layers composition towards the minimums on the equilibrium diagrams liquidus curves. In this case, the analysis of the equilibrium diagrams for iron-copper, boron-copper, silicon-copper [16] supposes that copper diffuses towards the outer surface. According to thermodynamic analysis, at 1873 K copper and boron vaporize. Consequently, the copper content reaches 2.4–3.0 at.% in the surface layers of F2, F3, F4 samples and on F1 alloys contact side; on F1 samples free side, the copper content is 1.3 at.%.

According to the XPS analysis, besides the principal elements, the alloys surface layers also contain oxygen (in oxides $Fe_3O_4$, FeO, $Fe_2O_3$, $Nb_2O_5$, NbO, $B_2O_3$, CuO, $Cu_2O$, $SiO_x$) and nitrogen in boron nitride. The concentration of oxygen and nitrogen in the surface layers of the free side for all the alloys is higher (30–40% at.) than on the contact side (8–30 at.%). The biggest content of oxygen is in the surface layer of F1 alloys free side (48 at.%).

The oxide film on F1, F2, F3, F4 alloys surface is almost similar. F2, F3 and F4 surfaces are enriched with boron and silicon: up to 14÷38 at.% of boron and 20÷25 at.% of silicon. The silicon content on both sides of F1 alloy is also increased up to 21 at.% on the contact side and 38 at.% on the free side. The boron concentration is reduced down to 9.4 at.% on the contact side and 8.4 at.% on the free side. Boron and silicon promote amorphous structure of metal alloys; however, in this case, the decisive role is played by copper; it is copper within the surface layers that promotes their crystallization. The amorphous state is seen in the surface layer of F1 alloys free side with the lowest content of copper. The surface layers of other alloys, despite the high content of the amortizing elements (boron and silicon), are crystallized.

Upon comparison of the concentration-profile shape and topography of the considered alloys surfaces (Table 11.1), it has been found that the concentration-profile shape depends on the mean square roughness and the size of the crystalline inclusions in the surface layer. With a mean square roughness $S_q \leq 1.5$ nm and inclusions $d \leq 40$ nm, the profiles shape is monotonous. For rougher surfaces and bigger inclusions, there are oscillations on the concentration profiles.

## 11.3.2    $FE_{70}CR_{15}B_{15}$ ALLOYS

In the papers [17, 18], we used atomic force microscopy, XPS and thermodynamic modeling for integrated study of the morphology of the surface layers of $Fe_{70}Cr_{15}B_{15}$ alloy. These studies demonstrate that the films free side is almost completely amorphous; the amorphous matrix has crystallized areas of 200–500 nm (with an average crystal size of 40 nm) and bigger hemispheric formations. In most surfaces if the alloys contact side is crystallized: the correlation between the areas of the crystallized and amorphous segments is approximately 3:1. The grain size is between 25 and 80 nm (Table 11.1).

The crystallized areas are small crystals of iron γ-phase. Iron and chromium borides, oxygen compounds of iron and chromium (oxides and spinel) form bigger inclusions. Inclusions of boron nitrides are also found [17, 18].

Figure 11.4 (a, b) shows the concentration profiles of the elements distribution throughout the depth of the alloys surface layers. Comparison of the profiles shape and the topography of these surfaces demonstrate that the amorphous side gives a monotonous shape and the crystallized side gives an oscillatory shape of profile. The decrease of the iron concentration is caused by the primary etching out of iron boride on the outer side of the crystalline inclusions. On the free side, the amount of boride inclusions is far less than on the contact side; therefore, the distribution profiles of iron and chromium are monotonous. The oscillatory profiles of boron distribution are connected to the formation of BN inclusions which amount was almost equal on both surfaces.

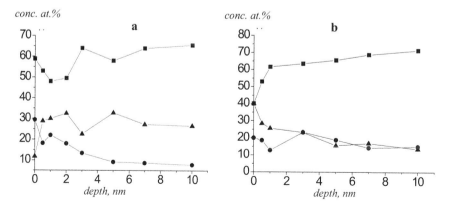

**FIGURE 11.4**  Concentration profiles of the principal elements distribution throughout $Fe_{70}Cr_{15}B_{15}$ alloys surface layers of contact (a) and free sides (b).

The crystallineness is caused by higher boron content and the drum surface acting as crystallization centers [19]. On the free side, there is less boron. According to the thermodynamic analysis, at 1873 K boron vaporizes. Obviously, this transition goes faster on the free side due to a less cooling rate than on the contact side. Therefore, the contact sides surface layer is depleted of boron and there are fewer amounts of borides.

## 11.4   CONCLUSION

There is a firm belief that the structural and phase transformations in the surface layers reflect similar processes in the bulk that cannot be registered with regular volume structural methods.

Still, it is necessary to note that the formation of the structural and phase composition of the thin outer surface layers is determined by the processes unusual for the bulk:

- thermostimulated redistribution of the alloy components which direction depends on the components concentration proportion and may be predicted from analysis of equilibrium diagrams [15];
- vaporization of the components;
- the components interaction with gas environment.

Atomic force microscopy provides information about the structural condition of the outer surface. The integral analysis with XPS and thermodynamics makes it possible to assess the structural condition of the thin surface layers about 10 nm thick; while their examination with traditional structural methods is problematic.

The shape of the concentration profiles of the elements distribution in the depth of the surface layers, as per XPS analysis, depends on the mean square roughness and the size of the crystalline inclusions in the surface layer. With a mean square roughness $S_q \leq 1.5$ nm and inclusions $d \leq 40$ nm, the profiles are of monotonous shape. On the concentration profiles obtained through X-ray electron spectroscopy with ion bombardment, there are oscillations for rougher surfaces and bigger inclusions.

The assessment of the structural condition of the surface layers through the concentration profiles shape is appropriate only for layers of 6–7 nm or thinner. The reason is that, besides sputtering, argon ion bombardment causes crystallization of the surface layer [13].

## KEYWORDS

- amorphous alloys
- concentration profiles
- surface layers
- thermodynamic analysis
- X-ray photoelectron spectroscopy`

## REFERENCES

1. Naguib, H. M., & Kelly, R. Criteria for bombardment-induced structural changes in non-metallic solids. Radiation effects. 1975, V. 25, 1–12.
2. Holloway, P. H., & Bhattacharia, P. Limitation of ion etching for interface analysis. Surf. Interface Anal. 1981, V.3. №3, 118–125.
3. Kelly, R. On the problem of whether mass of chemical bonding is more important to bombardment-induced compositional changes in alloys and oxide. Surf. Sci. 1980, V.100, P.85–107.
4. Wolf, G. K. Chemical effects of ion bombardment. Instr. Inorg. Chem. 1979, V.83, 1–34.
5. Hoffmann, S. Quantitative depth profiling in surface analysis: a review. Surf. Interface Anal. 1988, V.2. №4, 148–160.
6. Lomaev, I. L., & Lomaeva, S. F. Application of atomic force microscopy and x-ray photoelectron spectroscopy to measuring thickness of surface coating for nanostructured materials. Phys. Low-Dim. Struct. 2003, № 2/3, P. 175–182.
7. Kozhevnikov, V. I., Merzljakov, P. G., & Votjakov, V. A. Razrabotka fotojelektronnogo spektrometra na baze serijnogo spektrometra JeS-2401, Materialy XI Mezhd. nauchno-prakt. konf. "Fundamentalnye i prikladnye problemy priborostroenija i jekonomiki." 2008, 64–67.
8. Nefedov, V. I. Rentgenojelektronnaja spektroskopija himicheskih soedinenij: Spravochnik. M.: Himija. 1984, S. 28–31.
9. Povstugar, V. I., Shakov, A. A., Mihajlova, S. S., Voronina, E. V., & Elsukov, E. P. Razlozhenie slozhnyh rentgeno-fotojelektronnyh spektrov s pomoshhju bystrogo diskretnogo preobrazovanija Furye i uluchshennoj proceduroj shodimosti reshenija. Ocenka primenimosti metodiki. ZhAH. 1998, T.53. №8, S. 795–799.
10. Vatolin, N. A., Moiseev, G. K., & Trusov, B. G. Termodinamicheskoe modelirovanie v vysokotemperaturnyh neor-ganicheskih sistemah. M.: Metallurgija. 1994, 352c.
11. Goncharov, O. Ju. Termodinamicheskoe modelirovanie vysokotemperaturnogo okislenija splavov Fe-Cr na voz-duhe. Neorg. mat. 2004, T.40. №12, S.1476–1482.
12. Volkov, V. A., Ladyanov, V. I., & Cepelev, V. S. Osobennosti poverhnostnoj i ob#emnoj kristallizacii lent amorfnogo splava Fe76,1Si13,8V6,1Nb3Cu. Metally. 1998, №6, S.37–43.

13. Lomayeva, S. F., Kanunnikova, O. M., & Povstugar, V. I. AFM and XPS investigation of surface layers of nanocrystalline FeSiBNbCu. Phys. Low-Dim. Str. 2001, №3/4, C.271–276.

14. Teraoka, Y., & Komaki, M. Surface segregation and surface melting in segregating alloys. Surf. Sci. 1999, V.439, P. 1–13.

15. Gilmutdinov, F. Z., & Kanunnikova, O. M. Prognozirovanie izmenenij sostava poverhnosti mnogokomponent-nyh splavov pri termicheskih vozdejstvijah. FMM. 1997, T.84. Vyp.2. S. 78–88.

16. Diagrammy sostojanija dvojnyh i mnogokomponentnyh sistem na osnove zheleza. Spravochnik. Pod.red. Ban-nyh, O. A., Drica, M. E. M.: Metallurgija. 1986, 440s.

17. Goncharov, O. Ju., Kanunnikova, O. M., Lomaeva, S. F., & Shakov, A. A. Sostav poverhnostnyh sloev, obrazujushhihsja pri poluchenii amorfnogo splava Fe70Cr15B15. FMM. 2001, T.91. №6, S.64–71.

18. Kanunnikova, O. M., Lomaeva, S. F., & Gilmutdinov, F. Z. A comparative analysis of surface layers of amorphous Fe70Cr156B15 alloy by means of atomic force microscopy and x-ray electron spectroscopy. Phys. Low-Dim. Struct. 2001, № 3/4, P. 333–340.

19. Brovko, A. P., Maslo, V. V., & Paderno, D. Ju. i dr. O prirode kristallov na kontaktnoj poverhnosti amorfnyh lent splavov (Fe,Cr)85B15. Metallofizika. 1990, T.12. №4, S.116–119.

# CHAPTER 12

# SYNTHESIS OF ULTRAFINE METAL OXIDE PARTICLES IN THE POLYMER MATRIX OF ULTRA HIGH MOLECULAR WEIGHT POLYETHYLENE

A. M. NEMERYUK and M. M. LYLINA

*Research Institute of Chemical Reactants and Ultrapure Compounds (IREA), Moscow, Russia*

## CONTENTS

## 12.1 INTRODUCTION

Ultrahigh molecular weight polyethylene (UHMWPE) is a product of the polymerization of ethylene, carbon-chain polymers with a molecular weight reaches millions of units. This material is characterized by high strength and tribological properties, chemical resistance. Obtaining of materials based on UHMWPE containing modifying fillers in the form of ultrafine, nanoscale particles is a promising area of research. This can significantly improve the performance of the composites based on UHMWPE, affecting the character of the supramolecular structure of the polymer, which is of interest as a basis for the development of approaches to the creation of new structural and functional materials.

Modification of UHMWPE with small amounts (0.1–0.5%) of the traditional solid lubricants such as graphite and molybdenum disulfide increases the wear resistance of UHMWPE is 50–60%, whereas the administration of a low pressure polyethylene has no effect on the durability of materials. Traditionally, composite materials obtained by the mechanical mixing of components and subsequent processing into articles by various methods. Such an approach is not always possible to obtain composites having filler homogeneously distributed in the polymer matrix, as well as with nanoscale filler particles. To increase the compatibility of the filler particles with the matrix often use different methods of surface activation of polymer particles – mechanochemical process, low temperature plasma treatment, ionizing radiation, also chemical methods are used for activation.

X-ray diffraction, infrared spectroscopy and electron microscopy showed that the modification of the polymer mentioned nanofillers leads to the formation of an ordered (lamellar) supramolecular structure.

Of considerable interest to create effective new materials based on UHMWPE is the use of nano-dispersed fillers. Use of nanoparticles as fillers is relatively new and very effective method of modifying the structure and properties of polymers, allowing to obtain a unique combination of physical and mechanical properties.

Insertion of additives in nanoscale inorganic particles – aerosil, talc, alumina, accompanied by improved physical and mechanical properties of the polymer [1, 2]. This increases the resistance to abrasion, resistance to cracking, changing a number of other properties. For example, the insertion of aerosil as nanomodifier accompanied by increased crystallinity of UHMWPE, increased hardness, and reduction in friction wear. The composite material is based on UHMWPE and hydroxyapatite differs improved technical specifications [3, 4]. The formation of the functional properties of the composites is significantly affected by modification of UHMWPE by nanoparticles and nanofibers in low concentrations (0.1–0.5 wt.%) [5].

Strength characteristics depend on the method of modification, the type of inserted particle size and concentration. It was shown that the size of inserted particles should be significantly less than the UHMWPE particles [6]. For small sizes of inserted particles and concentrations up to 1%, an increase of the strength at break is visible. Further increases of inserted particles are accompanied by a decrease in tensile strength and impact resistance. Explain the results can be using the model of hardening thermoplastics containing ultrafine inorganic fillers, which is proposed in Ref. [7]. In this paper, the method of thermally stimulated depolarization shown the presence of filled thermoplastics spontaneous polarization charge. It is shown that the polarization field of the filler particles is induced dipole moment in the polymer, which provides an increase in strength due to the appearance of additional electrostatic interactions. Registered an increase in strength, change in crystallinity and melting point of the thermoplastic.

Typically, polymer composite materials based on UHMWPE prepared by mechanically mixing nanopowders and powdered UHMWPE under intense mechanical action. Wherein the mechanical activation of the binder and the filler powders provides uniform distribution of nano-powder in the binder and further improves the physical and mechanical and tribological characteristics of the composite. However, certain types of composite materials based on UHMWPE prepared using the nanoparticle suspension in organic solvents capable of dissolving at elevated temperatures, UHMWPE, such as decalin [8, 9]. Obtained by this method materials are perspective for use as dielectrics included in the capacitor of high performance.

The methods of making nanocomposite materials based on nanoparticles in processes for the production of polymer present in the solution in the reaction medium or the reaction medium are known, if used polymer melts. Nanoparticles formed during chemical reactions of precursor having high surface energies stabilized by the polymer molecules in solution that prevents formation of large clusters and promotes the homogeneous distribution of nanoparticles in the polymer matrix.

There is a great interest is the synthesis of nanoparticles directly into the polymer matrix. This eliminates the possibility of agglomeration of nanoparticles, as local voids in the polymer matrix are original nanoreactors limiting movement of nanoparticles.

## 12.2   THEORY (METHODS AND TECHNIQUES)

Based on the analysis of existing methods of modification of UHMWPE, methods for producing inorganic nanoparticles in non-aqueous media, and experimental studies the optimal way to obtain UHMWPE modified with metal oxide nanoparticles was determined. The method consists in the use of chemical transformations of metal compounds, mostly halides, resulting in the formation of metal oxide nanoparticles.

A method for obtaining nanoparticles of metal oxides by reacting halides, alkoxides or other compounds of metals with an oxygen donor, preferably alcohols and ketones, as well as ethers has been described in Ref. [10]. However, the temperature required for interactions metal alkoxides with alcohols (200–360°C) exceeds the limits of thermal stability of UHMWPE. When using metal halides formation of oxide nanoparticles occurs at temperatures of 80–100°C, below the melting temperature of UHMWPE (110–120°C). Conducting processes at temperatures below the melting temperature of the polymer avoids having high viscosity melt that obstructs the further processing of the polymer, in contrast to the powder.

The properties of transition metal halides, particularly their solubility and aggregative state are essential to the use of these compounds as precursors of oxide nanoparticles obtained directly into the polymer matrix. Thus, titanium tetrachloride is a liquid under normal conditions, well soluble in many organic solvents, while the chlorides of zirconium and hafnium are solids, the solubility of which is limited. This phenomenon is associated with the polymer structure of the halides of zirconium and hafnium. However, to obtain ultrafine oxide particles in a polymer matrix UHMWPE must use solutions of the starting compounds. It is known that tetrachlorides of zirconium and hafnium can form complexes with various organic compounds, such as aromatic hydrocarbons, ethers, and ketones. The resulting complexes are soluble in organic solvents and can be regarded as precursors for the preparation of oxide nanoparticles in a polymer matrix.

Among the possible complex compounds formed by tetrachlorides of zirconium and hafnium have been investigated complexes with tetrahydrofuran (THF), xylene and acetophenone. It has been found that the optimum combination of properties, including high solubility, boiling point and the stability of the resulting solution has acetophenone. Application THF limited low boiling point of this compound, and the use of xylene or other aromatic hydrocarbons are complicated by the formation of gels UHMWPE having a high viscosity.

Solvent power of acetophenone in relation to UHMWPE is insufficient to form gels at temperatures of 80–100°C, that allow to obtain UHMWPE containing ultrafine particles of transition metal oxides and suitable for further manufacture of compact samples of composites.

As donor compound by reacting oxygen with a transition metal halide, benzyl alcohol was used. The use of benzyl alcohol to obtain nanoparticulate oxides described in Refs. [11–13]. Previously, it was shown that UHMWPE partially swells among benzylalcohol, but gel formation is not observed, thus can be obtained UHMWPE powder suitable for further processing steps comprising benzyl alcohol.

## 12.3   EXPERIMENTAL RESEARCHES

### 12.3.1   GETTING UHMWPE IMPREGNATED WITH BENZYL ALCOHOL

In a round bottom flask equipped with a stirrer, reflux condenser and thermometer was charged with 50 g of UHMWPE powder and 200 mL of benzyl alcohol. The mixture was heated to 75–80°C and stirred for 10 hours, then cooled to room temperature, the UHMWPE was separated by filtration and washed with chloroform to remove excess benzyl alcohol. The product was dried under water pump vacuum at room temperature to constant weight.

## 12.3.2   GETTING UHMWPE CONTAINING NANOPARTICLES OF ZIRCONIUM (HAFNIUM) OXIDE

In a round bottom flask equipped with a stirrer, reflux condenser, thermometer, and charged with 50 mL of acetophenone and 4.6 g of zirconium tetrachloride or 6.2 g of hafnium tetrachloride (0.02 mol) and stirred at room temperature for one hour, then is heated at 60–80°C for 4 hours until complete dissolution. After complete dissolution of halide to the reaction mixture was added benzyl alcohol impregnated UHMWPE 10 g. The mixture was stirred for 10 hours, then UHMWPE is separated by filtration and repeatedly washed with chloroform and then continuously extracted with chloroform in Soxhlet extractor for 8 hours. UHMWPE containing ultrafine particles oxides air-dried to constant weight. The product is a white or light cream powder containing 0.05–0.1% zirconium or hafnium oxides.

## 12.4   RESULTS AND DISCUSSION

In the study of UHMWPE grains by electron microscopy have shown that interaction with benzyl alcohol changes the morphology of the UHMWPE grain. Figures 12.1 and 12.2 shows micrographs of grains UHMWPE starting and after treatment with benzyl alcohol.

Also, samples of UHMWPE treated with benzyl alcohol were studied by IR spectroscopy. Figures 12.3 and 12.4 shows the IR-spectrum UHMWPE both before and after treatment with benzyl alcohol.

The presence in the IR-spectrum of UHMWPE treated with benzyl alcohol, absorption bands in the range of 3500 cm$^{-1}$ confirms the presence of benzyl alcohol in the investigated sample.

Processing UHMWPE containing benzyl alcohol with acetophenone solutions of tetrachlorides of zirconium or hafnium leads to the penetration of the metal halide in the polymer matrix, where the interaction with the benzyl alcohol contained in UHMWPE, with the formation of ultrafine oxide particles. On the surface of the grains of UHMWPE in the study by electron microscopy, have relatively large agglomerates, but in the polymer matrix are ultrafine particles oxides of nanometric dimensions, as shown in Figure 12.5.

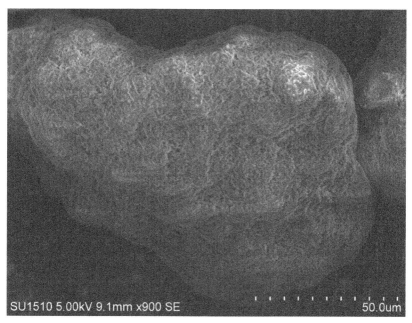

**FIGURE 12.1** Starting UHMWPE grain.

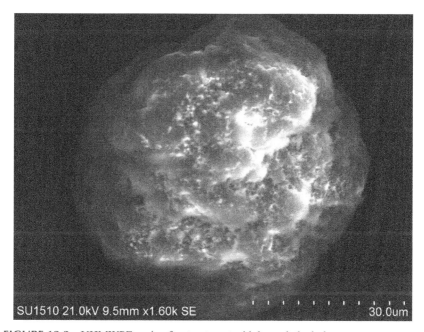

**FIGURE 12.2** UHMWPE grain after treatment with benzyl alcohol.

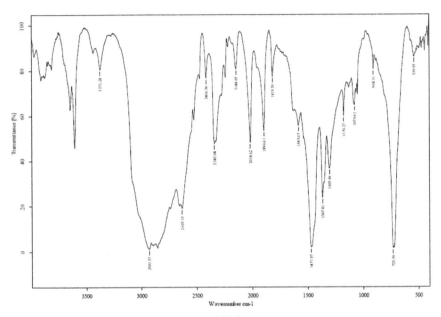

**FIGURE 12.3**　IR-spectrum Starting UHMWPE.

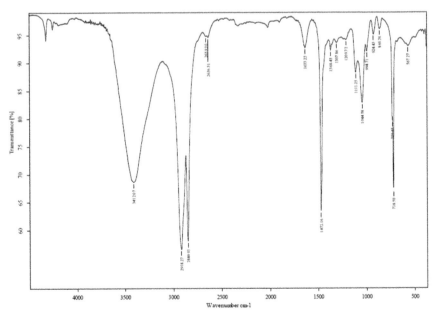

**FIGURE 12.4**　IR-spectrum of UHMWPE containing benzyl alcohol.

The results obtained in studies of UHMWPE specimens were examined by various methods, including using dielectric relaxation spectroscopy (DRS), and differential scanning calorimetry (DSC). It was found a significant influence of the introduction of small additions of oxides of zirconium and hafnium on the properties of UHMWPE.

The measurements were carried out using the device Broadband Dielectric Impedance Spectrometer BDC 80 and the DSC 8500.

In the study of matter is determined by the DRS dielectric loss tangent at certain temperatures. UHMWPE samples containing the metal oxides as well as starting UHMWPE were investigated by DRS at temperatures ranging from −170°C to −150°C.

In general, the spectra of DRS starting and the modified with oxides UHMWPE have similar features, but the dielectric loss tangent at temperatures from −80°C to −50°C is higher, reaching $0.01 \times 10^{-3}$ for samples containing hafnium oxide. Somewhat lower values of dielectric loss angle determined

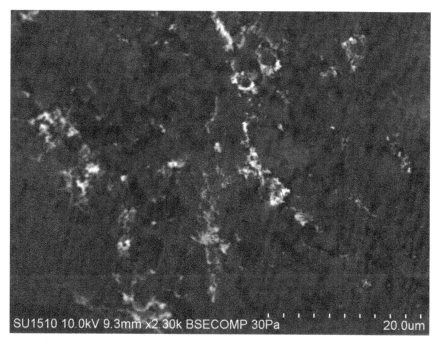

**FIGURE 12.5** Ultrafine particles and agglomerates of the oxides on the surface of the grain UHMWPE.

for samples nanoparticulate oxides of titanium and zirconium, whereas the starting UHMWPE dielectric loss tangent within this temperature ranges on the order less. Similar values of dielectric loss tangent obtained at temperatures of 60–90°C. At the moment the theoretical basis of the data insufficiently formulated and is being studied, but the increase of the scattering of electromagnetic energy in the environment of UHMWPE samples containing oxides of titanium, zirconium and hafnium indicates a greater polarizability of the medium, due to the presence of oxide particles. Also, the dependence of the dielectric loss tangent of the chemical nature of the filler, in the presence of hafnium oxide value of the dielectric loss tangent is higher than in the presence of oxides of zirconium and titanium, which is correlated with the data of DSC, demonstrating a greater impact on the hafnium oxide supramolecular polymer packaging. The latter relates to the degree of polarization of the surface of the filler particles. Thus, the data received by DRS confirm a theoretical model relating the degree of polarization of the surface of the filler particles with the influence on the properties of the polymer matrix.

During the investigations by DSC revealed phase transitions in materials based on UHMWPE containing nanoparticles of transition metal oxides, not typical for the starting UHMWPE. Thus it is confirmed previous theoretical position expressed about the impact of nano-sized additives on the supramolecular structure of UHMWPE. This is reflected in an increase in the degree of crystallinity, which has a beneficial effect on the properties of materials, as most of the polymer macromolecules ordering packaging promotes the values of the strength, abrasion resistance. Especially noteworthy the growth of the initial melting point of UHMWPE-based materials containing nanoparticles of oxides of zirconium and hafnium, and a more pronounced influence of hafnium oxide nanoparticles that can be attributed to a greater polarization of the surface of these particles. Table 12.1 shows the data obtained from studies based on material samples of UHMWPE containing oxide nanoparticles by DSC.

## 12.5   CONCLUSION

Thus, the approach is designed to produce nanomodified polymer composites, allowing to obtain oxides of zirconium and hafnium directly into a polymer matrix of UHMWPE using reactivity of complexes of zirconium and hafnium halides with acetophenone by reaction with benzyl spirit.

**TABLE 12.1** Research Results of Materials on the UHMWPE Basis by DSC

| Sample | Phase transition temperature 1, °C | Phase transition temperature 2, °C | Phase transition temperature 3, °C | Melting point, °C |
|---|---|---|---|---|
| UHMWPE | None | None | None | 110 |
| UHMWPE + $TiO_2$ 0.1% | 76 | 89 | 98 | 112 |
| UHMWPE + $ZrO_2$ 0.1% | None | 91 | 100 | 114 |
| UHMWPE + $HfO_2$ 0.1% | None | 93 | 101 | 116 |

## KEYWORDS

- hafnium oxide
- Nanoparticles
- nonaqueous media
- UHMWPE
- zirconium oxide

## REFERENCES

1. Rockenberger, J., Scher, E. C., & Alivisatos, A. P. A new nonhydrolytic single precursor approach to surfactant-caped nanocrystals of transition metal oxides. *J. Am. Chem. Soc.* 1999, vol. 121, no. 49, pp. 11595–11597. doi: 10.1021/ja993280v.
2. Shim, M., & Guyot-Sionnest, P. Organic-capped ZnO nanocrystals: Synthesis and n-type character. *J. Am. Chem. Soc.* 2001, vol. 123, pp. 11651–11654. doi: 10.1021/ja0163321.
3. Joo, J., Yu, T., Kim, Young W., Park, H. M., Wu, F., Zhang, J. Z., & Hyeon, T. Multi-gram scale synthesis and characterization of monodisperse tetragonal zirconia nanocrystals. *J. Am. Chem. Soc.* 2003, vol. 125, pp. 6553–6557. doi: 10.1021/ja034258b.
4. Jun, Y.-W., Casula, M. F., Sim, J.-H., Kim, S. Y., Cheon, J., & Alivisatos, A. P. Surfactant-Assisted Elimination of a High Energy Facet as a Means of Controlling the Shapes of $TiO_2$ Nanocrystals: *J. Am. Chem. Soc.* 2003, vol. 125 no. 51, pp. 15981–15985. doi: 10.1021/ja0369515.
5. Tang, J., Fabbri, J., Robinson, R. D., Zhu, Y., Herman, I. P., Steigerwald, M. L., & Brus, L. E. Solid-Solution Nanoparticles: Use of a Nonhydrolytic Sol–Gel Synthesis To Prepare $HfO_2$ and HfxZr1-xO2 Nanocrystals. *Chem Mater.* 2004, vol. 16, no. 7, pp. 1336–1342. doi: 10.1021/cm049945w.

6. Niederberger, M., Garnweitner, G., Krumeich, F., Nesper, R., Cölfen, H., & Antonietti, M. Tailoring the surface and Solubility Properties of Nanocrystalline Titania by a Nonaqueous in situ Functionalization Process. *Chem. Mater.* 2004, vol. 16, no. 7, pp. 1202–1208. doi: 10.1021/cm031108r.

7. Bakshi, S. R. The nanomechanical and nanoscratch properties of MWNT reinforced Ultrahigh-Molecular-Weight Polyethylene coatings. *J. Minerals. Metals and Materials Society*, 2007. vol. 59, no. 7. pp. 50–53.

8. Park, S.-J. Preparation and characterization of layered silicate-modified Ultrahigh-Molecular-Weight Polyethylene nanocomposites. *J. Ing. Eng. Chem.*, 2005. vol. 11 no. 4. pp. 561–566.

9. Peltzer, M. Thermal characterization of UHMWPE stabilized with natural antioxidants. *J. Thermal Analysis and Calorimetry,* 2007, vol. 87, no. 2, pp. 493–497. doi: 10.1007/s10973-006-7453-1.

10. Chae-Woong Cho, Sangkyu Lee, Sung-Churl Choi, Ungyu Paik, & Jea-Gun Park. Microstructural evolution of $BaTiO_3$/ultrahigh-molecular-weight poly(ethylene) (UHMWPE) cast body: influence of free organic additive in a nonaqueous medium. *Journal of Electroceramics*, 2006, Vol. 17. Issue 2–4. 345–349. doi: 10.1007/s10832-006-7725-y.

11. Nemeryuk, A., Sudarikova, E., Lylina, M., Romanova, E., Ramesh, A., & Omelchenko, A. Synthesis of ultrafine metal oxide particles in non-aqueous media as a method of modification of UHMWPE *Chemical Industry Today* 2013, vol. 12, 27–37.

12. Nemeryuk, A., Sudarikova, E., Lylina, M., Romanova, E., & Zhdanovich, O. Modification of ultrahigh molecular weight polyethylene oxide nanoparticles of transition metals. *Chemical industry today* 2013, vol. 11, 25–28.

13. Niederberger, M., & Garnweitner G. Organic reaction pathways in the nonaqueous synthesis of metal oxide nanoparticles. *Chem. Eur. J.* 2006, vol. 12, no. 28, pp. 7282–7302. doi: 10.1002/chem.200600313.

# CHAPTER 13

# POLYANILINES: TWO CASES STUDIES AND UPDATES

G. E. ZAIKOV

*Russian Academy of Sciences, Moscow, Russia*

## CONTENTS

## 13.1   FIRST CASE STUDY

Discovery of the polymers possessing by own electroconductivity due
to the presence of the system of alternant $\sigma$- and $\pi$-bonds results in the
occurrence and the development of new fundamental areas of researches
in chemistry, physics, materials science, etc. [1–3]. Electroconductive
polymers (ECP) combine high flexibility and plasticity, typical for the

polymers with high electroconductivity, whose value may approach to the conductivity of metals. This property of ECP has led that they are also often called as "synthetic metals." Among the large number of modern organic and inorganic materials the electroconductive polymers are valued as "strategic" materials, that have become by the objects of intense researches in the laboratories of the world's leading scientific centers and major industrial corporations [4].

Polyaniline (PAn) and polymers based on its derivatives occupy an especial place among the electroconductive polymers. Due to the special electronic spatial structure of aniline and its derivatives, their polymers have a wide range of physical and chemical properties, such as the ability to the reversible oxidation-reduction, long-continued resistance to not only moisture or air, but so much more aggressive media, etc. The combination of low prime cost with ease of synthesis makes such materials indispensable in molecular electronics, biochemistry, medicine, chemical current sources, etc. The color of PAn depends on the $pH$ value and on the applied potential, and can vary from light-yellow (at $pH = 1$ and the electrode potential $E = -0.2$ $V$) to purple-red ($pH = 4$ and $E = + 1.4$ $V$), passing through yellow, green, dark-blue, black, and many their tints [5, 6]. This property of PAn was the basis for its use as $pH$ indicators [5, 7], in electrochromic [6, 8–11] and electroluminescent [12, 13] devices. Polyaniline's layers possess by good corrosion properties during protection of steel [14–17], zinc [18], copper [19], aluminum and its alloys (see Chapter 8). Thin layers of PAn are used in molecular electronics [2, 20], as detectors and dosimeters of $\gamma$-radiation [21], in the manufacture of various types of sensor devices for determining of pH [22, 23], viruses [24], and also important analytes such as ammonia [25], $H_2O_2$ [26], $NO_2$ [27], CO [28], HCN [29], $H_2$ [30], $CH_4$ [3], glucose [32], carbamide [33], etc. Polyaniline and its derivatives are used as cathode materials in chemical sources of electrical energy [34, 35] and electrochemical supercapacitors [36–39].

The polymerization of aniline is very easily carried out chemically in aqueous acidic solutions using as oxidizing agents the peroxide compounds (in particular, peroxydisulfates [40–42], hydrogen peroxide [7, 43], benzoyl peroxide [44]), the ions in higher oxidation level ($Fe^{3+}$, $Ce^{4+}$, $Cr_2O_7^{2-}$, $IO_3^-$, $ClO_2^-$, $VO_3^-$) [45–50], the oxides of transition metals ($V_2O_5$ [51],

$WO_3$ [52], $MnO_2$ and $PbO_2$ [50]) and enzymes [53–56], etc. Polyaniline layers on the surface of metals or semiconductors were also obtained by the method of vacuum deposition [57, 58] or by the polymerization in plasma [59, 61]. However, often the method of electrochemical polymerization is used due to the opportunity to get thin polymeric layers on the surface of the conductive substrate. It has been reported about obtaining of PAn on the platinum electrode during its polarization under potentiodynamic [62] or galvanostatic [63] conditions both in the aqueous [62–64] and organic [65] media. In all cases, the structure and properties of the polymerizates are differed significantly, mainly due to varying degrees of protonation of nitrogen atoms in the polymeric chain. This difference leads to the volatility in the conductivity of the synthesized samples in a very wide range [66] from $10^{-15}$ to $10$ S·cm$^{-1}$.

With the use of *IR*, *UVis*, *ESR*, *NMR*, Raman and *X*-ray photoelectron spectroscopy methods, it was confirmed an existence of three forms of PAn, when the chain is a sequence: 1) only reduced (which contain only benzene rings)

dimer fragments, or, *so-called*, leucoemeraldine, **L**; 2) only oxidized (which contain only quinoid fragments)

dimer segments, or, *so-called*, pernigraniline, **P**; 3) reduced and oxidized dimer links, which are alternated, or, *so-called*, emeraldine, **E** [67−70]. In addition, the forms of PAn may be varied by degree of the protonation of nitrogen in the main chain.

MacDiarmid et al. [71], by measuring of the potential value of the platinum electrode during the chemical oxidative polymerization of aniline in aqueous acidic solutions $(NH_4)_2S_2O_8$, found that during the process it changes from +0.4 $V$ (initial value) to +0.66 $V$ and then to +0.75 $V$ (immediately and after 2 minutes after adding of the oxidant in the reaction mixture, respectively). After the 10 minutes period, the potential is decreased again to the +0.47 $V$. All fragments of polyaniline, which was formed at the potential +0.75 V, are oxidized, that is the polymer is in the form of pernigraniline, while the product formed at +0.43 V, is the emeraldine. This enabled the authors to the conclusion that the main product of the aniline polymerization is emeraldine that can be formed by at least in two ways: as a result of the pernigraniline reduction and oxidation of aniline by pernigraniline, respectively. According to experimental data, the more likely there is a second way.

Mechanism of the aniline formation has been the subject of many studies. Formation of the structure of polyaniline proceeds through the stages of the oxidation of aniline (an electron detachment) with the following recombination of radical particles-intermediates that leads to an elongation of the polymer chain. It is proposed, that the oxidation of aniline's molecule is single-electron process whereby the cation-radical is formed [72], that was confirmed by ESR spectra during the electrooxidative deposition of PAn in 1.0 $M$ $H_2SO_4$ aqueous solution at the platinum electrode under scanning potential range from −0.1 to +0.85 $V$ [73]. In this case, the strong dependence of the intensity of the ESR signal of polyaniline on electrode potential was observed in the range +(0.2–0.8) $V$. Possible ways to convert the aniline's cation-radicals were the subject of the discussion in many papers [5, 70, 74–77], but the proposed schemes of the transformations of aniline are imperfect since the existence of various tautomeric forms of particles-intermediates and the optimality of the proposed routes of further transformations were not confirmed. Moreover, with the exception of the *MacDiarmid's* and *A. Epstein's* article [78], the chemical mechanism of the initiating of aniline's polymerization was almost not considered. Therefore, the mechanism of oxidative polymerization of aniline and their derivatives in aqueous solutions, particularly in the presence of peroxydisulfate-anions it was analyzed by us in detail, based on the results of electro-chemical studies and quantum-mechanical calculations.

## 13.2 THE MECHANISM OF OXIDATIVE POLYMERIZATION OF ANILINE IN AQUEOUS SOLUTIONS

### 13.2.1 ANALYSIS OF THE INITIAL STAGES OF OXIDATION OF ANILINE

Cyclical voltammogram of aniline obtained in acidic aqueous solution at a platinum electrode is shown on Figure 13.1. The first maximum of oxidation current at +0.22 $V$, the height of which increases with each successive cycle of the potential scanning was attributed to the oxidation of neutral forms (i.e., $L$ and $E$) of polyaniline to cation-radicals, namely $L^{+\cdot}$ and $E^{+\cdot}$, respectively [79]. According to the results of Ref. [40], this maximum corresponds to the transition into the redox-couple "leucoemeraldine $\leftrightarrow$ emeraldine." The second maximum at +0.8 V corresponds to the transformation of emeraldine into pernigraniline, which is accompanied by the transfer of one electron and by the detachment of two ions [73] $H^+$. Transformation of emeraldine $\leftrightarrow$ pernigraniline in the acetonitrile can be also occurred as a result of chemical oxidation of emeraldine by pyridine or by superoxide-ion $O_2^{-\cdot}$ (in the presence of dissolved molecular oxygen in solution) [80]. The attempts of the interpretation of the third (middle)

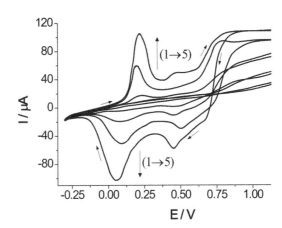

**FIGURE 13.1** Cyclical voltammogram of platinum electrode in 0.1 $M$ solution of aniline during the first five cycles of the potential scanning (base electrolyte is 1 $M$ HCl aqueous solution; $s_E = 20$ mV·c$^{-1}$) (Note: Here and hereinafter in this chapter, all electrode potentials are presented relatively saturated Ag/AgCl electrode).

peak were made in the Refs. [81, 82], under which it can be attributed to secondary transformations of polyaniline that result in the reduction of conductivity of polyaniline's layer on the electrode surface. In particular, such secondary transformations can be cross-linking of polymeric chains through nitrenium-cation [81], or the oxidation of products of partial degradation of PAn, which is formed at the polarization of the electrode under potentials over than $+1.2\ V$ [83]. Growth of the heights of maxima of oxidation current during each subsequent scan cycle of potential suggests that they are associated with the conversion of polymeric aniline, which is formed on the electrode.

According to the conventional scheme (*see* Figure 13.2), the oxidation of aromatic amines occurs through several stages, which are covering the transfer of two electrons and the detachment of two $H^+$ ions [72]. This scheme can be considered only preliminary, because it does not detail the mechanism of electron transfer, and does not estimate the likelihood of the proceeding of process accordingly with each two possible ways of its occurrence. There are two possibilities for the detachment of the second electron. The first is that possible direct electron transfer (the oxidation of cation-radical **II** to doubly-charged cation **III**), while in the second stage that precedes the deprotonation of cation-radical to the radical **IV**.

Moreover, given scheme ignores also the possibility of two ways of one-electron oxidation of original aniline [84]. According to the first, presented on the scheme, the electron detachment is occurred from $2s$ level of nitrogen atom

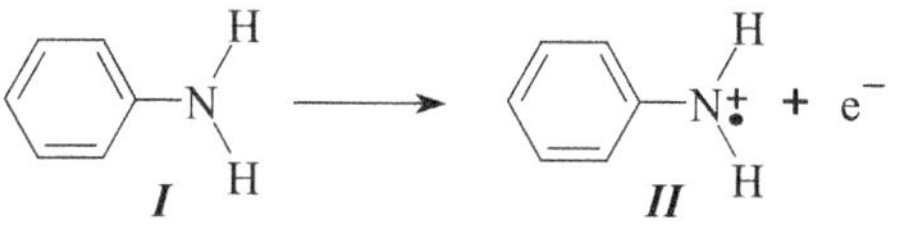

$$ \text{(structure I)} \longrightarrow \text{(structure II)} + e^- \qquad (1) $$

$$ C_6H_5-NH_2 \xrightarrow{-\bar{e}} C_6H_5-\overset{+\bullet}{N}H_2 \xrightarrow{-\bar{e}} C_6H_5-NH_2^{2+} $$
$$ \qquad I \qquad\qquad II \ \big|{-H^+} \qquad\qquad III \ \big|{-H^+} $$
$$ C_6H_5-\overset{\bullet}{N}H \xrightarrow{-\bar{e}} C_6H_5-NH^+ $$
$$ \qquad IV \qquad\qquad\qquad V $$

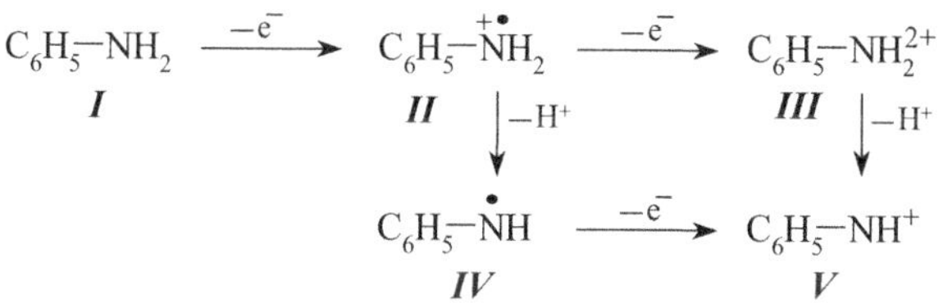

**FIGURE 13.2** The sequence of transformations of aniline during its oxidative polymerization according to Ref. [72].

leading to the formation of cation-radical $II$. At the same time, it is also possible one-electron oxidation of aniline's molecules, which will be accompanied by heterolytic breaking of N–H bond

$$\tag{2}$$

Also it is necessary to take into account the possibility of isomerization of cation-radical $II$ and radical $IV$, which proceed according to the following schemes:

$$\tag{3}$$

$$\tag{4}$$

The results of quantum chemical calculations, which have been done for the particles $I$–$VII$ [84], showed, an increase of the oxidation level in sequence aniline $\rightarrow$ cation-radical $\rightarrow$ dication is accompanied by an increase both of total energy $E_{tot}$ of respective particles (from $-103,518$ to $-92,653$ kJ·mol$^{-1}$) and its binding $E_{bin}$ (from $-6,210$ to $-4,131$ kJ·mol$^{-1}$) as well as electronic components $E_{el}$ (from $-421,384$ to $-403,648$ kJ·mol$^{-1}$). At this, the energy of the highest occupied molecular orbital (HOMO) $E_{HOMO}$ regularly is decreased for the same sequence, since as the oxidation of particles is complicated with increasing of their oxidation level. In addition, it was concluded about almost practically identical probability of the existence of benzenoid cation-radical $II$ and isomeric to it cation-radical of quinoid type $VI$, since they are characterized by practically almost identical energy characteristics.

Let's separately analyze the oxidation process of radical $IV$. Geniés and Łapkowski [72] claim that this neutral radical easily can be oxidized to the particle $V$, since as redox-potential of the process $IV \rightarrow V$ is lower compared with the transition $I \rightarrow II$. However, this statement is not agreed with the results of quantum-chemical calculations. Given the fact, that the electron transfer during the oxidation of the particles takes place from HOMO, then the comparison of values $E_{HOMO}$ for aniline and particle **IV** (−8.2135 against −9.2442 eV, respectively) shows, that the transformation **I** → **II** is more probable compared to the **IV** → **V** process [84]. Therefore, at the excess of aniline in the reaction mixture in the beginning its full primary oxidation takes place and then possible further conversion of monomeric particles that lead to the formation of the polymer.

Analysis of the electron occupancy of atomic orbitals (**AO**) of nitrogen atom, conducted in accordance with the Mulliken's method [85] shows the significant delocalization of electrons in a molecule of aniline and intermediates of its oxidation [84]. In particular, on the 2s **AO** of nitrogen into molecule of aniline (**I**) is 1.427 electrons instead of two, while on the $p_z$ **AO** is 1.841 electrons instead of one (*see* Table 13.1). This permits to consider the electronic structure as *sp*-hybridization, since 0.574 electrons move from 2s to 2p level. This transition takes place mainly on $2p_z$ level due to the same type of symmetry (axial) for s- and $p_z$ types of orbitals. It should be noted that the redistribution of the electron density in the molecule of

**TABLE 13.1** Electron Occupancy of Atomic Orbitals of the Nitrogen Atom

| Particle | | Atomic orbitals of the nitrogen | | | |
|---|---|---|---|---|---|
| | | 2s | $p_x$ | $p_y$ | $p_z$ |
| I | α | 1.427 | 1.103 | 1.041 | 1.841 |
| II | α | 0.580 | 0.513 | 0.485 | 0.938 |
| | β | 0.608 | 0.493 | 0.464 | 0.344 |
| III | α | 0.579 | 0.525 | 0.472 | 0.502 |
| | β | 0.586 | 0.527 | 0.474 | 0.545 |
| IV | α | 0.804 | 0.686 | 0.543 | 0.875 |
| | β | 0.799 | 0.673 | 0.512 | 0.170 |
| V | α | 0.760 | 0.741 | 0.523 | 0.346 |
| | β | 0.760 | 0.741 | 0.523 | 0.346 |
| VI | α | 0.605 | 0.532 | 0.485 | 0.918 |
| | β | 0.583 | 0.486 | 0.464 | 0.351 |
| VII | α | 0.809 | 0.692 | 0.549 | 0.867 |
| | β | 0.796 | 0.649 | 0.512 | 0.194 |

aniline causes the delocalization of electrons, so even the hydrogen atom in ortho-position of benzene ring carries an excess positive charge (to 0.108 of elementary charge). In the particles-intermediates of the oxidation process, the delocalization of the electron density is increased and the charge on the same hydrogen atom exceeds 0.199 of elementary charge. The strongest change of the charge is observed for the nitrogen atoms, in particular, from 0.051 (neutral molecule of aniline *I*) to 0.464 and 0.795 of elementary charge (cation-radical *II* and dication *III*, respectively).

## 13.2.2  ANALYSIS OF THE STAGE OF RECOMBINATION OF INITIAL RADICAL PARTICLES

The next stage of the transformations during an oxidative polymerization of aniline is the formation of dimers as the result of recombination of initial radical products of the aniline oxidation, namely particles *II*, *IV*, *VI* and *VII*. It should be noted, that the radical of imine type (*IV*), can take part in the chain transfer to another molecule of aniline as a result of hydrogen atom detachment from it, the most probable in *para*-position

$$\tag{5}$$

which can also takes part in recombination reactions. Moreover, it can be assumed, that the isomerization of initial radical particles (*see* equals Eqs. (3) and (4)) is a sequence of the processes analogous to the Eq. (5), with the following rearrangement of the formed particle into the intermediate of quinoid type.

Also it should be noted, that with the use of the methods of mass- and ion-spectrometry it was revealed the existence of two isomeric ions of aniline with a positive charges on the nitrogen or carbon atoms [86]. This means that the protonation of aniline can be occurred not only at the place of the nitrogen atom but also carbon atoms of the benzene ring. Quantum-chemical calculations performed on *DFT* and *MP4*-levels confirmed the possibility of the proton addition to the carbon atom in the *para*-position of the molecule of aniline [87]. Conducted at this analysis on the basis of the orbital *Fukui* indexes didn't give the clear answer to what is protonation of nucleophilic centers − carbon or nitrogen atom in the *para*-position prevails, because the probability of both processes are virtually identical. Therefore, in our opinion, the oxidation of aniline in theory is possible also in the place of carbon atom in the *para*-position of the benzene ring that will be occurred in accordance with the scheme [88]

$$\qquad\qquad\qquad\qquad\qquad\qquad\qquad\qquad\qquad\qquad\qquad + \ e^- + H^+ \qquad (6)$$

At this, in the scheme of the polymerization transformations of aniline is superfluous the stage of isomerization of nitrene-radical, and the formation of dimer particles will be occurred through direct recombination of nitrene and *para*-aminobenzene radicals. However, since the existence of the latter particle experimentally was not confirmed, the possibility of the process (1.6) we have not taken into account.

In order to standardize the possible processes of recombination can be classified as the interaction of type "head-to-head" (*h*−*h*), "tail-to-tail" (*t*−*t*) or cross-connection of type "head-to-tail" (*h*−*t*) and "tail-to-head" (*t*−*h*). In accordance with the general principles of thermodynamics the most probable will be the structures that have minimal energy. Therefore, the main condition for the formation of bond between the particles, starting from aniline and ending by polymer, is the less total energy of the forming particle compared with the sum of the energies of initial fragments, which, be combined, actually form the particle-product, that is,

$$E_s < \sum E_{f,i} \qquad (7)$$

where $E_s$ is the energy of the compound formation as a result of recombination; $E_{f,i}$ is the energy of initial fragments (radicals or cation-radicals).

According to the results presented in Table 13.2, the ratio (6) in the case of recombination of cation-radicals is not satisfied as to the total energy of the particle and binding energy regardless of the recombination type of particles with each other (direct, namely *h*−*h* and *t*−*t* or cross, type *h*−*t* and *t*−*h*). This result may be due to electrostatic repulsion between of the same charged particles. Exactly by this interaction can be explained the big values of total energy for dimers formed as a result of the interaction of "head-to-head." In the case of uncharged radical it is observed a clear dependence of $E_{tot}$ and $E_{bin}$ on the type of linkage of initial radicals: the highest values are observed for dimers formed by connection type "head-to-tail." Moreover, the energy of dimers of type "head-to-tail" and "tail-to-head", which are formed from the cross-connections, is also different, energetically more favorable is the latter structure, as evidenced by the corresponding values of $E_{tot}$, $E_{bin}$ and $E_{el}$. Therefore, based on quantum-chemical calculations it can be predicted, that the recombination will be easier took place in the case of initial radicals and their linkage will be occurred as a result of the cross-interaction. The recombination of the type "tail-to-head" is more energetically favorable compared with the interaction of "head-to-tail." Geniés and Łapkowski [70] suggested that the formation of PAn is the result of recombination of cation-radicals according to the type "head-to-tail" with the simultaneous deprotonation

$$2\,C_6H_5 - NH_2^{+\bullet} \longrightarrow C_6H_5 - NH - C_6H_4 - NH_2 + 2\,H^+ \qquad (8)$$

However, the minimal values of energy don't agree with the uncharged dimers (*see* Table 13.2). Therefore, the formation of structure

which is a product of recombination of cation-radicals (***II*** + ***VI***) of type "head-to-tail" and the structural unit of future polymeric chain looks more likely.

The proposed mechanism of oxidative polymerization of aniline describes only the main stages of this process, based on the fact that it is developed in accordance with the most likely scheme of the transformations, namely: monomer → dimer → tetramer → ... → oligomer → ..., etc. and with the formation, as a result, of the conductive polymer having the unbranched structure. It should be noted at once, that such a mechanism is implemented in highly acidic (pH < 2) medium. At higher *pH* values

**TABLE 13.2**   Energy Changes in the Result of the Recombination of Initial Radicals

| Initial particles | Total energy of starting particles, kJ·mol$^{-1}$ | | | Type of dimer | Energy of dimers, kJ·mol$^{-1}$ | | |
|---|---|---|---|---|---|---|---|
| | $-E_{tot}$, | $-E_{bin}$, | $-E_{el}$, | | $-E_{tot}$ | $-E_{bin}$ | $-E_{el}$ |
| II + **II** | 188,057 | 10,941 | 812,863 | $t-t$ (**VI** + **VI**) | 187,595 | 10,479 | 1,206,770 |
| | | | | $h-h$ (II + **II**) | 205,452 | 10,907 | 1,182,070 |
| | | | | $h-t$ (II + **VI**) | 205,402 | 10,781 | 1,204,690 |
| **IV** + **IV** | 186,349 | 11,759 | 777,434 | $h-h$ (IV + **IV**) | 186,417 | 11,828 | 1,153,140 |
| | | | | $t-t$ (**VII** + **VII**) | 186,530 | 11,858 | 1,110,970 |
| | | | | $h-t$ (IV + **VII**) | 204,334 | 11,916 | 1,152,380 |
| **II** + **IV** | 187,203 | 11,350 | 795,148 | $h-t$ (IV + **VII**) | 205,038 | 11,518 | 1,174,790 |
| | | | | $t-h$ (**VI** + II) | 205,042 | 11,527 | 1,182,140 |

the interaction between the monomeric and di- or trimeric intermediates, as theoretically was shown by stejskal et al. [89], could lead to the formation in a reaction mixture along with the polymer also low molecular oligomeric products, as evidenced by the fact that the yield of the polymeric product is always much lower than 100%. In addition, as a result of intramolecular oxidative cyclization it is possible also the formation of branched tetramers of aniline such as pseudomauveine [89] which is well-known by-product of the oxidation of aniline [90], which, however, may also participate in the polyaniline's chain propagation [91]. The formation of these and other similar products, which is occurred at $pH = 2.5-10.0$ and/or weak oxidant is analyzed in detail in Refs. [92, 93].

### 13.2.3   AN INTERACTION ANILINE – PEROXYDISULFATE

Chemical method of the PAn's synthesis, particularly as a result of oxidation of the initial monomer by peroxydisulfate in acidic aqueous solutions is probably the most common. Moreover, since the processes of chemical and electrochemical oxidation of aniline are occurred almost identically, and the obtained products are neared both in structure and physical-chemical properties, it is likely that the radical intermediates, which are formed during the electrochemical and chemical processes are also identical. Therefore, the chemical oxidation of aniline with peroxydisulfate ions has been analyzed separately [84].

It is well know, that the aromatic amines in combination with the peroxides form a redox system. The first stage of the process is the formation of complex "PAn − peroxydisulfate anion," the decomposition of which further leads to the generation into a system of radical nature particles, which respectively initiate further polymerization [94]. The possibility of the formation of complex was investigated by us using the method of cyclic voltammetry. In particular, the influence of the peroxydisulfate's admixtures on the process of electrochemical oxidation of aniline has been analyzed. According to the cyclical voltammograms of aqueous solutions of aniline, which are shown on Figure 13.3, after adding of $(NH_4)_2S_2O_8$ the position of the first maximum of oxidation [−(0.2−0.24) $V$] virtually is unchanged compared to the cyclical voltammograms in the absence of peroxydisulfate ions. The height of this maximum increases with the increase of the concentration of peroxydisulfate because it accelerates the formation of PAn, which was recorded visually.

At the same time, an impact of the aniline's additives on the reducing process of $S_2O_8^{2-}$-anions is much appreciable. In this case (*see* Figure 13.4) for the process it was observed the shift of potential towards more positive potentials. In particular, at the addition of 0.01 $M$ of aniline solution to 0.01 $M$ $(NH_4)_2S_2O_8$ solution the reduction potential of $S_2O_8^{2-}$-anions is shifted from −1.47 to −1.05 $V$. But, an increase of the concentration of aniline to 0.02 $M$ does not lead to further significant shift of the reduction

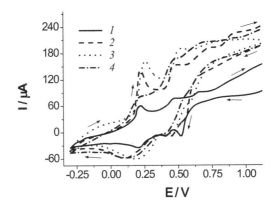

**FIGURE 13.3**   Cyclical voltammograms of platinum electrode in 0.1 $M$ solution of aniline + 1 $M$ aqueous solution of HCl in the presence of $(NH_4)_2S_2O_8$ at the concentration, $M$: 0.0 (1); 0.01 (2); 0.02 (3) and 0.04 (4) ($s_E$=20 $mV \cdot s^{-1}$).

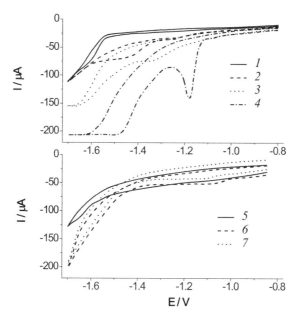

**FIGURE 13.4** Cyclical voltammograms of platinum electrode in aqueous solution of aniline + 0.01 $M$ $(NH_4)_2S_2O_8$ + 1 $M$ KCl at the concentration of aniline, $M$: 0.0 (1); $1\times10^{-3}$ (2); $2\times10^{-3}$ (3); $5\times10^{-3}$ (4); $8\times10^{-3}$ (5); $1\times10^{-2}$; (6) та $2\times10^{-2}$; (7) ($s_E$= 20 $mV·s^{-1}$).

potential of the $S_2O_8^{2-}$-anions. The highest rate of reduction was observed at addition of $2\times10^{-3}$ $M$ aniline solution when the magnitude of the oxidation current reaches a value 41 µA. This effect may be due to the catalytic influence of small amounts of aniline, while a further increase of its content can inhibit the active centers on the surface of the electrode.

Maximal shift of the potential is observed at equimolar (1:1) ratio of aniline and peroxydisulfate, so, obviously that this is the stoichiometric ratio of the components in the intermediate reactive complex. For such systems the donor-acceptor nature of this complex is the most likely. It is obvious that in a couple "aniline-peroxydisulfate" the donor is aniline, while the acceptor of the electrons is peroxydisulfate (its peroxide group). The calculated by us values of total energy and binding energy for the complex **An**–$S_2O_8^{2-}$ were lower on 89.25 and 89.41 kJ·mol⁻¹ respectively, compared with isolated initial particles, and the value of the enthalpy of complex formation[84] $\Delta_f H$ was –93.65 kJ·mol⁻¹. Since the values of the all energies are decreased with the formation of the complex, it shows its energy advantage.

The decomposition of complex $An–S_2O_8{}^{2-}$

$$(9)$$

leads to the formation of aniline's cation-radical, $SO_4{}^{\cdot-}$ anion-radical and $SO_4{}^{2-}$ anion. Further, the particles of radical nature can enter into the secondary chemical transformations. In our opinion, the anion-radicals can recombine both with the cation-radicals of aniline

$$(10)$$

and also to oxidize the neutral aniline to radical *IV*

$$C_6H_5 - NH_2 + SO_4^{\cdot-} \longrightarrow C_6H_5 - \overset{\cdot}{N}H + HSO_4^{-} \qquad (11)$$

At the same time, it is possible the simultaneous occurrence of these two processes, resulting in the formation the dianiline sulfate, which next will be decomposed into aniline's radical and hydrosulfoaniline:

$$(12)$$

In accordance with the Glarum and Marshall [73] during the electrochemical oxidation of aniline in the range of electrode potentials $+(0.05–2.00)$ $V$ it was found the greatest intensity of the ESR signal, which indicates the presence of paramagnetic particles. In this case, the ESR spectrum contained only one absorption line with the splitting factor equal to $2.00270 \pm 0.00005$. It was assumed that this signal corresponds to the product of one-electron oxidation of aniline. Therefore, the assumption about the formation of particle *IV* during the chemical oxidation of aniline is quite reasonable.

## 13.3 AN IMPACT OF THE SUBSTITUENT'S NATURE ON MECHANISM OF THE SYNTHESIS, STRUCTURE, AND PROPERTIES OF POLYANILINES

In view of the above-proposed mechanism of polymerization, the significant effect on the rate of the process and the structure of products will have the nature and position of the substituent in benzene ring of aniline. Therefore, we have studied the process of polymerization of nitro- [84, 94] and oxymethyl [95, 96] derivatives of aniline. This choice was caused primarily by the difference in electronic properties of the substituents, since as $NO_2$-group has strong electron-acceptor properties, whereas the group $CH_3O$-group possesses by electron-donor properties as well as by voluminosity of the substituents and, therefore, by possible creation of spatial difficulties during the polymerization process at the stage of the radicals recombination.

### 13.3.1 COPOLYMERS OF ANILINE AND NITROANILINES

An attempt of chemical synthesis of polymeric nitroanilines was performed in 1 *M* HCl aqueous solutions, using the $(NH_4)_2S_2O_8$ as oxidant [94]. Despite the fact that the color change of reaction mixture clearly pointed to the interaction between the components of the system, but the polymer formation even after 24 hours of stirring the reaction mixture was not observed. In view of the solubility of the products of interaction, it was concluded that the polymerization process is stopped at the stage of the formation of dimers. Moreover, it was found that the introduction of *ortho*-nitroaniline in the reaction mixture during the chemical synthesis of aniline significantly reduces the yield of polymeric product. At this, the results presented in Table 13.3 show, that the yield of polymeric product the smaller the higher is content of the nitro-derivative in the initial polymerization mixture and thus nitro-aniline derivatives inhibit the radical chain process. It can be assumed, that in this case the initial cation-radical of nitroaniline reacts with the same particle on the place of oxygen atom

**TABLE 13.3**  The Yield of PAn and of Copolymer Aniline:o-Nitroaniline During Their Chemical Synthesis Depending on Molar Ratio of Monomers ($r$) into Reactive Mix at the Use of $(NH_4)_2S_2O_8$ as Oxidant (the Duration of the Synthesis was 24 hours), and Also the Conductivity of the Obtained Samples

| $r$(aniline: o-nitroaniline) | 10:0 (PAn) | 9:1 | 8:2 | 7:3 | 6:4 | 5:5 |
|---|---|---|---|---|---|---|
| Yield of the product/% | 75.2 | 66.9 | 63.6 | 53.3 | 50.3 | 51.1 |
| Conductivity/ $S{\cdot}cm^{-1}$ | $3.02{\times}10^{-3}$ | $1.06{\times}10^{-3}$ | $1.39{\times}10^{-3}$ | $3.35{\times}10^{-4}$ | $2.25{\times}10^{-4}$ | $1.09{\times}10^{-5}$ |

of nitro-group. As a result, poorly reactive dimer of nitroaniline is formed, unable to further polyaddition reactions, and therefore the polymerization process terminates at the stage of the dimers formation.

In order to interpretate further of the obtained results, the copolymerization of aniline with various nitroanilines was studied by cyclic voltammetry. It turned out that unlike to aniline (see Figure 13.1) for which there are three peaks of currents oxidation, for nitroaniline only one weak peak is characteristic in the vicinity of +0.2 $V$ (see Figure 13.5), and then the working electrode scanning in the range of potentials (−0.3) − (+1.3) does not result in the formation of polynitroanilines on the electrode surface. At the same time, the significant changes was not observed on the cyclical voltammograms of acidic aqueous solutions of aniline at the introduction into them of additives of o-, m- or p-nitroanilines (see Figure 13.6), which may be evidence of the similarity of chemical or electrochemical oxidation of aniline and nitroanilines.

### 13.3.1.1  Spectral Properties and the Structure of Polyaniline and Copolymers of Aniline With Its Nitro-Derivatives

The structure of the synthesized samples of polyaniline and copolymers of aniline with o-nitroaniline was analyzed on the basis of their spectral properties [94]. FTIR-spectra of the investigated samples, pressed into tablets with KBr, were recorded in the spectral range 4400–400 $cm^{-1}$ using the spectrophotometer BRUKER IFS 66. For the obtaining of Raman spectra it was used the spectrophotometer BRUKER FRA106 with cooled liquid nitrogen, MCT detector and Nd: YAG laser ($\lambda = 1\,064$ $nm$) as the excitation source.

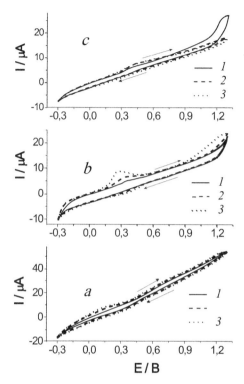

**FIGURE 13.5** Cyclical voltammograms of platinum electrode in 0.01 $M$ solutions of *ortho-* (*a*), *meta-* (*b*) and *para-*nitroaniline (*c*) + 1 $M$ aqueous solution of HCl during the first (1), second (2) and tenth (3) cycles of the potential scanning at the rate of scanning $s_E = 20 \, mV \cdot s^{-1}$.

Analysis of IR spectrum of polyaniline obtained by chemical method (*see* Figure 13.7, *a*), testified that it represents by itself the oxidized form of emeraldine. This conclusion is based on the presence of absorption bands in the spectrum corresponding to the valence stretching vibrations of bonds C=N (bands at 3434 and 1659 cm$^{-1}$), $^+$N–H (2921 and 1291–1238 cm$^{-1}$) and C–N= (1103 and 4963 cm$^{-1}$) [97], typical for this modification of PAn. The presence of >C=N$^+$–group in the structure of PAn was determined based on the Raman spectrum (*see* Figure 13.8), where there is a clear splitted band at 2200–2100 cm$^{-1}$. Wide band in the range of 3500–2300 cm$^{-1}$ (*see* Figure 13.8) can be attributed to stretching vibration of unoxidized group >C=N–. In addition, the band in the same area can generate the vibrations of C–N–group, and also quaternized

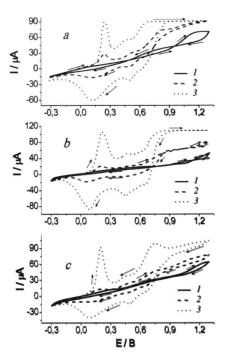

**FIGURE 13.6** Cyclical voltammograms of platinum electrode in 0.01 $M$ solutions of aniline + 1 $M$ aqueous solution of HCl with the additives of 0.01 $M$ solutions of *ortho-* (*a*), *meta-* (*b*) and *para*-nitroaniline (*c*) during the first (1), third (2) and fifth (3) cycles of the potential scanning ($s_E = 20 \ mV \cdot s^{-1}$).

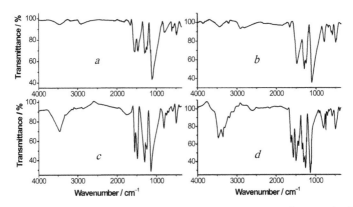

**FIGURE 13.7** FTIR spectra of samples of polyaniline (*a*) and copolymers of aniline and o-nitroaniline at molar ratio of the monomers 9:1 (*b*), 8:2 (*c*) and 7:3 (*d*).

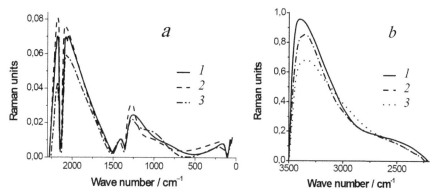

**FIGURE 13.8**  Raman spectra of polyaniline (1) and copolymers of aniline and *o*-nitroaniline at molar ratio of monomers 8:2 (2) and 7:3 (3) in the spectral areas 3500–2250 (*a*) and 2250–2270 *cm*$^{-1}$ (*b*), respectively.

nitrogen atom, whose structure is similar to unoxidized electrically neutral atom. This allows us to propose the following approximate formula for the synthesized sample of polyaniline

which is formed when the oxidation of initial molecule of aniline proceeds as the result of electron transfer from the 2*s* sublevel of the nitrogen atom.

Analysis of the spectra of copolymerization products obtained at different molar ratios (*r*) aniline: *o*-nitroaniline in the reaction mixture (*see* Figure 13.7, *b–d*), is complicated due to partial overlapping of the absorption bands corresponding to nitro-group ($v_{as}$ = 1650–1500 cm$^{-1}$ and $v_s$ = 1370–1250 cm$^{-1}$), with the bands corresponding to valence stretching vibrations of C–N bond (at 1360–1250 cm$^{-1}$) and deformation vibrations of (–HN$^+$–) group (1600–1500 cm$^{-1}$). The main differences in the spectra of copolymerization products are observed in the range of wave numbers from 1800 to 1200 cm$^{-1}$ and in the high-frequency region 3000–2750 cm$^{-1}$. Polymerization of the initial mixture at *r* = 7:3 leads to the appearance in spectrum of the product the band at 1626 cm$^{-1}$, which can be attributed to asymmetric stretching vibrations of NO$_2$-group. At the same time, for the all investigated ratios of the components in the spectrum is evident band at 616–582 cm$^{-1}$ associated with the presence of $SO_4^{2-}$-ions in the product. The form and the intensities ratio of the adsorption bands at 3442–3476 and 2900–300 cm$^{-1}$ are varied depending on the initial composition of the reaction mixture. So, for the ratio of 7:3 the band at ~3400 cm$^{-1}$ is splitted into two maxima at 3476.4 and 3353.9 cm$^{-1}$. Since the frequencies of the – HN– ta –HN$^+$–groups vibrations correspond to these bands, it indicates the presence both of oxidized and unoxidized links in a polymer chain.

Comparison of Raman spectra (*see* Figure 13.8) allows us to follow the trend of the changes in the oxidation level of polymer on place of N=C– group, using a decrease in the intensity of the absorption band at 2173 cm$^{-1}$ and increase of the bandwidth with a maximum at 2060 cm$^{-1}$. At the same time, the intensity of the band at 1282 cm$^{-1}$, which corresponds to asymmetric vibrations of >C–N<–group increases, suggesting the possibility of some decrease of oxidation level of products of compatible polymerization of aniline and *o*-nitroaniline compared to homopolymer of aniline. In spectra both of PAn and copolymer the band at 1432–1488 cm$^{-1}$ is observed, which is associated with the vibrations of N=N–group [67]. The presence of N–N bonds in the structure of polymer indicates the possibility of recombination of the type "head-to-head" at the stage of the dimers formation. The possibility of such units connection in the polymer chain was also considered in Ref. [67]. Thus, according to IR spectroscopic studies the fragment of polymeric chain containing the nitroaniline link can be described as follows:

$$\text{(1.12a)}$$

Exactly this structure was used by us to interpret the data of element analysis for the calculation of theoretical composition of copolymer.

It is well known, that the neutral polyaniline, no matter in which of the three forms it is existed, is the typical insulator with a size of an conductivity ~10$^{-10}$ *S*·cm$^{-1}$, and its transformation into the high-conductive state (doping) can be made by chemical or electrochemical ways, otherwise acidic or oxidative [2]. With the use of the method of impedance spectroscopy it was found two components of the conductivity of polyanilines, namely electronic and ionic [98]. In order to explain the electroconductivity in terms of electronic component when the charge is transferred by the electrons of the conduction band of PAn, the notion of *solitons* and *polarons* is used [1]. However, in our view, the spread of the soliton hypothesis proposed for the processes for polyacetylene doping, on the other, including the polyaniline, conductive polymers, are not sufficiently justified, as this virtual structures is the result of the *Schrödinger* equation solution. At the same time, the polarons and bipolarons, whose formation is a high probable at the initial stages of the polymerization process, can be interpreted as a cation-radical center or positively charged center in the singlet state in the polymer chain with a conjugated system of $\pi$-bonds, which are often used in chemistry. The increase of electronic conductivity of polyaniline, which is determined by the number of paramagnetic centers on the chain (polarons or bipolarons) can be achieved during its oxidation. On the other hand, the conductivity of PAn increases also as a result of the processing of polyaniline base with strong acids, (so-called acidic doping with the formation of the salt of appropriate forms of PAn) due to the quaternization of nitrogen atom of imine group as a result of donor-acceptor interaction with $H^+$ ions. In this case, the charge will be transferred mainly by counterions. In turn, ionic component of the conductivity associated with such macrocharacteristics of polymer as the porosity, crystallinity degree, orientation of macrochains, globularity or fibrillation of the structure of polymer. Moreover, when the result of oxidation of PAn is the formation of polarons or bipolarons, then the charge on them is compensated by the ions, which are transferred from the volume of the electrolyte solution through the pores of the surface or due to the deprotonation of imine groups of the polymer and, thus two components of the conductivity of polymers with the system of conjugated $\pi$–bonds are interrelated.

An increase of the number of unpaired electrons result in the growth of electronic component of the conductivity for the all forms of polyaniline. At the same time, the ion component increases with the increase of the number of charges on the chain, which are localized on the quaternized atoms of nitrogen. However, given the possibility of acidic doping and the fact that the polarons is too short-lived particles, it must be concluded that the structure (12) cannot exist long. Moreover, according to data presented in Table 13.3 the conductivity of synthesized samples not reaches the values $1-5$ $S \cdot cm^{-1}$ that corresponds to $50^{-th}$ % degree of the acidic doping [99]. Therefore, the structure of synthesized samples likely can be described by the formula

$$(13)$$

which corresponds to the $H^+$-doped form of emeraldine, namely emeraldine salt with the doping level of polymer 25%, resulting from the values of the conductivity of obtained polycondensates.

### 13.3.1.2 Element Analysis and Thermogravimetry of the Synthesized Copolymers

The results of the element analysis of the obtained samples are represented in Table 13.4. It should be noted at once, that the experimental values of the carbon content were on 1−6 *mass. %* lesser compared with the theoretical values calculated in accordance with the proposed above structure. The content of chlorine and nitrogen was undervalued compared with hydrogen

**TABLE 13.4** The Results of Element Analysis of Synthesized Samples of Polyaniline and Its Copolymers with *o*-Nitroaniline (Analyzer *GA* 1108, *Carlo Erbo*)

| Sample | | Composition, mass. % | | | | | |
|---|---|---|---|---|---|---|---|
| | | C | H | N | S | $SO_4^{2-}$ | Σ |
| Theoretical data | | | | | | | |
| Polyaniline (emeraldine form) | | 62.74 | 4.13 | 12.20 | 6.97 | 20.91 | 99.98 |
| Copolymer | 9.1 | 57.01 | 3.76 | 13.86 | 6.33 | 19.00 | 99.97 |
| (molar ratio | 8.2 | 57.01 | 3.76 | 13.86 | 6.33 | 19.00 | 99.97 |
| aniline: | 7.3 | 57.01 | 3.76 | 13.86 | 6.33 | 19.00 | 99.97 |
| *o*-nitroaniline) | 6.4 | 57.01 | 3.76 | 13.86 | 6.33 | 19.00 | 99.97 |
| Experimental data | | | | | | | |
| Polyaniline (emeraldine form) | | 53.05 | 4.46 | 10.23 | 4.35 | 13.04 | 85.13 |
| Copolymer | 9.1 | 56.15 | 4.32 | 11.58 | 3.97 | 11.82 | 83.87 |
| (molar ratio | 8.2 | 56.00 | 4.19 | 12.77 | 3.81 | 11.43 | 84.59 |
| aniline: | 7.3 | 54.23 | 4.16 | 13.77 | 4.16 | 10.52 | 82.67 |
| *o*-nitroaniline) | 6.4 | 55.75 | 4.09 | 14.21 | 2.71 | 8.13 | 82.18 |

the content of which, in turn, was somewhat overvalued. In addition, it was found, that the total content of the elements (C, H and N) in the sample was only 85 and 95% for the copolymer and PAn, respectively. However, similar results also were obtained by other researchers [100, 101]. According to Kobayashi [100] these experimental data can be explained by the presence of water in the polymer even after its drying under vacuum, the content of which can be ranged from 2 up to 17 *mass. %*. So, if to carry out of the PAn's synthesis under special conditions, then total content of the elements can reach 97%[102]. In addition, the results presented in Table 13.4 show, that both the polyaniline and the all samples of copolymers contain the sulfur. As a result of electrostatic binding of $HSO_4^{2-}$-anions (the product of the reduction of peroxydisulfate, which was used for the synthesis as an oxidizing agent), to remove these anions from the samples by repeated washing with water is practically impossible. So, the anions $HSO_4^{2-}$ thus have the function of doping agent (counter-ion) of PAn. On the other hand, it is also possible chemical binding of the sulfur in the structure of the polymer. As it was shown above, at the initial stages of chemical synthesis of PAn after the destruction of the $An-S_2O_8^{2-}$ complex, which is accompanied by the transfer of an electron from the aniline's molecule to peroxide group, possible the recombination of radical intermediates, namely cation-radicals of aniline and sulfate-anion radicals (*see* eq. (6.11)) with the formation of N—O-bond. The possibility of N—O-bond formation confirms the presence into IR-spectrum of the polymerization product the absorption bands in the range 1302–1238 $cm^{-1}$.

So, if to take into account the analytically determined sulfur, then the total content of the components reaches about ~100%. At this, the excess of hydrogen can be attributed to the presence in synthesized samples of ammonium ions and also water, which was confirmed by the results of the thermogravimetry studies. As shown on Figure 13.9, a slight decrease in mass of the sample (~1.5 *mass. %*) is observed in the temperature range 45–105°C and related, apparently, with the removal of physically bound water. These data are in good agreement with the results of differential scanning calorimetry [103, 104], according to which the endothermic peak at 30–120°C corresponds exactly to this process (desorption of water). During heating of the sample from 30 to 500°C the most significant decrease of mass (~35 *mass. %*) was observed in the temperature range 105–250°C. Spectral analysis of gaseous products of destruction

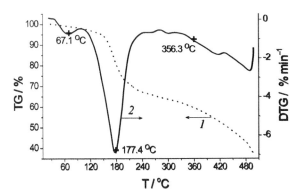

**FIGURE 13.9** Integral (1) and differential (2) derivatogram of the sample of copolymer of aniline and *o*-nitroaniline ($r = 7:3$) (microbalance NETZSCH TG29; atmosphere is Ar; scanning rate of the temperature 15 K·min⁻¹).

of polymer sample with $r = 7:3$ showed that for the basic product characteristic are the absorption bands with maxima at 2359 and 2344 cm⁻¹. Somewhat less intense bands are observed at 668 and 1370–1320 cm⁻¹ (*see* Figure 13.10). If the first two bands can be attributed to stretching vibration of N=C and C=O-groups, the following two asymmetric valence vibrations correspond to $NO_2$-group and an amino-group in aromatic amines [97]. Also somewhat intense band at 668 cm⁻¹ can also be associated with the deformation vibrations of N–H–group.

Electrochemical destruction of polyaniline was studied in detail by Stilwell and Park [83]. They confirmed fixed earlier [100] fact of the

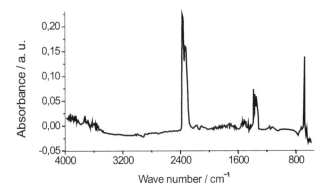

**FIGURE 13.10** FTIR absorption spectrum of gaseous products of thermolysis of the sample of copolymer of aniline and *o*-nitroaniline ($r = 7:3$) after 1364 *s* after the start of thermolysis.

benzoquinone accumulation during the destruction of PAn. At the same time, the final product of the destruction depends on the conditions of the process and the nature of the initial sample. In the presence of water the oxidative hydrolysis of polyaniline takes place, the main products of which are the quinone and $NH_3$:

$$(14)$$

The oxidative degradation of the copolymer, in contrast to electro-chemical destruction, passing through another sequence of transformations, which are different than the proposed in Ref. [83]. First, we took into account the presence of oxidant's particles in the sample. In our opinion, during the thermolysis the particles of radical nature may be generated due to the thermal decomposition of adduct, which is formed in reaction (10):

$$(15)$$

Sulfate-anion radicals may further oxidize partially oxidized polymer (emeraldine form) to pernigraniline:

$$(16)$$

If the emeraldine is oxidized by cation-radicals of aniline, then in addition to pernigraniline also the aniline is formed:

$$(17)$$

## 13.3.2 SYNTHESIS AND PHYSICAL-CHEMICAL PROPERTIES OF POLYANISIDINES

The researches of various aniline's derivatives are far from the completion. For example, among the all methoxyanilines the polymerization process only of its *ortho*-isomer is sufficiently thoroughly investigated [105–107]. Therefore, to identify the relationships between the conditions of synthesis and the properties of polymers, on the one hand, and the electronic structure of initial monomers—on the other hand, we have investigated the oxidative polymerization of isomeric anisidines and the structure of obtained polymers [96].

### 13.3.2.1 Cyclical Voltammetry of Isomeric Anisidines

Quantify to estimate the processes of the electron transfer from the electroactive particle to electrode allows the cyclic voltammetry. As shown on Figure 13.11, among the three isomeric anisidines the most easily is

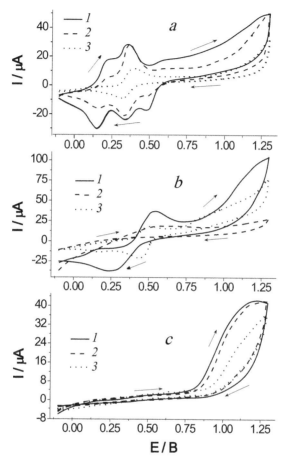

**FIGURE 13.11**    Cyclical voltammograms of 0.1 (1), 0.05 (2), and 0.01 $M$ (3) solutions of *ortho-* (*a*), *para-* (*b*), and *meta*-anisidines (*c*) in 0.2 $M$ aqueous solution of HCl during the first cycle of the potential scanning ($s_E$= 70 $mV \cdot s^{-1}$).

oxidized *ortho*-isomer, while the oxidation of *meta*-derivative is substantially difficult: the potential corresponding to the first oxidation wave is shifted from +0.21 $V$ for *ortho*-anisidine to +0.52 $V$ for *meta*-anisidine. Such impact of the position of substituent is true for different concentrations of electroactive monomers, and for different scanning rates of the potential electrode. Starting from 0.01 $M$ initial concentration of monomer, the oxidation of *ortho*-anisidine takes place in several stages, as shown by the presence of three peaks of the oxidation current on the cyclic voltammogram (*see* Figure 13.11, *a*), similarly as in the case of aniline (*see* Figure 13.1). Therefore, the first maximum of oxidation current at +0.21 $V$ corresponds to the transition "leucoemeraldine → emeraldine", the second oxidation current at +0.58 $V$ corresponds to oxidation to the pernigraniline form, and the middle oxidation current is associated with the oxidation of products of partial destruction of the polymer or with the cross-linking of the polymer chains [80]. With each next cycle of the potential scanning the values of redox currents increase, which indicates the accumulation of the polymer and its subsequent conversion on the electrode surface [70], so, the anisidne unlike nitro-derivatives of aniline capable to form of homopolymers. Moreover, an oxidation of poly-*ortho*-anisidine is reversible, since the difference between the maxima highs and reduction of polymer oxidation is negligible, which is evidence of the formation of electroconductive polymer with the system of conjugated $\pi$–bonds.

In accordance with the value of the first potential of oxidation (*see* Table 13.5) the isomeric anisidines can be placed in the following consequence:

*meta*-anisidine > *para*-anisidine > *ortho*-anisidine

At the same time, electron-donor properties of anisidine depend on the localization of electron density on the nitrogen atom of electrically amine-group. The results of the calculations of energy characteristics and the electronic structure of isomeric anisidines and aniline (*see* Table 13.5)

**TABLE 13.5**  An Impact of the Electronic Structure on the Position of the First Wave of Oxidation and Energy Characteristics of Aniline and Isomeric Anisidines

| Compound | $E_{ox}^{(I)}$ / $V$ | $-E_{bin}$/kJ·mol$^{-1}$ | $-E_{el}$/kJ·mol$^{-1}$ |
|---|---|---|---|
| Aniline | 0.18 | 6,185.87 | 407,436.01 |
| *ortho*-Anisidine | 0.21 | 7,757.39 | 645,210.92 |
| *para*-Anisidine | 0.44 | 7,763.21 | 645,553.56 |
| *meta*-Anisidine | 0.52 | 7,682.19 | 647,565.70 |
| Compound | $E_{HOMO}$/eV | $\Delta_f H$/kJ·mol$^{-1}$ | Charge on atom $N/e$ |
| Aniline | −8.0678 | −107.10 | 0.051 |
| *ortho*-Anisidine | −8.5662 | −63.26 | 0.077 |
| *para*-Anisidine | −8.3841 | −69.08 | 0.073 |
| *meta*-Anisidine | −8.6422 | −71.85 | 0.072 |

conducted using the MOPAC 93 (semiempirical PM3 level) showed, that the values of electronic energy and the energy of molecules binding as well as the enthalpy of formation for the consequence *meta* < *para* < *ortho*-isomer are decreased. In other words, the potential of oxidation increases with the increasing of thermodynamic stability of the anisidine's isomers. At the same time, the increasing of HOMO energy and, therefore, reducing of the negative charge on the nitrogen atom of the amino-group indicates the difficulty of the electron transfer process.

However, an unambiguous relationship between the value of oxidative potential of anisidines and the position of oxymethyl substituent in the aromatic ring of the initial monomer isn't observed. Therefore, it can be concluded, that the difference between the electrochemical properties of anisidines is determined not only by the redistribution of electron density between the substituents and benzene ring, but also the orientation of molecules in the adsorption layer that is associated with the geometry of the monomer's molecule. Obviously, the electroactive $NH_2$–groups of *ortho*-isomer of anisidine are oriented profitable compared with *meta*- and *para*-isomers, primarily because of the possibility of interaction with the surface also oxygen of oxymethyl group that promotes the electron transfer from the adsorbed particles on the electrode.

### 13.3.2.2 The Structure of Polymeric Anisidines

In accordance with discussed above mechanism of aniline's oxidation [84, 88] the first stage of the process is the transfer of an electron from 2*s*-orbital of the nitrogen atom with the formation of a cation-radical, which in the presence of base *B* can be accompanied by the deprotonation of amine-groups:

$$H_3C-O \underset{\overset{}{\phantom{}}}{\bigcirc} -N^{+\cdot}_H H + B \rightleftharpoons H_3C-O \underset{\overset{}{\phantom{}}}{\bigcirc} -N^{\cdot}_H + BH^+ \qquad (18)$$

Presented in the Table 13.6 results of quantum-chemical calculations of energy parameters of initial radical particles that arise as a result of the oxidation of isomeric anisidines, allow to conclude that accordingly to the Eq. (18) the formation of nitrenium-radicals is advantageous compared

**TABLE 13.6**   An enthalpy of formation, an energy of binding and the electron energy of intermediates of the anisidines oxidation

| Particle | $\Delta_f H/kJ \cdot mol^{-1}$ | $-E_{bin}/kJ \cdot mol^{-1}$ | $-E_{el}/kJ \cdot mol^{-1}$ |
|---|---|---|---|
| $o\text{-}CH_3O - C_6H_4 - \overset{+}{N}H$ | 101.53 | 302.96 | 22,993.87 |
| $p\text{-}CH_3O - C_6H_4 - \overset{+}{N}H$ | 105.76 | 324.02 | 21,250.69 |
| $m\text{-}CH_3O - C_6H_4 - \overset{+}{N}H$ | 109.99 | 330.25 | 20,647.92 |
| $o\text{-}CH_3O - C_6H_4 - \overset{+\bullet}{N}H$ | 708.28 | 706.90 | 1,190.73 |
| $p\text{-}CH_3O - C_6H_4 - \overset{+\bullet}{N}H$ | 712.47 | 712.51 | 613.03 |
| $m\text{-}CH_3O - C_6H_4 - \overset{+\bullet}{N}H$ | 732.73 | 732.73 | 17.96 |

with the existence of the corresponding to them cation-radicals. In particular, for the all anisidines heat of formation, an electron energy and the energy of binding of nitrenium-radicals are lower compared to cation-radicals, that is an equilibrium of the reaction (18) is shifted to the right.

Further recombination of nitrenium-radicals leads to the formation of N–N bond, which must be present in the structure of the polymer. This fact is confirmed by the results of IR-spectroscopy (*see* Table 13.7) of chemically synthesized in 1 $M$ aqueous HCl solutions samples of polyanisidines $(NH_4)_2S_2O_8$ was used as an oxidant). The band at 1461 $cm^{-1}$ which formerly was observed by Patil et al. [107] for electrochemically-synthesized samples of poly-*ortho*-anisidine corresponds to the stretching vibrations of N–N-bond. In addition, the band at ~1490 $cm^{-1}$ which is associated with the existence of quinoid structure indicates the deeper oxidation of poly-*ortho*- and poly-*meta*-anisidine's chains during the chemical synthesis of the polymer.

The results, which are shown in Table 13.7, demonstrate the similarity of IR-spectra of isomeric polyanizsidines. However, for the spectra of poly-*para*-anisidine typical are some significant differences. In particular, only for poly-*para*-anisidine the band at 1347 cm$^{-1}$ is observed, showing the formation of polar on C–N$^{\bullet+}$, as it was shown for PAn, doped with the sulfate-ions [108]. On the other hand, the spectrum of poly-*para*-anisidine

**TABLE 13.7**   Analysis of the Absorption Bands of *IR*-Spectra of Isomeric Anisidines

| Poly-*para*-anisidine | | Poly-*ortho*-anisidine | | Poly-*meta*-anisidine | |
|---|---|---|---|---|---|
| $\omega/cm^{-1}$ | **Vibrations** | $\omega/cm^{-1}$ | **Vibrations** | $\omega/cm^{-1}$ | **Vibrations** |
| 3,403 | $\nu_{N-H}$ | 3,448 | $\nu_{N-H}$ | 3,424 | $\nu_{N-H}$ |
| 3,220 | $\nu_{N-H}$ | 3,204 | $\nu_{N-H}$ | 2,937 | $\nu_{-CH_3}; \nu_{-CH_2-}$ |
| 2,038 | $\nu_{-NH^+}$ | 1,579 | $\nu_{N=N}$ | 2,834 | $\nu_{-O-CH_3}$ |
| 1,608 | $\nu_{N=N}; \nu_{-N-H}$ | 1,488 | $\nu_{C-N^+}$ | 2,038 | $\nu_{-NH^+}$ |
| 1,564 | $\nu_{C=C}; \nu_{N-H}$ | 1,461 | $\nu_{N-N}$ | 1,606 | $\nu_{N=N}; \gamma_{-N-H}$ |
| 1,510 | $\gamma_{-N-H}$ | 1,255 | $\nu_{-C-O-}; \nu_{-C-O-CH_3}$ | 1,578 | $\nu_{C=C}; \nu_{N=N}$ |
| 1,347 | $\nu_{-C-N^{\cdot+}}$ | 1,205 | $\nu_{C-N=}$ | 1,493 | $\nu_{C=N-}; \nu_{N-H}$ |
| 1,289 | $\gamma_{N-N}$ | 1,172 | $\nu_{C-N}$ | 1,464 | $\nu_{N-H}$ |
| 1,251 | $\nu_{-C-O-}; \nu_{-C-O-CH_3}$ | 1,179 | $\nu_{C-N}; \nu_{N-N}$ | 1,254 | $\nu_{-C-O-}; \nu_{-C-O-CH_3}$ |
| 1,172 | $\nu_{-C-N}; \nu_{N-N}$ | 1,046 | $\nu_{C-N}$ | 1,207 | $\nu_{C-N=}$ |
| 1,030 | $\nu_{-C-N<}$ | 1,017 | $\gamma_{C-C}; \nu_{N=N}$ | 1,158 | $\nu_{C-N}; \nu_{N-N}$ |
| 826 | $\gamma_{=C-H}$ | 828 | $\gamma_{=C-H}$ | 1,039 | $\nu_{-C-N<}$ |
| 520 | $\gamma_{C-H}; \gamma_{N-N}; \gamma_{C-N=}$ | 583 | $\nu_{-C-S}$ | 832 | $\gamma_{=C-H}$ |
| | | 547 | $\gamma_{C-H}; \nu_{N^+-H};$ | 688 | $\gamma_{C-C}; \gamma_{C-H}; \gamma_{C-N}$ |
| | | 451 | $\nu_{C-N}$ | 580 | $\nu_{-C-S}$ |

is somewhat simpler. So, it is not observed of the absorption bands in the range of wave numbers 1493–1461 cm⁻¹ indicating the absence of the N–N-bonds in this polymer.

The spectra of proton magnetic resonance (NMR) of synthesized samples of polyanisidines are also quite similar to each other (*see* Figure 13.12). However, unlike aniline [109], the signal from the protons of benzene ring at 7.4–7.7 ppm is weak, while the proton signal of the oxymethyl group at 2.3–2.7 ppm is enough intense. However, this fact was also recorded for the polymer of toluidine [110].

The reason for differences in spectral characteristics of polyanisidines is a different mechanism of their formation. The linkage of initial radicals of the type "head-to-tail" can occur only when the isomerization of nitrenium-radical into quinoid structure takes place. If the isomerization of *ortho*- and *meta*-anisidines easy proceeds due to the stabilization of the quinoid structure at the expense of the conjugation of π-bond of the cycle with the unpaired electron pairs of oxygen of the CH₃O-group

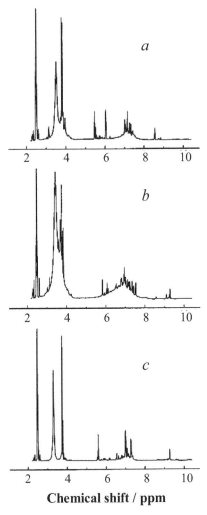

**FIGURE 13.12** NMR-spectra of synthesized samples of polymeric *ortho*- (*a*), *para*- (*b*) and *meta*-anisidine (*c*) (400 *MHz*; spectrometer Varian Whity 500 Plus; solvent is deuterated DMSO; external standard is Si(CH$_3$)$_4$).

$$\text{(19)}$$

then for *para*-isomer this process

$$(20)$$

is complicated, and therefore, the formation of the dimer by the interaction of the type "head-to-head" with the formation of N–N–bond, makes further growth of the chain more likely impossible. Moreover, in the case of recombination of "head-to-tail" type

$$(21)$$

the conjugation in the main chain is broken and thus the conductivity of the polymer is decreased and becomes impossible the reisomerization of dimer. Therefore, in our opinion the isomerization of the nitrenium-radical of *para*-anisidine will be occurred differently

$$(22)$$

which removes the all previous disagreements. Therefore, the formation of polymer after interaction of initial and isomerized radicals

$$(23)$$

is possible due to the reisomerization of the formed particle into the structure of benzenoid type with the following its oxidation (according to Eq. (1)) and further interaction with another particle of the radical nature. For example, the trimer formation can be as a result of the recombination of oxidized dimeric particle with isomerized initial radical of anisidine,

$$\text{(chemical structure with } O-CH_3 \text{ groups)} \tag{24}$$

and tetramer as a result of the recombination of two oxidized dimer particles or trimer and initial radicals and *etc.* similarly as in the case of oxidation of aniline or nitro-anilines [84, 94]. This propagation of polymer chain is the result of recombination of oligomeric and monomeric particles, leading to a rapid depletion of the monomer as it was revealed during the oxidation of aniline with peroxydisulfate in aqueous solutions [111]. Leucoemeraldine (*L*), which is formed in the first cycle of the potential scanning next more consistently is oxidized to emeraldine salt (*ES*) and pernigraniline (*P*):

$$\text{(chemical structures for } (L), (ES), (P)\text{)} \tag{25}$$

As it can be seen from the data presented in Table 13.7, experimentally found values of the hydrogen content in the synthesized sample of poly-*ortho*-anisidine exceed the theoretical value and *vice versa* the content of carbon is understated, which may be evidence that chemically synthesized poly-*ortho*-anisidine is not as emeraldine base but emeraldine salt (*ES*) or pernigraniline. Such transformations were observed, in particular, for the photoelectrochemical response of PAn intercalated in the film of the ethylphosphoric acid [112]. However, there is also another possibility, related with the transformation of initial radicals without the reisomerization of the products of the recombination. In this case the quinoid structure is remained unchanged and, as a result, the structure of the oxazyne type is formed

$$(26)$$

The possibility of such structure formation was showed by Widera et al. [106] for the oxidation of *ortho*-anisidine on platinum electrode. The content of carbon in the structure of oxazyne type is less than the leucoemeraldine form of poly-*ortho*-anisidine explaining the reduced value of the experimentally found carbon content in the polymer sample (56.03 *mass. %, see* Table 13.7a) compared with the theoretical (58.21 *mass. %*).

## 13.4 CHEMICAL SYNTHESIS OF POLYLUMINOL

The solubility of PAn in water and organic solvents is very low, causing some difficulties during the research of these materials as well as in the design of different kinds of devices on their basis. Therefore, the

**TABLE 13.7A** The Results of Element Analysis of Synthesized Samples of Polyanisidines (Analyzer GA 1108, Carlo Erbo)

| Sample | Composition/*mass. %* | | | | | |
|---|---|---|---|---|---|---|
| | C | N | H | $SO_4^{2-}$ | O | Σ |
| Theoretical data (emeraldine form) | | | | | | |
| Polyanisidine | 58.21 | 9.70 | 4.64 | 16.46 | 11.05 | 100.00 |
| Experimental data | | | | | | |
| Poly-*ortho*-anisidine | 56.03 | 8.00 | 5.01 | 13.00 | 11.05 | 93.09 |
| Poly-*meta*-anisidine | 58.40 | 7.90 | 4.45 | 12.10 | 11.05 | 93.90 |
| Poly-*para*-anisidine | 62.49 | 8.22 | 4.49 | 11.70 | 11.05 | 97.95 |

search for new materials based on polyaniline with improved physical and chemical properties continues. Two the most important directions of the researches today are the obtaining of polymers based on derivatives of aniline, considered above, and the synthesis of copolymers of aniline with heterocyclic compounds, such as pyrrol [113], luminol [114, 115], etc. From this perspective, the luminol, thanks to its structure as heterocycle and benzene ring, bound to the amino group, corresponds to both criteria. However, despite the wide range of the researches of polyluminol (PLm), many questions regarding the reaction mechanism with its participation remained unclear. In particular, the polymerization of luminol was studied only in terms of its electrochemical oxidation when the polymer was delayed on the electrode in the form of a thin layer.

The electrochemical obtaining of the polyluminol has been studied in detail for the first time by Zhang and Chen [116]. They obtained the polymer in acidic aqueous solutions in potentiodynamic mode, by 50-fold potential scanning of gold electrode ranging from $-0.2$ to $+1.2$ $V$. Today, in addition to the gold electrode [114–119], the electropolymerization of luminol was made on the traditional electrode materials such as ITO [117, 120], glass-carbon [116–119, 121], platinum [186, 122]. Depending on the electrode material the polyluminol can be obtained by scanning of the electrode potential also in a narrower range of potentials (e.g., $0.0–(+0.6)$ $V$ on glass-carbon in aqueous $H_2SO_4$ solution with $pH \sim 1.5$) [121]. At the same time it is showed that the graphite electrode can be modified with a layer of electrically PLm also in weakly acidic medium at $pH = 6$ [121] which has significant advantages during the immobilization of enzymes on such a surface to create the biosensors.

Since the luminol can be considered as a derivative of aniline, then the proposed mechanism of its electrochemical polymerization [116] is similar to the mechanism of the synthesis of PAn. This fact was confirmed by enhanced surface of Raman spectra of the PLm, which was by evidence of the transformation of the amine-groups into imide ones during oxidative polymerization of luminol, while N–N-group of hydrazide ring remained intact [117], or were deprotonated passing into N=N-fragment [116]. Therefore, by analogy with PAn, the structure of the reduced form of PLm can be represented as a set of dimer fragments (excluding possible protonation of imine-groups):

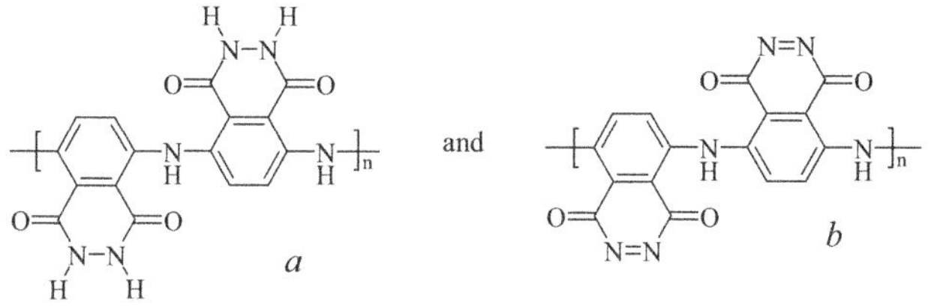

(27)

while the oxidized form of which is as follows:

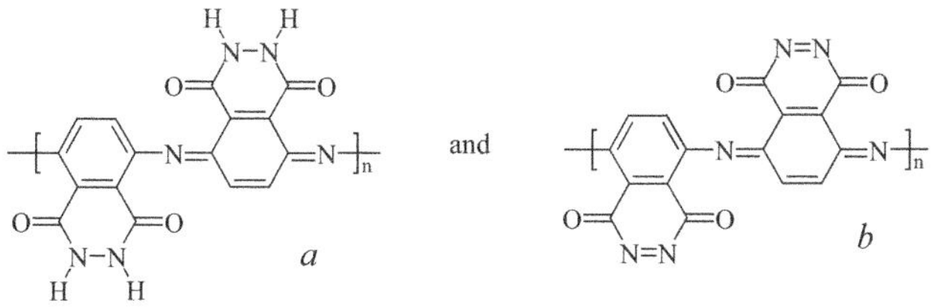

(28)

Another confirmation of the structure of polyluminol were the results of chemiluminescent researches. Since the synthesized polymer retained the fluorescent properties, it was by evidence that the heteroatomic cycle of luminol was kept during the oxidative polymerization. Moreover, it was shown that via copolymerization of luminol and aniline (molar ratio of monomers 1:(40−60)) the polymer which, unlike PAn, due to the presence in the structure of the polymer hydrazide group of luminol is electroactive and stable in alkaline medium [114, 115].

As it was already noted, the synthesis of PLm was carried out only electrochemically. However, given the similarity of PLm and PAn, logically to expect that the polyluminol can be also obtained chemically, which was conducted using a mixed aqueous-organic medium and suitable oxidant.

### 13.4.1 CYCLICAL VOLTAMMETRY OF 5-AMINE-2,3-DIHYDRO-1,4-PHTHALAZINEDIONE

Due to low solubility of inorganic oxidants in organic solvents, and luminol in water, chemical synthesis of PLm was conducted only in mixed aqueous-organic solvents. To estimate the possibility of polyluminol obtaining in such media, where organic component was **DMSO**, we have used the method of cyclic voltammometry [123]. As shown on Figure 13.13, at

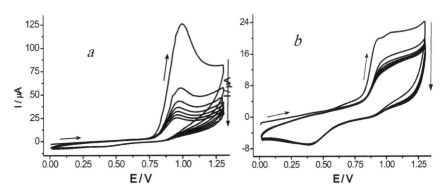

**FIGURE 13.13**   Cyclical voltammograms of platinum electrode in 0.1 ($a$) and 0.001 $M$ ($b$) solutions of luminol in mixed water-DMSO ($r_V$ = 1 : 9) solvent (base electrolyte is 0.2 $M$ $H_2SO_4$; $s_E$ = 30 $mV \cdot s^{-1}$).

electrode potential scanning in the range 0–(+1.3) $V$ the maximum of current on cyclical voltammogram is observed at potentials +(0.92–1.04) $V$. This maximum of current obviously is referred to the oxidation of luminol upon place of the amine-group since as potential of the aniline's oxidation consists of +1.13 $V$ [124]. Similar results were also obtained by the authors of Refs. [114, 115] who have noted that as a result of lower potential of the luminol's oxidation versus aniline the oxidation process of last is suppressed during the electrochemical copolymerization of these monomers at their molar ratio from 1:1 to 1:4. Each successive scanning of the potential leads to the depression of the oxidation currents due to the screening of the surface of electrode by the layer precipitated on its surface of insoluble product, indicating the non-conductivity of the formed film of polyluminol. Another evidence of the fact that the formed polymer represents by itself the dielectric is absence on the cyclical voltammograms of current peaks responsible for the formation of polarons and bipolarons due to electrochemical doping (for PAn at +0.2 and +0.8 $V$) [73]. The height of the current maxima is increased with the speed sweep potential increasing (*see* Figure 13.14). This result points to the fact that the speed of the process is limited by the diffusion of particles of electroactive material.

Studying the electrochemical oxidation of luminol (8.4×10⁻⁴ $M$) in 0.5 $M$ aqueous solution of $H_2SO_4$ on the glass-carbon electrode Zhang and Chen [116] received several other results, due to using of another electrolyte and electrode material. They observed two additional mild anodic peaks at +332 and +552 $mV$ and the corresponding to them two

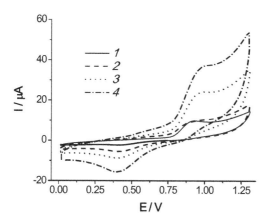

**FIGURE 13.14**   Cyclical voltammograms of platinum electrode in 0.05 M solution of luminol in mixed water-DMSO solvent ($r_V = 1:9$) at the values $s_E$, $mV \cdot s^{-1}$: 10 (1); 30 (2); 50 (3) та 100 (4) (base electrolyte is 0.2 M $H_2SO_4$).

cathode current peaks at +223 and +467 $mV$, the heights of which gradually increase with the increasing of number of cycles. These peaks corresponded, according to the authors' opinion, to the oxidation and reduction of the formed polyluminol's film. However, the attributing of both these current highs to the redox-processes of polyluminol film is seemed to us somewhat incorrect. In particular, the second maximum the most plausible is associated with redox processes involving the imide groups of luminol, as indicated by cyclic voltammogram of luminol in aqueous alkaline solutions (*see* Figure 13.15).

## 13.4.2   CHEMICAL POLYMERIZATION OF LUMINOL

Based on electrochemical measurements it can be concluded that at low values of the electrode potential the luminol is oxidized by hydrazine group, while at the high at the place of the amine-group. So, for the chemical synthesis two oxidants, namely $(NH_4)_2S_2O_8$, the standard values of the redox potential of which is 2.05 $V$, and $KIO_3$, the standard potential of which is twice smaller and is 1.085 $V$ were selected. It can be hoped, that the oxidation of various functional groups differently are displayed on the process of the condensation of luminol rings and the structure of the condensation product. As the standard values of the redox-potentials are given relatively to the hydrogen electrode, and the oxidation potentials on cyclic

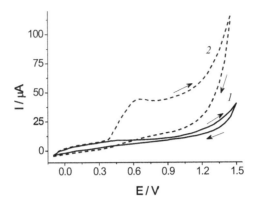

**FIGURE 13.15**  Cyclical voltammogram of platinum electrode in borate buffer solution ($pH$ = 10.0) in the absence of (1) and after addition (2) of $5 \times 10^{-5}$ $M$ solution of luminol ($s_E$ =10 $mV \cdot s^{-1}$).

voltammograms relatively saturated Ag/AgCl electrode ($E$ = +0.26 $V$), it is clear that the redox potential of $KIO_3$ is insufficient for the oxidation of luminol in the place of amine-group.

The possibility of luminol polymerization by chemical oxidation with the elimination of the product of reaction as separate phase has been investigated by us in a mixed aqueous-organic medium, when the dimethylsulfoxide (DMSO), dimethylformamide (DMFA) or N-methylpyrrolidone (MPD) were used as organic components. It was found, that the optimal volume ratio ($r_V$) of the components in a mixed solvent organic component: water was 9:1. At this, the yield of the product of polymerization of luminol does not depend on the nature of the solvent, and is determined by the ratio monomer: solvent (*see* Table 13.8). The lowest yield of the polycondensate was observed at the use of $KIO_3$ as an oxidant. An increase of the excess of $(NH_4)_2S_2O_8$ over luminol actually does not affect the yield, while an increase of excess amounts of monomer leads to a significant increase of the quantity of the obtained product.

### 13.4.3  SPECTRAL PROPERTIES OF POLYCONDENSATES OF LUMINOL

Color of the obtained final products of oxidative polymerization of luminol depends on the nature of oxidant. At the use of $KIO_3$ the black polymer

**TABLE 13.8** Dependence of the Yield of the Luminol's Oxidation Products on the Conditions Of Synthesis [$r_V$ (water:organic component) = 1:9; the Temperature of Reacting Mix is 20°C; the Duration of the Reaction 24 hours] [123]

| Organic solvent | C(Lum)/$_M$ | C(oxidant)/M | Yield/% |
|---|---|---|---|
| PLm I (Oxidant is KIO$_3$) | | | |
| DMSO | 0.22 | 0.11 | 5.7 |
| PLm II (Oxidant is (NH$_4$)$_2$S$_2$O$_8$) | | | |
| MPD | 0.22 | 0.11 | 26.5 |
| DMFA | 0.22 | 0.11 | 26.9 |
| DMSO | 0.22 | 0.11 | 26.9 |
| DMSO | 0.22 | 0.22 | 27.5 |
| DMSO | 0.22 | 0.33 | 27.8 |
| DMSO | 0.44 | 0.22 | 41.6 |
| DMSO | 0.55 | 0.22 | 42.5 |

(PLm I) is formed, and in the presence of (NH$_4$)$_2$S$_2$O$_8$ the obtained polymer had a green color (PLm II) like to polyaniline, which, along with the results of spectral studies is an evidence of significant differences in polymerization mechanism and product's structure depending on the nature of the used oxidant.

IR-spectra of luminol and samples of synthesized polyluminols are shown on Figure 13.16. Spectra of PLm (spectra $b-d$) similar to the spectra of luminol (spectrum $a$), and the most essential changes are observed in the high-frequency (2900–3500 cm$^{-1}$) area, where there are the bands of amine-group: the bands at 3331, 3420 and 3473 cm$^{-1}$ correspond to asymmetric and symmetric vibrations of amine-groups of aromatic amines [125]. The first and the third from these bands are characteristic for the spectra of the samples PLm II, synthesized in DMSO and MPD (*see* Figure 13.16, the spectra $b$ and $c$). The band at 3420 cm$^{-1}$ which corresponds to the vibration of N–H-bonds is absent in the spectra of samples PLm II, indicating the oxidation of the amine-group with peroxydisulfate during the synthesis. Available in a spectrum of luminol the bands of deformation and torsion vibrations of NH$_2$-group at 1628 and 492 cm$^{-1}$, respectively, are also absent in the spectrum of the polymer, which once again confirms the oxidation of luminol in a place of

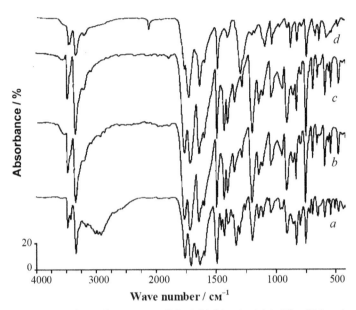

**FIGURE 13.16**   IR absorption spectra of the initial luminol (*a*), PLm II (synthesized in mixed water-DMSO (*b*) and water-MPD (*c*) solvent) and PLm I (*d*).

amine-group. In the spectra of polyluminols also available the azogroup –N=N–, the vibration frequency of which centered in the vicinity of 1588 cm$^{-1}$. Valence vibrations vc–н of benzene ring are observed at 3164–2911 and 3167–2920 cm$^{-1}$ in monomer and polymer, respectively. Three bands of vibrations in the area 1627–1754 cm$^{-1}$ correspond to the vibrations of amide-group [97]. An absence of the bands in the area 1610–1550 and at 1400 cm$^{-1}$ is agreed with the assumption that the carboxylates are absent in the structure of polyluminol.

Replacement of oxidant $(NH_4)_2S_2O_8$ on $KIO_3$ leads to the change of the structure of product that is displayed on the *IR*–spectrum of polyluminol obtained under these conditions. The form of absorption bands in the range 3200–3500 cm$^{-1}$ (*see* Figure 13.16, the spectrum *d*) which is similar to luminol (the spectrum *a*) indicates the tenure of the amine-group during the oxidation of luminol by potassium iodate. At the same time, the new absorption band at 2123 and 1289 cm$^{-1}$ can be attributed to the vibrations of –N=C=O group and amine-group of aromatic amines, respectively.

Luminol and PLm can be represented as analogs of aniline and polymeric anilines. However, unlike PAn, in the molecule of polyluminol no bands at 1500 and 1600 cm$^{-1}$ which were assigned by the authors of Ref. [26] to structural fragments

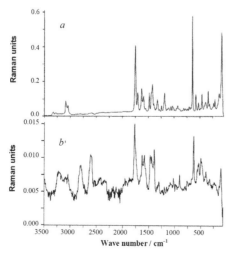

and

In addition, in spectra of the all synthesized samples of polyluminols absent the absorption bands, the vibration frequencies of which are in the vicinity of 1610 and 1400 cm$^{-1}$. Therefore, it can be assumed, that the final product of the luminol's oxidation both with $(NH_4)_2S_2O_8$ and $KIO_3$ is not aminophthalate (emitter of chemiluminescence), and, therefore, heteroatomic ring of luminol is not oxidized during the polymerization.

Initial luminol (*see* Figure 13.17*a*) and synthesized samples of PLm (*see* Figures 13.17*b* and Figure 13.18) were also characterized by Raman spectroscopy. It was found that the intensity of the Raman bands for the samples PLm II on two orders and for the sample PLm I on three orders lower compared of with the spectrum of luminol. This is obviously due to the strong absorption of dark green and black synthesized polymer samples (PLm I and PLm II, respectively). Compared with luminol the band at 3100 *cm*$^{-1}$ which corresponds to the valence vibrations of N–H–bond of the aromatic amines significantly is expanded, although the position of

**FIGURE 13.17**   Raman spectra of the starting luminol (*a*) and PLm I (*b*)

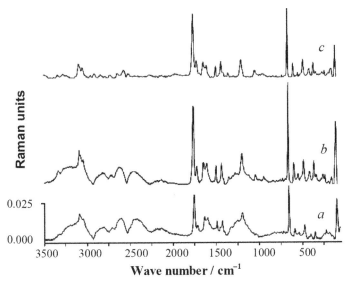

**FIGURE 13.18**   Raman spectra of the PLm II synthesized in mixed water-organic ($r_V$ = 1:9, organic component: DMSO (*a*), DMFA (*b*) and MPD (*c*)) solvent.

maximum (3089 cm$^{-1}$) remained unchanged. This not can be said for the spectrum of PLm I, where this maximum stretched in the range from 3046 to 3250 cm$^{-1}$ and reminds the band for the polyaniline (*see* Figure 13.8). Position of doublet band at 1749 (intensive) and 1705 cm$^{-1}$ (low-intensive) corresponding to valence vibrations of carbonyl group is little varied during the transition from luminol to PLm II. At the same time, for polyluminol obtained in the presence of KIO$_3$, $v_{c=o}$ is shifted to 1775 cm$^{-1}$ due to the different surroundings of the functional group in both cases. The bands at 1628, 1589 and 647 cm$^{-1}$ (intense) corresponding to the vibrations of benzene ring are observed in the spectra of the all synthesized samples.

References data on the Raman spectra of electroconducting polymers in general and PLm in particular are rather poor. The Raman spectra enhanced with the surface of electrochemically synthesized on Au-electrode polyluminol were of low quality [116], but major changes were recorded for the band of the stretching vibration vN–H, the intensity of which for polyluminol is much higher than similar intensity for the same band of luminol. High-quality spectra of (1′,2′-dicarboxy)ethylbenzotriazole on the copper surface were obtained by the authors of Ref. [127]. Here existing distinct bands of the stretching vibrations of the carboxyl group, benzene ring and a clear band of deformation vibrations $\delta N-H$ at 1162 cm$^{-1}$, which into obtained by us spectra was centered at 1174 cm$^{-1}$ for luminol, ranging from 1185 to 1187 cm$^{-1}$ for PLm II, and was absent in the case of PLm I. In the Raman spectrum of luminol and polyluminols available low-intensive band at 1245 cm$^{-1}$, which can be attributed, by analogy with PAn [128], to the deformation vibrations $\delta C-N$. In spectrum of polyluminols the bands that correspond to the presence in the structure of the C=N bonds were not found. Obviously, this is related to the absence of quinoid type structures, the formation of which is possible due to oxidation of polyluminol. This is a serious argument in favor of unparticipation of polyluminol chains in deep oxidation with the formation of structures such as emeraldine or pernigraniline, as it takes place in the case of chemical condensation of anilines [129].

## 13.4.4 THE STRUCTURE OF POLYCONDENSATES OF LUMINOL

The data of element analysis of polyluminols are shown in Table 13.9. Characteristically, that the content of carbon, nitrogen and hydrogen is little differed for the luminols obtained by oxidation with $(NH_4)_2S_2O_8$, while at the use of $KIO_3$ the element composition of samples is significantly

**TABLE 13.9** Data of Element Analysis of Luminol and Synthesized Samples of Polyluminol (Analyzer GA 1108, Carlo Erbo, Italy)

| Sample | | Composition, % | | | |
|---|---|---|---|---|---|
| | | C | H | N | O |
| Theoretical | | | | | |
| Luminol (Lum) | | 54.24 | 3.96 | 23.73 | 18.08 |
| Experimental | | | | | |
| Lum | | 54.24 | 3.94 | 23.61 | 18.22* |
| PLm I | | 54.25 | 3.88 | 23.66 | 18.21* |
| PLm II | DMSO; C(Lum)/C(Ox) = 1:1 | 53.80 | 3.84 | 23.40 | 18.96* |
| | DMSO; C(Lum)/C(Ox) = 2:1 | 54.17 | 3.83 | 23.48 | 18.52* |
| | MPD; C(Lum)/C(Ox) = 2:1 | 47.39 | 3.08 | 18.38 | 31.19* |

*Note: the values calculated based on found experimentally total content C, H and N.

different. The interpretation of these differences in chemical composition based on various possible mechanisms of oxidation in these two cases. Oxidation of luminol with $KIO_3$ takes place on amide groups, which are easier oxidated. The fact that the content of nitrogen is on 5% less, it can be postulated the loss of nitrogen atom by each monomer link. Probably accept, that at the use of $KIO_3$ as oxidant agent, the dimer of type is formed.

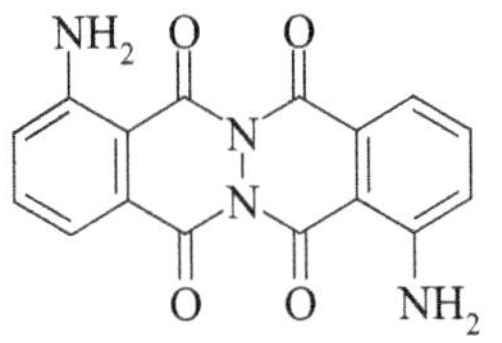

At this, the N–N bond is formed by connection of two residues of the luminol's molecules, each containing one nitrogen atom. The calculation of element composition for this structure gives the following results: 59.62% C; 17.39% N; 3.10% H and 19.88% O. A good agreement of the results of analysis with the proposed above structure takes place for nitrogen and hydrogen, however the carbon content exceeded on 5%. The products of $KIO_3$ reduction may be the iodine atoms, so they can be part of the product of luminol oxidation. On the possibility of formation of complex compound of iodine with nitrates or quaternary ammonium salts, particularly was pointed by Wells [130]. The structure of the product then can be represented by the following formula

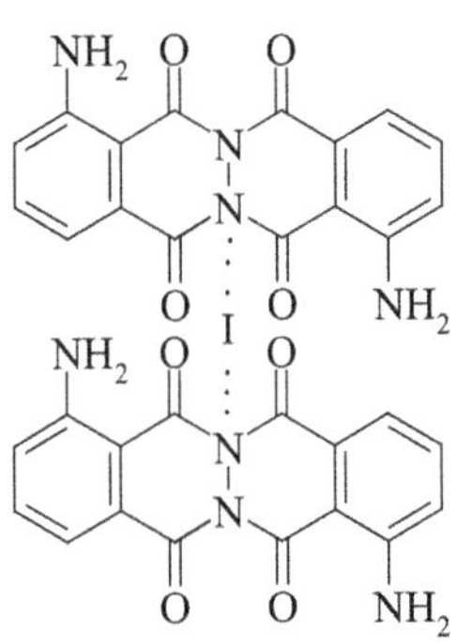

(29)

Calculation of chemical composition based on the structure (29) gave the following results: 49.74% C; 14.51% N; 2.85% H; 16.58% O; 16.32% I, and, so element composition in near to the experimentally found (*see* Table 13.9). In favor of this structure indicates the presence of the absorption band at 567 cm$^{-1}$ and IR bands at 526 and 581 cm$^{-1}$ in Raman spectra. Therefore, we can assume that the most likely structure of the luminol oxidation with $KIO_3$ (PLm I) product corresponds to depicted formula (29).

Oxidation of luminol with peroxydisulfate takes place on the place of the amino-group of luminol and the structure of the formed polycondensate contains the monomer links that are almost indistinguishable from the original structure of luminol. In favor of this there are data of the element analysis, IR and Raman spectra of the products of oxidation. By analogy with the polyaniline for polyluminol synthesized in strong acidic medium it was proposed the structure (27, *a*), which corresponds to fully reduced form of polyluminol [116]. However, it was not considered that the amide groups can be also oxidized, indirect evidence of which is, for example, the oxidation of luminol in the presence of montmorillonite [131, 132] containing the $Fe^{3+}$ ions.

$$(30)$$

Further electrochemical oxidation of the intermediate radical leads to the formation of azoquinone [133]. Studying the electrochemical oxidation of an aqueous solution of luminol in an acidic medium, it was suggested that in the area of low potential reversibly is formed exactly this compound [116]

$$(31)$$

and the oxidation takes place on amine-group at higher potentials. So, again similar to the analogy with the condensation of aniline it can be suggested that chemically synthesized polyluminol doesn't contain the hydrazine $-NH-NH-$ group, but azo-group $-N=N-$ in heteroatomic ring and the structure of the link will be correspond to the formula (27, *b*).

The hydrogen content in such a structure should be much smaller. Calculated for this structure content of the hydrogen (2.11%) is much smaller compared with the experimental values 3.83–3.88% for polymer and 3.95% for luminol. If to assume that the amide group in the luminol is not changed during the oxidation with $(NH_4)_2S_2O_8$, and bridge atoms of the nitrogen are protonated, then the content of hydrogen will be near to the experimentally determined. Because it can be argued that the oxidation of hydrazine groups of luminol in the presence of $(NH_4)_2S_2O_8$ doesn't proceed and then the polymer structure is similar (27, $a$). In the presented formula of the structural fragment of polyluminol, the authors of Ref. [116] does not exclude the possibility of an existence of protonated atoms of nitrogen.

An indirect confirmation of the conclusions about the mechanism of the formation of polymerizates, namely that the oxidation of luminol in mixed aqueous-organic solvents proceeds primarily not by hydrazine group (resulting in the formation of the emitter radiation 3-aminophtalate), but by amine-group with the formation of the polymer, can be the results of chemiluminescent researches in reaction systems, in which the polycondensate of luminol was obtained. In systems $0.22\,M$ Lum $+\,0.11\,M$ $(NH_4)_2S_2O_8$ slight luminescence was observed only in a mixed water-MPD solvent, while in the water-DMSO and water-DMFA solutions the generation of luminescence was not fixed. Adding to the mix besides $(NH_4)_2S_2O_8$ of hydrogen peroxide leads to the intensification of luminescence in the presence of MPD and before its appearance in the presence of DMSO and DMFA. However, the intensity of radiation even in the presence of $H_2O_2$ is very small (on the 2–3 orders lower compared to the intensity of chemiluminescence of luminol under optimal conditions), that is more obvious, if we take into account on 1–2 order higher concentration of luminol during the synthesis of its polymeric form compared to concentrations, that are used in chemiluminescent techniques.

### 13.4.5  THERMAL STABILITY OF CONDENSATES OF LUMINOL

We have also studied the thermal stability of polyluminols synthesized under different conditions and compared to the properties of the original luminol. As it can be seen from the integral and differential curves of thermal analysis shown on Figure 13.19, the less thermally stable is monomeric luminol, for the sample of which the first maximum on DTG-curve,

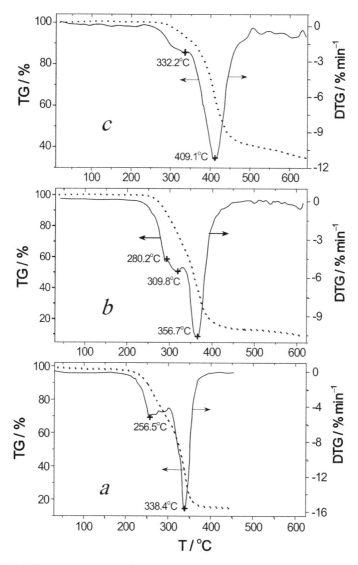

**FIGURE 13.19**  Integral and differential derivatograms of initial luminol (*a*) and PLm II synthesized in mixed water-DMSO (*b*) and water-MPD (*c*) solvent (atmosphere is Ar; microbalance NETZSCH TG29; $s_T = 10°C·min^{-1}$).

which is associated with destruction on the hydrazine group with the nitrogen elimination is observed at the destruction temperature ($T_D$) 256.5°C, while for the most thermally stable PLm II (synthesized in MPD) $T_D = 334.2°C$ (against 280.2°C for PLm II synthesized in DMSO).

Another derivatogram has been obtained for polyluminol synthesized with the use of $KIO_3$ as oxidant (*see* Figure 13.20). Substantial differences in the nature of DTG curves of PLm I and PLm II once again affirm the significant difference in the structure of the obtained samples. Differential thermogram of sample PLm I (*see* Figure 13.20) is much "richer" as to the number of available highs, indicating not only the much more complex mechanism of thermolysis of the structure (29), but the entire difference between the products of PLm I and PLm II. In particular, if the final losses of mass both of luminol and samples PLm II (DMSO) make up more than 85%, for the samples of PLm II (MPD) and PLm I this value consists of only ~55%.

Sample PLm I is characterized by significantly lower thermal stability even compared to monomeric luminol. In particular, the first maximum on the DTG–curver related, obviously, with the loss of mass by the iodine sample, is observed at 62.8°C and a maximal rate of the loss of mass for the sample PLm I is observed at $T_{D,max}$ = 282.5°C. In comparison, for luminol and samples PLm II synthesized in DMSO and MPD, $T_{D,max}$ is 338.4°C, 356.7°C and 409.1°C, respectively.

## 13.5   SECOND CASE STUDY

Electroconductive polymers (ECP) are relatively new class of polymers[1-5], the most famous representative among which is the

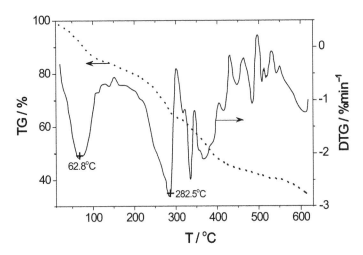

**FIGURE 13.20**   Integral and differential derivatograms of the sample PLm I.

polyaniline (PAn). Easy methods of synthesis, the possibility of acquiring through the mechanism of doping/dedoping of various forms, unique physicochemical properties among which important are high electrical conductivity, chemical sensitivity, multicolored electrochromism, catalytic properties, limited solubility, chemical and thermal resistance, high adhesiveness to the surfaces of different nature, etc., make the PAn and composites on its basis by important materials in modern technologies [1, 2, 4–9].

The most common usable after chemical synthesis are the electrochemical methods of PAn obtaining, namely galvanostatic method (**GS**), potentiostatic method (PS) and potentiodynamic method (PD) [10–14]. Galvanostatic polymerization of aniline (An) makes easy to control the properties of the PAn's films, namely molecular weight of polymer and the thickness of deposited coatings. Potentiostatic mode of polymerization in turn permits to control by the reactivity of electrochemically active intermediates of the starting monomer's oxidation during reaction. Potentiodynamic polymerization (cyclic voltammetry), which is carried out under cyclic scanning of the potential electrode allows to control in real time both of An oxidation and redox conversions of deposited PAn. Cyclic voltammetry is actively used for the determination of mechanisms of An oxidation, redox reactions of PAn [15], mechanisms of ion exchange in films of PAn [16], as well as for the researches of the film's stability [17], dispersions and PAn's composites [18], capacitive characteristics of PAn's films [19–23], electrical activity of PAn's films [23] and others.

Electrochemical methods are used for obtaining of polymeric coatings free from the oxidants directly on the metals surface. In addition to noble metals (Pd, Pt, Au), which are used the most often, by the working electrodes (WE) can serve also active metals Mg, Al, Ti, Cr, Fe, Ni, Cu, Zn, Ag, In, Pb) [17, 25–38], the alloys based on Fe including also stainless steels, such as trademark SS 304 [39, 40], SS 316L [41, 42], etc., the aluminum alloys (AA), including polycrystalline AA 1100 [43], AA 2024 43–T3 [44, 45], AA 2024–T6 [46], AA 3004 [47], AA 5182 [48], AA 7075 [49] and amorphous alloys [50–53], magnesium alloy by trademark AZ91D [54], the alloy $Co_{67}Cr_{29}W_4$ [55] and others. Sometimes the films of PAn are synthesized on metal oxide surfaces such as IrOx [56], $SnO_2$ [57], $PbO_2$ [58], $In_2O_3 \times SnO_2$ (ITO) [59].

By choosing the electrochemical method and the conditions of the aniline's oxidation on the surface of WE it can be obtained the PAn's film with different structure of macromolecular chains (from linear to branched), different morphology from micro- or nanowires to micro- or nanotubes and also different topography of the surface from smooth films to the layers with developed surface [60–63]. The morphology of electrochemically deposited PAn's films on the surfaces of electrodes is influenced by various factors, including the method of electrochemical oxidation of An [10–13, 47, 64], current density [61], velocity potential sweep in the first and subsequent cycles [65], the boundary of the potential sweep [39] and the value of applied potential [39], duration of polymerization and conductivity of electrode [66], nature of anion of acid-electrolyte [67], nature [12, 68], temperature [68], pH of medium [64], the presence of oxide film on the surface of WE [69], others. Films of PAn, deposited by potentiodynamic method have more developed surface than obtained by other electrochemical methods [49, 70, 71].

An important feature of ECP, which makes their use, is a phenomenon of the proceeding of redox transformations in macromolecules of ECP under the action of potential, current or chemical factors. Another feature of films of ECP, which is important in their application, is their morphology and structure. These characteristics come to the fore, when PAn is used in chemo- and biosensors (since determine the size of response of PAn films at detecting of various substances-analytes) [72], in the manufacture of supercapacitors (increasing of their capacity) [19, 73, 74], electrochemical energy sources (reducing the polarization losses) [23, 75], in catalytic coatings (increasing the effectiveness of catalysis) [76–80], in nano- and microdrives (artificial muscles) (increasing an angle of their deformation) [81, 82], more. The morphology of PAn's layers is also crucial under their application in modern electronic technologies. The large number of active centers per unit of mass of the polymer accessible to oxidation/reduction reaction as well as the porosity of the PAn's structures allow rapid diffusion of ions in polymer networks that are the most important requirements for the construction of a high-energy batteries, supercapacitors, transducers of chemosensors and biosensors, nano- and microdrives (artificial muscles) [23], etc. However, smooth and adhesive, chemically and thermally stable, impermeable to $H_2O$ and small ions such $Cl^-$, films of PAn also have broad prospects for their use, including a protective anticorrosion coatings of active metals and alloys on their basis [36, 49, 83].

Surface condition of WE, especially from the active metals, has an important significance for the deposition of PAn's film. Usually, in literature data it is not described the condition of WE surfaces, such as cleanness of the surface treatment, the presence of oxide films, the composition and structure of alloys of different nature, but there are only way of the surface preparation WE. The clarification of an impact of the condition and composition of oxide films on potentiodynamic oxidation of An is an important task because it allows the use of active metals without additional costs for surface preparation, and in many cases enhance the adhesion of electrochemically deposited films of ECP. Suitable materials for WE, which are characterized with listed above properties, can serve the samples of amorphous metal alloys based on aluminum [50–52]. Therefore, important is the study of potentiodynamic deposition of PAn's films on amorphous metal surfaces of WE due to the possibilities of their future use.

## 13.6 EXPERIMENTAL

### 13.6.1 MATERIALS

Aniline (99.5%) of the "*Aldrich*" company was distilled in a vacuum. Solutions of sulfuric acid ($H_2SO_4$) were prepared from the standard titrimetric substance of Cherkasy State Plant of Chemicals production. Distilled water was used as the solvent. Ethanol was distilled under normal conditions.

Samples of AMA based on aluminum of composition $Al_{87}Ni_8Y_5$ (AlNiY–electrode), $Al_{87}Ni_8Ce_5$ (AlNiCe-electrode), $Al_{87}Ni_8Gd_5$ (AlNiGd-electrode), $Al_{87}Ni_8Dy_5$ (AlNiDy-electrode), as well as polycrystalline aluminum (Al-electrode, purity 99.995%) in the form of plates with the thickness ~40 μm, size ~2.0×0.2 $cm^2$ and active surface ~0.2 $cm^2$ were used as the working electrodes (WE).

### 13.6.2 ELECTROCHEMICAL DEPOSITION OF POLYANILINE'S FILMS

In the ribbon AMA obtained by the flow turning method there are two sides, namely contact (adjacent to the cooling drum) with developed (defective) surface and the external (which is in contact with the atmosphere of helium) having a smooth surface [51]. In the presented work, the films of PAn were deposited simultaneously on both sides of the AMA samples, which were used as the working electrodes. Working electrodes were previously washed with ethanol and were air-dried for 5 min. Electrodeposition of PAn was performed from air-free argon for 10 min. 0.25 $M$ aqueous solution of An in 0.5 $M$ $H_2SO_4$ simultaneously on both sides of the electrodes at the speed of a potential sweep 50 $mV/s$ within $-200-1200$ $mV$. The films of PAn on the working electrodes were formed for 75 cycles of the sweep potential. Electrodes with the coated films of PAn were washed with distilled water and dried at room temperature.

### 13.6.3 INSTRUMENTAL METHODS

The deposition of PAn's films was conducted by potentiodynamic method at the facility for electrochemical and electrochemiluminescent researches $CVA$-1 accordingly to three-electrodic scheme with Ag/AgCl reference electrode by $EVL$-1$M$4 mark. All values of the electrode potentials are regarded as for this reference electrode. Cyclic voltammograms were recorded on a personal computer. Platinum plate (99.9%) with the size of 1×1 cm was used as the antielectrode.

The structure of the synthesized PAn films was studied using X-ray diffraction ($XRD$) and Fourier-transform infrared spectral ($FTIR$) analysis. The diffractograms of the samples were received with the use of diffractometer by $DRON$-3 mark ($Cu$-$K\alpha$ radiation, $\lambda = 1.54060$ Å). $FTIR$ spectra of the samples were recorded with the use of spectrophotometer $NICOLET$ $IS$ 10 in reflection mode. $XRD$ and $FTIR$ analysis of PAn's films was performed directly on the surface of WE.

To study the morphology of obtained PAn's layers ($SEM$-images) on WE and energy dispersive X-ray ($EDX$) analysis of composition of films it was used the scanning electron microscope microanalyzer $SELMA$ by $PEMMA$-102–02 mark.

### 13.7 RESULTS AND DISCUSSION

### 13.7.1 POTENTIODYNAMIC OXIDATION OF ANILINE

Analysis of potentiodynamic curves of An's oxidation and redox transformations of PAn on the surface of AlNiY-electrode [50] showed the difference in a form of a peak for oxidation of An and cyclic voltammograms on a peak of An's oxidation and curves of redox transformations of PAn on Pt, Au, Al and other electrodes [17, 49, 70]. Therefore, to characterize and understand the processes occurring under potentiodynamic oxidation of An on the electrodes with the Al-based AMA we have conducted researches on potentiodynamic oxidation of An on polycrystalline Al-electrode. Cyclical voltammograms depicted on Figure 13.21 obtained by oxidation of An in its 0.25 $M$ aqueous solution of 0.5 $M$ $H_2SO_4$. The first branch of the anode cycle within potentials $0 \div 1200$ $mV$ is slightly dissolving of Al electrode [36]. As shown on Figure 13.21a, characteristic high-current peak, just as on Pt-electrode [17, 50, 52], which is associated with the electrochemical oxidation of An, is not observed on the cyclical voltammograms due to the presence of oxide film $(Al_2O_3)$ on the surface of Al-electrode, which prevents the oxidation of An. At the prolonged (for 25 cycles) potential ($E$) scanning of Al-electrode, gradually dissolving of the oxide film or the formation of pittings in it takes place, where the oxidation of a small number of An is occurred [50]. Clearly to establish the potential of An's oxidation is difficult due to very small current. However, at detailed study of cyclical voltammograms it can be said that this takes place at E $\approx$ 1000 $mV$. For 25 cycle on the cathodic branch of cyclical voltammogram at $E \approx 460$ $mV$ there is a low-current peak of the reduction of PAn, or rather transformation of its pernigraniline form into emeraldine, which at $E \approx 200$ $mV$ is reduced deeper, namely to leucoemeraldine form of PAn. On the next anodic branch of cyclical voltammogram a peak

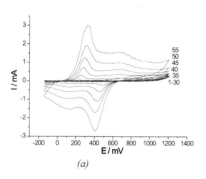
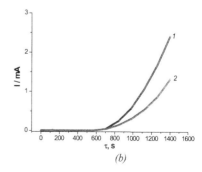

           (a)                                     (b)

**FIGURE 13.21**  Cyclical voltammograms of Al-electrode in 0.25 M aqueous solutions of An + 0.5 M $H_2SO_4$ (the numbers of cycles depicted on figure): a are 1–55 cycles; b are kinetic curves of the formation of anodic peaks: the first (**1**) and the second (**2**).

at $E \approx 250\ mV$ is observed, which corresponds to oxidative transformation leucoemeraldine/emeraldine and as well as a peak at $E \approx 650\ mV$, which corresponds to the oxidation of emeraldine to pernigraniline (see Figure 13.21a) [15, 36, 84]. In subsequent cycles the values of anodic and cathodic current peaks are increased (see Figure 13.21). At the anode potential sweep to ~1000 $mV$ on the WE the PAn is formed as the PNAn salt (PNAnS), which is one of the main forms of polyaniline. And at the cathodic potential sweep to ~ –200 $mV$ the PAn is formed as LEm (LEmS), which is also a form of PAn [1–2, 4–5, 15, 36].

Increase of the number of cycles leads to the displacement of the potential of the first anodic peak (LEm/Em) in the region of higher potentials (see Figure 13.21a), which is a sign of some growing of the electrical resistance of PAn's film. The difference between the values of potentials of the first anodic peaks of the 30-th and the 75-th cycles is ~250 $mV$. Figure 13.21b demonstrates the presence of the induction period (parallel to the abscissas plot of the curve) with the duration ~650 $s^{17}$. The duration of the induction period was determined only taking into account the time of anodic potential sweep. An availability of the induction period is caused by the dissolution of the oxide film on the surface of Al-electrode. The ascending branch of the curve shows that the growth of the film of PAn on the Al-electrode begins actively to occur after the end of the induction period. The potentials of the first anodic peak grow stronger than the potentials of the second anodic peak (see Figure 2.1, b). Overall, the form of cyclical voltammograms of the redox transformations of PAn on Al-electrode is similar to that described in Ref. [36].

On Figure 13.22 there are cyclical voltammograms of AlNiY-electrode in 0.25 $M$ solution of An in 0.5 $M\ H_2SO_4$. At the anodic branch of the first cycle within the potentials $220 \div 470\ mV$ the low-current peak at $E = 280\ mV$ of AlNiY-electrode dissolution is observed. Then the passivation of WE takes place. Peak, which is responsible for oxidation of An is not observed on the anodic branches of 1–40 cycles (see Figure 13.22, a). At the cathodic branch of the ~ 40 cycle of the potential sweep at $E \approx 300\ mV$ the cathodic peak is observed that as in the case of Al-electrode, corresponds to the reduction of PNAn to Em form of PAn. After this peak at $E \approx 50\ mV$ the reduction process of Em to LEm is started. At the anodic potential scanning at $E \approx 400\ mV$ the peak is formed, which corresponds to the oxidation of LEm in Em

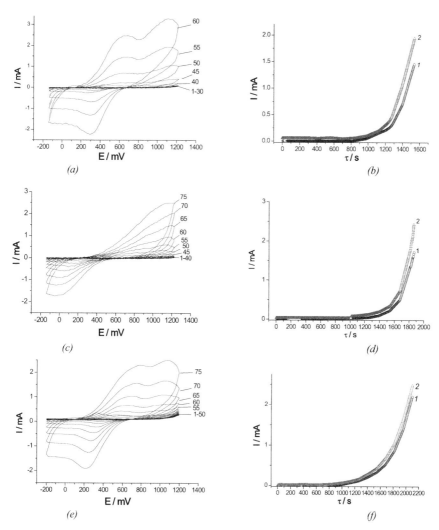

**FIGURE 13.22** Cyclical voltammograms of working electrodes in 0.25 M aqueous solutions of An + 0.5 M H$_2$SO$_4$ (the numbers of cycles depicted on figure): a is AlNiY, c is AlNiCe, $e$ is AlNiGd, $g$ is AlNiDy and kinetic curves of the formation of anodic peaks on the electrodes: $b$ is AlNiY, $d$ is AlNiCe, $f$ is AlNiGd, $i$ is AlNiDy.

(*see* Figure 13.22, $a$). Active formation of PAn on AlNiY-electrode is started after ~800 $s$ of anodic scanning potential and is described by a sharp rise on kinetic curve (*see* Figure 13.22, $b$). Currents of the second anodic peak grow more intense than the currents of the first anodic peak (*see* Figure 13.22, $b$).

Figure 13.22, *c* shows the cyclical voltammograms of AlNiCe-electrode. For 45 scanning cycles of the potential the obvious signs of An's oxidation are not traced. An analysis of the cathodic branch of cyclical voltammogram of the 45 cycle shows, that the low-current peak at $E \approx 220 \; mV$ corresponds to reduction of PNAn. Already in the 46 cycle on the cyclical voltammogram two peaks of PAn oxidation at $E \approx 620 \; mV$ and 1050 $mV$ are traced. Form of the red/ox peaks on cyclical voltammograms is differed from the peaks on the cyclical voltammograms of Al-electrode (*see* Figure 13.22, *a*). Cathode peak at $E \approx 200 \; mV$ corresponds to the reduction of PAn obtained in the anode process. This peak with increasing of the number of cycles is shifted to the cathode side and in 75 cycles its potential is equal to 25 $mV$. Continuous scanning of the potential leads to higher anodic and cathodic currents peaks (*see* Figure 13.22, *d*). The character of anodic branches of cyclical voltammograms shows that the oxidation of LEm into Em and Em into PNAn is similar to this process on other WE. The difference between the values of the potentials ($\Delta E$) of the first anodic peaks of 51 and 75 cycles consists of ~301 $mV$. For the kinetic curve (*see* Figure 13.22, *d*) inherent an existence of the induction period. Active formation of PAn begins after achievement of 1100 *s*; then, sharp rise is appeared on the curves. The currents of the first anodic peak are increased stronger than the currents of the second peak (*see* Figure 13.22, *d*).

Figure 13.22, *e* shows the cyclical voltammograms of AlNiGd-electrode. For 50 cycles on cyclical voltammogram at $E \approx 450 \; mV$ the peak of PAn reduction is observed (*see* Figure 13.22, *e*). At the anode branch of 51 cycle two peaks at potentials ~580 and ~1080 $mV$ are observed. Continuous scanning of the potentials leads to higher currents redox peaks of PAn and the shift of their anode potential in the anode side and cathode potential in the cathode side. The difference between the values of potentials ($\Delta E$) of the first anodic peaks of 51 and 75 cycles consists of ~897 $mV$. Potential of the cathodic peak of 75 cycles is 130 $mV$. Kinetic curve of the anodic processes proceeding on AlNiGd-electrode is shown on Figure 13.22, *e*. Apparently, the sharp rise on the kinetic curve is observed after 1500 *s* and is a sign of the start of active formation of PAn on the surface of WE. Currents of anodic peaks increase with the same intensity (*see* Figure 13.22, *f*).

Figure 13.22, *g* presents the cyclical voltammograms of AlNiDy-electrode. At the anodic branch of the first cycle within the potentials

$100 \div 300\ mV$ the low-current peak is observed at $E = 280\ mV$, which corresponds to the dissolution of AlNiDy-electrode. Then the passivation of WE takes place. The pick, which corresponds to the An oxidation on the anodic branches of 1–54 cycles is not observed (see Figure 13.22, g). At long potential sweep the cathodic peak at $E \approx 190\ mV$ is formed on the 54 cycles. At the anodic branch of 55 cycles two distinct peaks with anodic potentials ~790 and ~1080 $mV$ are formed. These red/ox peaks correspond to conversion of PAn like to Al-electrode. As shown on Figure 13.22, i the start of active formation of PAn's film begin after 1150 s of the completion of the induction period. The currents of the first anodic peak grow stronger than the currents of the second peak (see Figure 13.22, i).

Different duration of the induction period, typical for the beginning of active formation of PAn's film on each of the investigated WE, caused by the different properties of the passivation surface oxide films and the duration of their dissolution [17, 85]. There is an obvious fact that the An's molecules on the surface of WE do not inhibit the process of surface oxide films dissolution [86].

For red/ox transformations of PAn on the Al-electrode is characteristic higher reversibility (small difference of the potentials $\Delta E'$ between conjugated anodic and cathodic peaks on the cyclical voltammograms) compared with AMA-electrodes (see Table 13.10). At the transition from AlNiY, AlNiGd to AlNiDy electrode the reversibility of the red/ox process conversion of PAn is increased, the value $\Delta E$ is decreased (see Table 13.10).

**TABLE 13.10** Parameters of Potentiodynamic Oxidation of Aniline and Redox Transformations of Polyaniline on the Investigated Electrodes (the 45 cycles)

| WE | Potentials ± 1.0/$mV$ | | | | | $\Delta E/mV$ |
|----|------|------|------|------|------|--------|
| | **Oxidation An** | **Redox transformations of PAn** | | | $\Delta E/mV$ | |
| | | **Anodic peaks** | | **Cathode peak** | | |
| Al | ~1000 | 270 | 660 | 82 | — | 280 | 250 |
| AlNiY | ~1050 | 700 | 1100 | 220 | — | 880 | 350 |
| AlNiCe | ~1050 | 620 | 1185 | 210 | — | 887 | 301 |
| AlNiGd | ~1050 | 750 | 1083 | 190 | — | 897 | 224 |
| AlNiDy | ~1050 | 780 | 1080 | 180 | — | 900 | 161 |

Note: *$\Delta E$ (reversibility) was determined as the difference between of the first anodic and the second cathode peaks. **$\Delta E'$ was determined as the difference between the potentials of 50 and 75 first anodic peaks.

Exclusion has the AlNiCe-electrode: the reversibility red/ox conversion in macromolecule of PAn is significantly lower than on other electrodes of the AMA.

Analysis of the results of potentiodynamic oxidation of An and red/ox transformations of PAn on the electrodes of Al and AMA shows that the existing difference in peaks' form of red/ox transformations of PAn, as well as in the kinetics of PAn's deposition are due to surface condition of WE, which is determined by the influence amorphic component (Y, Ce, Gd and Dy), which, obviously affects the higher resistance of the oxide film on the surface of WE. With an increasing of number of scanning cycles of potential the difference between the potential of the first anodic peak of 50 and 75 cycles is decreased in the series (*see* Table 13.11):

$$AlNiY > AlNiCe > AlNiGd > AlNiDy$$

The obtained films of PAn on WE had dark green/black color characteristic for PAn in oxidized form Em/PNAn [36], so in general, electrochemical oxidation reaction of An at the anode (WE) can be described by the scheme [17, 87]:

$$C_6H_5NH_2 \xrightarrow{-ne^-,-nH^+} -(C_6H_4 - NH)_n -$$

The main peaks of electrochemical conversion of PAn find the anodic peaks at potential $\sim 300$ $mV$ and $\sim 750$ $mV$, which correspond to the oxidation of LEm to partially oxidized Em form, and the second peak corresponds to the oxidation of Em to fully oxidized PNAn form [23].

**TABLE 13.11**   Structural Parameters of PAn in the Films on the Surface of WE

| Sample | Crystallinity degree/% at | | The ratio of intensities | |
|---|---|---|---|---|
| | 2. = 24.1 | 2. = 20.4 | $2\theta_{24.1}/2\theta_{20.4}$ | $I_{1484}/I_{1570}$ |
| Al | 1.3 | 13.8 | 1.2 | 1.3 |
| AlNiY | 1.0 | 13.9 | 1.1 | 1.2 |
| AlNiCe | 2.3 | 18.1 | 2.0 | 1.3 |
| AlNiDy | 1.8 | 8.2 | 2.6 | 1.3 |
| AlNiGd | 1.3 | 16.8 | 3.0 | 1.2 |

## 13.7.2 STRUCTURAL STUDIES

### 13.7.2.1 X-Ray Diffraction Analysis of PAn Products on Working Electrodes

Figure 13.23 shows the diffractograms of PAn, potentiodynamically deposited on Al and AMA electrodes in the form of films. As shown from Figure 13.23, on the all diffractograms against the background of halo within $2\theta = 17 \div 28^0$ there are two characteristic peaks of PAn at $2\theta = 20.4^0$, which corresponds to EmB and at $2\theta = 24.1^0$, which corresponds to EmS [53, 88, 89]. The peak centered at $2\theta = 20.4^0$ can be ascribed to the periodicity parallel to the polymer chain, while the peak at $2\theta = 24.1^0$ may be caused by the periodicity perpendicular to the polymer chains of PAn [75, 90, 91].

The intensities of diffraction peaks of the PAn's sample, formed on the Al-electrode (*see* Figure 13.23, curve 1) are the lowest, and the intensities of diffraction peaks of PAn on the AlNiY–electrode are almost proportionate. The intensity of the diffraction peak of PAn at $2\theta = \sim24.1^0$ formed on the surface of AlNiCe, AlNiGd and AlNDy electrodes, is higher than the intensity of peak at $2\theta = \sim20.4^0$ certifying the change ratio EmB EmS

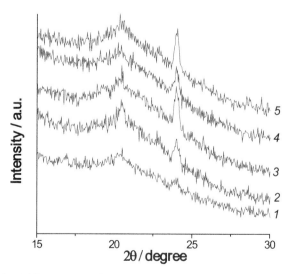

**FIGURE 13.23**  Diffractograms of PAn on the surface of working electrodes: 1 is Al; 2 is AlNiY; 3 is AlNiCe; 4 is AlNiGd; 5 is AlNDy.

toward EmS and its higher crystallinity (sharp and narrow peak). Characteristic peaks at $2\theta = \sim 24.1^0$ on diffractograms of PAn's samples on the AlNiCe, AlNiGd and AlNDy electrodes (*see* Figure 13.23, diffractograms 3, 4 and 5) is higher than the peaks on the Al and AlNiY electrodes (*see* Figure 13.23, diffractograms 1 and 2), which is indicative of a higher degree of crystallinity of PAn in films (*see* Table 13.11). The presence of two diffraction peaks in the background amorphous halo is a sign that the formed polyaniline has amorphous-crystalline structure in which coexist crystallites of EmS (metalloids) with the crystallites of unprotonated form of EmB [36, 92]. High intensity of the diffraction peaks at $2\theta = \sim 24.1^0$ compared with the peaks at $2\theta = \sim 20.2^0$ shows that the proportion of metalloid crystallites is too higher (*see* Table 13.11).

### 13.7.2.2   FTIR Spectra of PAn Products on Working Electrodes

The FTIR spectra of reflection of PAn's films deposited by potentiodynamic method on the surface of WE are shown on Figure 13.24 and the band assignments are collected in Table 13.23. FTIR spectra of PAn's films on the surface of WE are, mainly, similar to the FTIR spectra of PAn synthesized in chemical [93−97] and electrochemical [44, 47, 90, 98−102] ways.

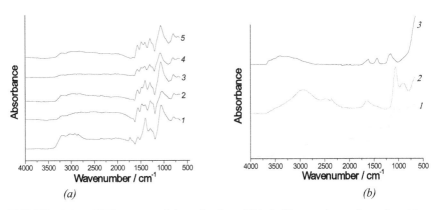

**FIGURE 13.24**   FTIR spectra of the reflection of PAn's films on the surface of working electrodes: **1** is Al; **2** is AlNiY; **3** is AlNiCe; **4** is AlNiGd; **5** is AlNiDy; **1** is the sample PAn-$H_2SO_4$ (chemically synthesized (a); **2** is salt $Al_2(SO_4)\times 18H_2O$ and **3** is film of $Al_2O_3$ (*b*) on Al-electrode.

**TABLE 13.12**  The Assignment and Frequency of the Peaks in the Region FTIR Spectra 1600–650 cm$^{-1}$ of PAn Samples on Figure 13.24

| Sample | Interatomic bond/vibrational frequency, cm$^{-1}$ | | | | | | | |
|---|---|---|---|---|---|---|---|---|
| | C=C quinoid ring | C=C benzoid ring | C–N | C–N–C | C=C quinoid ring | SO$_4^{2-}$ | B–NH–Q or B–NH–B | C–H |
| Al | 1564 | 1481 | 1296 | 1244 | 1136 | 1065 | 973 | 796 |
| AlNiY | 1564 | 1487 | 1301 | 1246 | 1149 | 1046 | 970 | 799 |
| AlNiCe | 1574 | 1487 | 1302 | 1247 | 1149 | 1070 | 972 | 799 |
| AlNiGd | 1574 | 1488 | 1301 | 1245 | 1149 | 1077 | 986 | 799 |
| AlNiDy | 1574 | 1489 | 1302 | 1244 | 1150 | 1073 | 972 | 799 |

The main characteristic bands characteristic for PAn are the following 3 443, 3 200, 3 050, 2 920, 2 846, 1 576, 1 490, 1 301–1 299, 1 150–1 136 i 808–783 cm$^{-1}$. The locations of these characteristic peaks present a good agreement with the literature [44, 47, 90, 93–102]. The peaks positions of various bonds obtained in these spectra (*see* Figure 13.24) along with their bonds are given in Table 13.12.

Deformation bands at 3400–3200 cm$^{-1}$ and 2950–2830 cm$^{-1}$ correspond to valence vibrations of N–H groups of aromatic amines, and to valence vibrations of C–H groups of aromatic ring [68, 103, 104]. Absorption bands within 3400–2800 cm$^{-1}$, commonly are called "H-peaks", also attributed to the hydrogen bonds between the regularly placed PAn chains that form the amino (–NH–) and attaching proton imino (–NH$^+$–) groups [94, 89]. Bands within 3270 and 3200 cm$^{-1}$ are also referred to intra- and intermolecular hydrogen-bonded N–H stretching vibrations of the secondary amines [105]. The presence of a weak band at 2920 cm$^{-1}$, which corresponds to N–H modes may be a sign of crosslinking (structuring) of polymer chains [106] or the formation of fully reduced LEm and partially oxidized Em bonds at 2922 and 2923 cm$^{-1}$ correspond to N–H modes with sharp intense bond edges and fully oxidized pernigraniline (PNAn) form of bond at 2922 cm$^{-1}$ with low intense bond edge as shown on Figure 13.24 [102].

A weak peak on FTIR spectra of PAn on Al-electrode at 1733 cm$^{-1}$ (*see* Figure 13.24, the spectrum 1) obviously can be attributed to the absorptions of C = O group, which indicates that PAn in films is slightly overoxidized during the growth process of electropolymerization [107]. A similar band on the FTIR spectra of PAn's films on the AMA electrodes is not observed (see Figure 13.24, the spectra 2–5). The weak band at ~1665 and an intense band at ~1407 cm$^{-1}$ on FTIR spectra of PAn on AMA electrodes are referred to cross-linking of PAn's macromolecules [108]. The combination of observed on the FTIR spectra of the absorption bands at 1665, 1407 and 687 cm$^{-1}$ was attributed by the authors of Ref. [109] to the first oligomeric products of oxidation of aniline.

The characteristic peaks at about 1570 and 1485 cm$^{-1}$ correspond to the stretching vibrations of N=$Q$=N ring, N–$B$–N ring, respectively (where $Q$ refers to the quinonic type rings and $B$ refers to the benzenic-type rings) [44, 47, 68, 89, 93–103, 105–107]. An intensity of the "benzoid" band exceeds an intensity of the "quinoid" band, which is a sign of dominance of the benzoid groups in the structure of macromolecular chain.

On FTIR spectra of the samples at 1404–1410 cm$^{-1}$ an intensive band exists (see Figure 13.24, a). The intensity of this band on the spectrum of PAn on Al and AlNiCe-electrodes is high enough (see Figure 13.24, a, the spectra 1 and 3), and on the other AMA electrodes (see Figure 13.24, a, the spectra 2, 4, 5) is commensurate with the intensity of the band at ~1490 cm$^{-1}$. Nature of band at 1404 – 1410 cm$^{-1}$ (see Figure 13.24, a) requires a detailed study. In the works, in which there are FTIR spectra of PAn on the aluminum-containing surfaces, particularly AA 2024 T6 (sharp band at 1380 cm$^{-1}$)[46] and PAn/pumise (acute low-intensity band at 1415 cm$^{-1}$) [110] not identified. On FTIR spectra of PAn on the not aluminum-containing surfaces, such as, PAn/activated carbon (band at 1400 cm$^{-1}$) [111] or PAn/dodecylbenzenesulfonic acid (weak sharp band at 1410 cm$^{-1}$) [112] the nature of this band is also not determined.

The weak band on FTIR spectra of PAn at 1404 cm$^{-1}$ authors of Ref. [113] attributed to the stretching vibrations of –C–N and $Q$–$B$–$Q$ group of semiquinoid ring of PAn. The peaks at 1403 and 1490 cm$^{-1}$ authors of Ref. [114] attributed to the C=C stretching vibrations of benzinoid group. Small peaks appeared at 1400 cm$^{-1}$ authors of Refs. [79, 105, 115] attributed to the aromatic C–C stretching vibrations of phenazine-like segments or branched structures in PAn oligomers. The pair of bands at 1443 cm$^{-1}$ and 1041 cm$^{-1}$ on FTIR spectra of PAn authors of Ref. [116, 117] attributed to the N–N single bond, which is the defect of the polymer chains. The intense sharp bands on Raman spectra of PAn at about 1651 and 1400 cm$^{-1}$

authors of Ref. [108] attributed to the cross-linking of PAn. Weak peak at 1400 cm$^{-1}$ the authors of Ref. [118] attributed to the C–N stretching in the neighborhood of a quinonoid ring. The intensity of the peak at 1404 – 1410 cm$^{-1}$, also can correspond to the oxidized units (quinoid rings), and it very low due to low polymerization potential [119].

The band at 1404 – 1410 cm$^{-1}$ on FTIR spectra of PAn the authors of Refs. [120, 121] attributed to stretching vibration of Al–O bond or stretching vibrations –C–N and Q–B–Q groups of semiquinoid ring of PAn, or which were associated with phenazine-like or branched structures in PAn oligomers [120–122].

For identification of band at ~1407 cm$^{-1}$ on FTIR spectra of PAn's samples on WE, we have conducted an analysis of FTIR spectra of samples, namely, chemically synthesized sample in 0.5 $M$ solution of H$_2$SO$_4$ (PAn-H$_2$SO$_4$), salt Al$_2$(SO$_4$)$\times$18H$_2$O and film of Al$_2$O$_3$ on Al-electrode (*see* Figure 13.24, *b*). As shown on Figure 13.24, *b*, on the FTIR spectra of the sample PAn-H$_2$SO$_4$ in the spectrum **1** there are all characteristic bands peculiar for chemically synthesized PAn in the presence of H$_2$SO$_4$ [100, 115, 123, 124]. On FTIR spectra of Al$_2$(SO$_4$)$\times$18H$_2$O the sharp intense peak at 1062 cm$^{-1}$ can correspond to SO$_4^{2-}$ groups (*see* Figure 13.24, *b*, the spectrum 2) [100, 113, 123]. On FTIR spectra of Al$_2$O$_3$ within the 1700–900 cm$^{-1}$ there are three characteristic peaks (*see* Figure 13.24, *b*, the spectrum 3). The most intense is the peak at 1153 cm$^{-1}$, which apparently corresponds to the valence vibrations of Al–O bond [121].

High intensity of the peak at 1407 cm$^{-1}$ on FTIR spectra of the PAn's sample, deposited on the Al-electrode (*see* Figure 13.24, *a*, the spectrum **1**) may be caused by overlapping of the bands inherent to Al$_2$(SO$_4$)$_3$, which may be present in the PAn's film, by the formation of branched structures or semiquinoid ring in macromolecules of PAn, or which were associated with phenazine-like structures in PAn oligomers. Obviously, the presence of the peak at 1407 cm$^{-1}$ on FTIR spectra of PAn's samples on WE may correspond to the above factors together [114–117, 125, 129].

The intensity peak at about ~1299 cm$^{-1}$ and shoulder at ~1247 cm$^{-1}$ are attributed to C–N stretching vibration corresponding to –NH$^+$– in protonic acid doped [95, 104, 115]. The intensity of the band at about ~ 1299 cm$^{-1}$ indicative of the degree of PAn's doping, i.e., the presence of EmS providing the conductivity of PAn. It's well-known, that the doping-protonation of PAn is occurred on imine nitrogen atoms of the macromolecules with the formation of semi-quinoid cation-radicals, which are so-called polyarons [94]. The form of band within $\sim$1210–890 cm$^{-1}$ with intensive peak at 1068 cm$^{-1}$ on all FTIR spectra is complex and obviously has all three bands, namely band (shoulder) at $\sim$1135 cm$^{-1}$, band at $\sim$1068 cm$^{-1}$ and band (shoulder) at $\sim$982 cm$^{-1}$ (*see* Figure 13.24, *a*) [126]. The weak bands at $\sim$1135 and $\sim$982 cm$^{-1}$ can correspond to B–NH–Q or B–NH–B groups that are formed during the doping reaction [104, 127]. These characteristic absorption peaks at $\sim$1135 cm$^{-1}$ (corresponding to electron delocalization degree) and $\sim$1244 cm$^{-1}$ (corresponding to –NH$^+$– in protonic acid doped PAn) (*see* Figure 13.24, *a*) became lower in intensity with decreasing acidity of the preparation media [115]. The authors of Ref. [128] attributed of the intense peak at $\sim$1144 cm$^{-1}$ to the plane vibrations of C–H groups of the pernigraniline. At $\sim$1135 cm$^{-1}$ it can be occurred the band of Me–O bond in composites PAn/metal oxides [129] or C–O stretching vibrations can be found [130] at 1132 cm$^{-1}$. The bands at $\sim$1135 and $\sim$796 cm$^{-1}$ can also correspond to the plane and out of plane deformation vibrations of C–H bond in benzene cycle, respectively, confirming the para-replacement in the structure of the molecules of synthesized PAn [131–133]. Low intensity band at $\sim$796 cm$^{-1}$ which can be traced on the FTIR spectra of samples is a unique key to identify the type of joining molecules of An in macromolecular chain of PAn [95, 99, 103, 113, 134, 135]. The presence of this band indicates 1.4-linkage of An's molecules in macromolecules of PAn [99, 131–133, 136].

After the finish of potentiodynamic deposition of PAn's film on WE in 0.5 $mol·l^{-1}$ H$_2$SO$_4$, PAn is in doped SO$_4^{2-}$ state, namely in the form of EmS/PNAnS. This confirms also the peak at $\sim$1068 cm$^{-1}$, which is a sign of the absorptions of SO$_4^{2-}$ groups [100, 113, 123].

On FTIR spectra of electrochemically deposited PAn's films on WE from AMA it is observed the shift of the main characteristic bands (*see* Figure 13.24, *a* and Table 13.13) regarding to PAn's film on Al-electrode ($\sim$1564 cm$^{-1}$) in the short-wave area of the spectrum for PAn on AlNiY–electrode to $\sim$1564 cm$^{-1}$ and 1485 cm$^{-1}$, while the main characteristic bands of PAn in films on AlNiCe, AlNiGd and AlNiDy-electrodes are "quinoid" bands at 1574 cm$^{-1}$, whereas for the "benzoid" band at $\sim$1487 cm$^{-1}$ a small shift is observed. More significant in relation to characteristic band at 1065 cm$^{-1}$ of PAn on the Al-electrode is bathochromic displacement of this characteristic band for PAn on AlNiY–electrode (~1046 cm⁻¹) and hypsochrome displacement for the PAn's films on AlNiCe (~1070 cm⁻¹), AlNiDy (~1073 cm⁻¹) and AlNiGd-electrode (~1076 cm⁻¹).

Based on diffractograms (*see* Figure 13.23) it were calculated the crystallinity degrees of PAn in films on WE, and by FTIR spectra (*see* Figure 13.24, *a*) it were determined the ratio of the intensities of "quinoid" (~1484 cm⁻¹) and "benzoid" (~ 1570 cm⁻¹) peaks (*see* Table 13.12). As shown in Table 13.22, the crystallinity degree of PAn in the form of Em**B** (2θ = 24.1⁰) is low, and the crystallinity degree of PAn in the form of Em**S** (2θ = 20.4⁰) is much higher. During the electrochemical oxidation of An on WE the Em**S** and PNAn**S** are formed to a greater extent than the Em**B**. The ratio of the "quinoid" (~1484 cm⁻¹) and "benzoid" (~1570 cm⁻¹) intensities is greater than 1 confirming the formation predominantly of benzoid groups in the composition of PAn's film on the investigated WE.

### 13.7.3   MORPHOLOGICAL PROPERTIES OF PAN'S FILMS ON WORKING ELECTRODES

The surface of WE and PAn's films on these electrodes were analyzed by scanning electron microscope-microanalyzer (SEM–images). For the analysis the electrodes with the PAn's films deposited at 75 cycles of the potential scanning were used. The analysis of the PAn's films deposited only on the contact surfaces of WE has been done. Figure 13.25 shows the SEM-images of the contact surfaces of Al and, as an example of the AMA' surfaces, the surface of AlNiY-electrode.

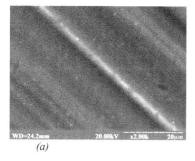 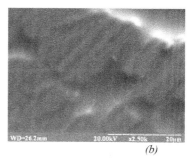

*(a)*                                            *(b)*

**FIGURE 13.25**   SEM-images of the contact side of Al (*a*) and AlNiY (*b*) electrodes.

The surface of Al-electrode is smoother than the surface of AlNiY-electrode. The surface of AlNiY–electrode, as well as other AMA-electrodes are more developed due to the topography of the drum on which the tapes of AMA were formed. Ledges on the surface of AMA-electrodes are rough and can serve as defects, and the deepenings are smoother (*see* Figure 13.25, *b*). For oxide films on Al and on alloys on its basis inherent an inhomogeneity in thickness [17, 43, 69] to facilitate the formation of pittings in areas with thinner oxide film or in areas with the defects.

Figure 13.26 illustrates the SEM images of the surfaces of PAn's films with different magnifications, deposited from 0.25 $M$ solution of An in 0.5 $M$ $H_2SO_4$ on WE. From the SEM images (*see* Figure 13.26, *a, c, i, g, h*) it is shown, that against the background of spongy, branched polymeric units on Al-electrode dark areas (deepenings) are traced (*see* Figure 13.26, *a*), and on the surface of WE from AMA, namely AlNiY, AlNiGd and AlNiDy the cracks exist (*see* Figure 13.26, *c, g, h*). SEM image on Figure 13.26, *i* certifies uneven deposition of PAn on AlNiCe-electrode. A similar morphology of the film was obtained by the authors of Ref. [137] at the anode polymerization of aniline on ITO and Au-electrodes.

As shown on Figure 13.26, *a* the PAn's film on Al-electrode has the surround character and is more porous. SEM images of higher magnifications (*see* Figure 13.26, *b, d, f, e, j* show that the PAn's film is uneven on the thickness formation. For the film formation characteristic is the formation of spatial pungent structures. The PAn's films deposited via potentiodynamic oxidation of An from the solutions of $H_2SO_4$ on Al and AMA-electrodes have porous and spongy (branched) structure, which is typical for electrodeposited PAn's films on different electrode materials [13, 23, 39, 42, 45, 67, 138]. Spongy morphology of the PAn's films on WE caused by the fact that the formed macromolecular PAn represents by itself the original nano- or microelectrodes for the initiation of new polymer formation, namely sprout of various shapes, such as a thread or rod.

### 13.7.4   ENERGY-DISPERSIVE X-RAY (EDX) SPECTROMETRY

Figures 13.27 and 13.28 show the EDX-spectra of the area of PAn's film on the surface of Al and AlNiDy (as an example of the spectrum obtained

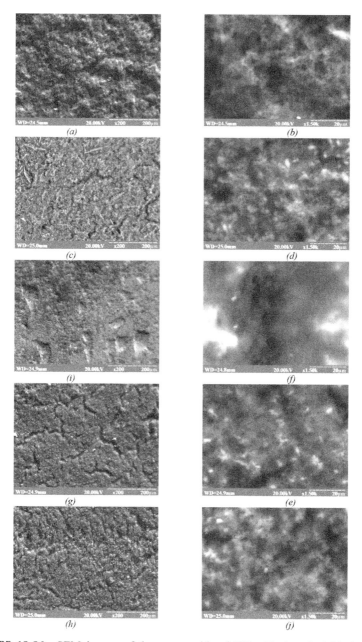

**FIGURE 13.26**  SEM–images of the contact side of WE with deposited PAn's films: $a, b$ – Al, $c, d$ – AlNiY, $i, f$ – AlNiCe, $g, e$ – AlNiGd, $h, j$ – AlNiDy. $a, c, i, g, h$ – magnification in 200 times. $b, d, f,$ e, $j,$ – magnification in 1500 times.

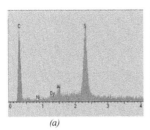 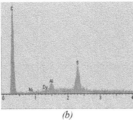 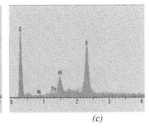

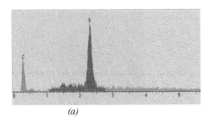 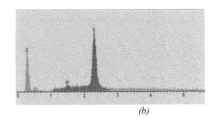

(a)                                    (b)

**FIGURE 13.27**    EDX-spectra of the area of PAn's film on Al-electrode (*see* Figure 13.26, *a, b*): *a* is gray area, *b* is light area.

(a)                    (b)                    (c)

**FIGURE 13.28**    EDX-spectra of the area of PAn's film on AlNiDy-electrode (*see* Figure 13.26, *h, j*): *a* is dark area, *b* is gray area, *c* is light area.

for PAn obtained on the AMA-electrodes) electrodes (*see* Figure 13.26). The peak at 0.52 *keV* corresponds to the nitrogen atom (N). The intense peak at 2.3 *keV* confirms the presence of sulfur (S), which is part of the ions $HSO_4^-$ or $SO_4^{2-}$ that are doping component of polyaniline. Low-intensity peak at ~1.5 *keV* corresponds to aluminum (Al), which due to the dissolution of the oxide film is as an impurity in the form of $Al_2(SO_4)_3$ in the film of PAn.

On the EDX-spectra of PAn's films areas (*see* Figures 13.26–13.28) on the surface of AlNiDy-electrode there are weak intensive peaks of Al, Ni and Dy. Evidently, such elements present in the PAn's film as the sulfate additives.

Analysis of EDX-spectra of different areas of PAn's film confirms that PAn (availability C) covers the entire surface of the electrode. But the PAn's films in dark areas obviously have a less-developed surface and are denser. The presence of atoms of metals Al, Ni and Dy in these areas of PAn's film caused by the formation of sulfate salts such $Al_2SO_4$, confirming also the existence of S atoms. However, identification of the metal atoms in the films of PAn can also be caused by the porosity of the films of polymer.

## KEYWORDS

- aniline
- aqueous solutions
- polyanilines
- polyluminol
- polymerization
- synthesis

## REFERENCES

1. Heeger, A. J. Semiconducting and metallic polymers: the fourth generation of polymeric materials. Curr. Appl. Phys. 2001, 1, 247–267.
2. MacDiarmid, A. G. "Synthetic metals": a novel role of organic polymers. Curr. Appl. Phys. 2001, 1, 269–279.
3. Shirakawa, H. The discovery of polyacetylene film – the dawning of an era of conducting polymers. Curr. Appl. Phys. 2001, 1, 281–286.
4. Croce, F., Panero, S., Passerini, S., & Scrosati, B. The role of conductive polymers in advanced electrochemical technology. Electrochim. Acta 1994, 39, 255–263.
5. Batich, C. D., Laitinen, H. A., & Zhou, H. C. Chromatic changes in polyaniline films. J. Electrochem. Soc. 1990, 137, 883- 885.
6. Mortimer, R. J., Dyer, A. L., & Reynolds, J. R. Electrochromic organic and polymeric materials for display applications. Displays 2006, 27, 2–18.
7. Dutta, D., Sarma, T. K., Chowdhury, D., & Chattopadhyay, A. A polyaniline-containing filter paper that acts as a sensor, acid, base, and endpoint indicator and also filters acids and bases. J. Colloid Interf. Sci. 2005, 283, 153–159.
8. Jelle, B. P., & Hagen, G. Performance of an electrochromic window based on polyaniline, Prussian blue and tungsten oxide. Solar Energy Mater. Solar Cells 1999, 58, 277–286.
9. somani, P., Mandale, A. B., & Radhakrishnan, S. Study and development of conducting polymer-based electrochromic display devices. Acta mater. 2000, 48, 2859–2871.
10. Mortimer, R. J. Electrochromic materials. Chem. Soc. Rev. 1997, 26, 147–156.
11. Lacroix, J. C., Kanazawa, K. K., & Diaz, A. Polyaniline: a very fast electrochromic material. J. Electrochem. Soc. 1989, 136, 1308–1313.

12. Carpi, F., & De Rossi, D. Colors from electroactive polymers: electrochromic, electroluminescent and laser devices based on organic materials. Opt. Laser Technol. 2006, 38, 292–305.

13. Yang, Y., & Heeger, A. J. Polyaniline as a transparent electrode for polymer light-emitting diodes: lower operating voltage and higher efficiency. Appl. Phys. Lett. 1994, 64, 1245–1247.

14. Sazou, D., Kourouzidou, M., & Pavlidou, E. Potentiodynamic and potentiostatic deposition of polyaniline on stainless steel: Electrochemical and structural studies for a potential application to corrosion control. Electrochim. Acta. 2007, 52, 4385-4397.

15. Nicho, m. e., hu, H., González-Rodriguez, J. G., & salinas-bravo, V. M. Protection of stainless steel by polyaniline films against corrosion in aqueous environments. J. Appl. Electrochem. 2006, 36, 153–160.

16. Cook, A., Gabriel, A., Siew, D., & Laycock, N. Corrosion protection of low carbon steel with polyaniline: passivation or inhibition? Curr. Appl. Phys. 2004, 4, 133–136.

17. Santos, J. R. Jr., Mattoso, L. H. C., & Motheo, A. J. Investigation of corrosion protection of steel by polyaniline films. Electrochim. Acta 1998, 43, 309-313.

18. Williams, G., Holness, R. J., Worsley, D. A., & McMurray, H. N. Inhibition of corrosion-driven organic coating delamination on zinc by polyaniline. Electrochem. Commun. 2004, 6, 549-555.

19. Özyılmaz, A. T., Tüken, T., Yazıcı, B., & Erbil, M. The electrochemical synthesis and corrosion performance of polyaniline on copper. Prog. Org. Coat. 2005, 52, 92-97.

20. Saxena, V., Malhotra, B. D. Prospects of conducting polymers in molecular electronics. Curr. Appl. Phys. 2003, 3, 293–305.

21. Laranjeira, J. M. G., Khoury, H. J., de Azevedo, W. M., et al. Polyaniline nanofilms as a monitoring label and dosimetric device for gamma radiation. Mater. Charact. 2003, 50, 127–130.

22. Grummt, U.-W., Pron, A., Zagorska, M., & Lefrant, S. Polyaniline based optical pH sensor. Anal. Chim. Acta 1997, 357, 253-259.

23. Talaie, A. Conducting polymer based pH detector: a new outlook to pH sensing technology. Polymer 1997, 38, 1145- 1150.

24. Tahir, Z. M., Alocilja, E. C., & Grooms, D. L. Polyaniline synthesis and its biosensor application. Biosens. Bioelectron. 2005, 20, 1690-1695.

25. Timmer, B., Olthuis, W., & van den Berg, A. Ammonia sensors and their applications – a review. Sensor. Actuat. B-Chem. 2005, 107, 666-677.

26. Michira, I., Akinyeye, R., Somerset, V., et al. Synthesis, characterization of novel polyaniline nanomaterials and application in amperometric biosensors. Macromol. Symp. 2007, 255, 57-69.

27. Do, J.-S., & Chang, W.-B. Amperometric nitrogen dioxide gas sensor based on PAn/Au/Nafion® prepared by constant current and cyclic voltammetry methods. Sensor. Actuat. B-Chem. 2004, 101, 97-106.

28. Ram, M. K., Yavuz, Ö., Lahsangah, V., & Aldissi, M. CO gas sensing from ultra-thin nano-composite conducting polymer film. Sensor. Actuat. B-Chem. 2005, 106, 750-757.

29. Li, J., Petelenz, D., & Janata, J. Suspended gate field-effect transistor sensitive to gaseous hydrogen cyanide. Electroanalysis 1993, 5, 791-794.

30. Conn, C., Sestak, S., Baker, A. T., & Unsworth, J. A polyaniline-based selective hydrogen sensor. Electroanalysis 1998, 10, 1137-1141.

31. Campos, M., Bulhões, L. O. S., & Lindino, C. A. Gas-sensitive characteristics of metal/semiconductor polymer Schottky device. Sensor. Actuat. A-Phys. 2000, 87, 67-71.

32. Pan, X., Kan, J., & Yuan, L. Polyaniline glucose oxidase biosensor prepared with template process. Sensor. Actuat. B-Chem. 2004, 102, 325-330.

33. Luo, Y.-C., & Do, J.-S. Urea biosensor based on PANi(urease)-Nafion®/Au composite electrode. Biosens. Bioelectron. 2004, 20, 15-23.

34. MacDiarmid, A. G., Yang, L. S., Huang, W. S., & Humphrey, B. D. Polyaniline: electrochemistry and application to rechargeable batteries. Synth. Met. 1987, 18, 393–398.

35. Novák, P., Müller, K., Santhanam, K. S. V., & Haas, O. Electrochemically active polymers for rechargeable batteries. Chem. Rev. 1997, 97, 207-281.

36. Fan, L.-Z., Hu, Y.-S., Maier J., et al. High electroactivity of polyaniline in supercapacitors by using a hierarchically porous carbon monolith as a support. Adv. Funct. Mater. 2007, 17, 3083-3087.

37. Nadagouda, M. N., & Varma, R. S. Green approach to bulk and template-free synthesis of thermally stable reduced polyaniline nanofibers for capacitor applications. Green Chem. 2007, 9, 632-637.

38. Jang, J., Bae J., Choi, M., & Yoon, S.-H. Fabrication and characterization of polyaniline coated carbon nanofiber for supercapacitor. Carbon 2005, 43, 2730-2736.

39. Ryu, K. S., Lee, Y., Han, K.-S., et al. Electrochemical supercapacitor based on polyaniline doped with lithium salt and active carbon electrodes. Solid State Ionics 2004, 175, 765-768.

40. MacDiarmid, A. G., & Epstein, A. J. Polyanilines: a novel class of conducting polymers. Faraday Discuss. Chem. Soc. 1989, 88, 317-332.

41. Chiang, J.-C., MacDiarmid, A. G. 'Polyaniline': Protonic acid doping of the emeraldine form to the metallic regime. Synth. Met. 1986, 13, 193-205.

42. MacDiarmid, A. G., Chiang, J. C., Richter, A. F., & Epstein, A. J. Polyaniline: a new concept in conducting polymers. Synth. Met. 1987, 18, 285-290.

43. Sun, Z., Geng, Y., Li, J., et al. Chemical polymerization of aniline with hydrogen peroxide as oxidant. Synth. Met. 1997, 84, 99-100.

44. Ram, M. S., & Palaniappan, S. Benzoyl peroxide oxidation route to polyaniline salt and its use as catalyst in the esterification reaction. J. Mol. Catal. A-Chem. 2003, 201, 289–296.

45. Armes, S. P., Gottesfeld, S., Beery, J. G., et al. Conducting polymer-colloidal silica composites. Polymer 1991, 32, 2325- 2330.

46. Yan, H., & Toshima, N. Chemical preparation of polyaniline and its derivatives by using cerium(IV) sulfate. Synth. Met. 1995, 69, 151-152.

47. Ayad, M. M., & Shenashin, M. A. Polyaniline film deposition from the oxidative polymerization of aniline using $K_2Cr_2O_7$. Euro. Polym. J. 2004, 40, 197-202.

48. Chowdhury, P., & Saha, B. Potassium iodate-initiated polymerization of aniline. J. Appl. Polym. Sci. 2007, 103, 1626- 1631.

49. Li, X.-X., & Li, X.-W. Oxidative polymerization of aniline using $NaClO_2$ as an oxidant. Mater. Lett. 2007, 61, 2011-2014.

50. Ballav, N. High-conducting polyaniline via oxidative polymerization of aniline by $MnO_2$, $PbO_2$ and $NH_4VO_3$. Mater. Lett. 2004, 58, 3257-3260.
51. Kuwabata, S., Idzu, T., Martin, R. C., & Yoneyama, H. Charge-discharge properties of composite films of polyaniline and crystalline $V_2O_5$ particles. J. Electrochem. Soc. 1998, 145, 2707-2710.
52. Bernard, M.-C., Hugot-Le Goff, A., & Zeng, W. Elaboration and study of a PANI/ PAMPS/$WO_3$ all solid-state electrochromic device. Electrochim. Acta 1998, 44, 781-796.
53. Cruz-Silva, R., Romero-García, J., Angulo-Sánchez, J. L., et al. Comparative study of polyaniline cast films prepared from enzymatically and chemically synthesized polyaniline. Polymer. 2004, 45, 4711-4717.
54. Cruz-Silva, R., Romero-García, J., Angulo-Sánchez, J. L., et al. Template-free enzymatic synthesis of electrically conducting polyaniline using soybean peroxidase. Euro. Polym. J. 2005, 41, 1129- 1135.
55. Shen, Y., Sun, J., Wu, J., & Zhou, Q. Synthesis and characterization of water-soluble conducting polyaniline by enzyme catalysis. J. Appl. Polym. Sci. 2005, 96, 814-817.
56. Rumbau, V., Pomposo, J. A., Alduncin, J. A., et al. A new bifunctional template for the enzymatic synthesis of conducting polyaniline. Enzyme Microb. Tech. 2007, 40, 1412-1421.
57. Traore, M. K., Stevenson, W. T. K., McCormick, B. J., et al. Thermal analysis of polyaniline. Part I. Thermal degradation of HCl-doped emeraldine base. Synth. Met. 1991, 40, 137-153.
58. Ivanov, V. F., Gribkova, O. L., Nekrasov, A. A., & Vannikov, A. V. Comparative spectroelectrochemical investigation of vacuum evaporated and electrochemically synthesized electrochromic polyaniline films AgI. J. Electroanal. Chem. 1994, 372, 57-61.
59. Karakişla, M., Saçak, M., & Akbulut, U. Conductive polyaniline/poly(methyl meth-acrylate) films obtained by electropolymerization. J. Appl. Polym. Sci. 1996, 59, 1347-1354.
60. Lin, Y., & Yasuda, H. Effect of plasma polymer deposition methods on copper corrosion protection. J. Appl. Polym. Sci. 1996, 60, 543-555.
61. Gong, X., Dai, L., Mau, A. W. H., & Griesser, H. J. Plasma-polymerized polyaniline films: Synthesis and characterization. J. Polym. Sci. Pol. Chem. 1998, 36, 633-643.
62. Boschi, T., Di Vona, M. L., Tagliatesta, P., & Pistoia, G. Behavior of polyaniline electrodes in aqueous and organic solutions. J. Power Sources 1988, 24, 185-193.
63. Taguchi, S., & Tanaka, T. Fibrous polyaniline as positive active material in lithium secondary batteries. J. Power Sources 1987, 20, 249- 252.
64. Diaz, A. F., & Logan, J. A. Electroactive polyaniline films. J. Electroanal. Chem. 1980, 111, 111-114.
65. Osaka, T., Ogano, S., & Naoi, K. Electroactive polyaniline deposit from a nonaqueous solution. J. Electrochem. Soc. 1988, 135, 539-540.
66. Geniès, E. M., Boyle, A., Łapkowski, M., & Tsintavis, C. Polyaniline: A historical survey. Synth. Met. 1990, 36, 139-182.
67. Ohsaka, T., Ohnuki, Y., Oyama, N., et al. IR absorption spectroscopic identification of electroactive and electroinactive polyaniline films prepared by the electrochemical polymerization of aniline. J. Electroanal. Chem. 1984, 161, 399-405.

68. Syed, A. A., & Dinesan, M. K. Review: Polyaniline – a novel polymeric material. Talanta 1991, 38, 815-837.
69. Kaplan, S., Conwell, E. M., Richter, A. F., & MacDiarmid, A. G. Ring flips as a probe of the structure of polyanilines. Macromolecules 1989, 22, 1669-1675.
70. Łapkowski, M., & Geniés, E. M. Evidence of two kinds of spin in polyaniline from in situ EPR and electrochemistry: Influence of the electrolyte composition. J. Electroanal. Chem. 1990, 279, 157-168.
71. Manohar, S. K., MacDiarmid, A. G., & Epstein, A. J. Polyaniline: Pernigraniline, an isolable intermediate in the conventional chemical synthesis of emeraldine. Synth. Met. 1991, 41, 711-714.
72. Łapkowski, M., & Geniés, E. M. Spectroelectrochemical evidence for an intermediate in the electropolymerization of aniline. J. Electroanal. Chem. 1987, 236, 189-197.
73. Glarum, S. H., & Marshall, J. H. The in situ esr and electrochemical behavior of poly(aniline) electrode films. J. Electrochem. Soc. 1987, 134, 2160-2165.
74. La Croix, J.-C., Diaz, A. F. Electrolyte effects on the switching reaction of polyaniline. J. Electrochem. Soc. 1988, 135, 1457-1463.
75. Rudzinski, W. E., Lozano, L., & Walker, M. The effects of ph on the polyaniline switching reaction. J. Electrochem. Soc. 1990, 137, 3132-3136.
76. Habib, M. A., & Maheswari, S. P. Electrochromism of polyaniline: An in situ FTIR study. J. Electrochem. Soc. 1989, 136, 1050-1053.
77. Shim, Y.-B., Won, M.-S., & Park, S.-M. Electrochemistry of conductive polymers VIII. J. Electrochem. Soc. 1990, 137, 538-544.
78. MacDiarmid, A. G., & Epstein, A. J. The polyanilines: Potential technology based on new chemistry and new properties. In: W.R. Salaneck, David Thomas Clark, Emil J. Samuelsen (Eds.). Science and Applications of Conducting Polymers: Proceedings of Sixth Europhysics Industrial Workshop, Lofthus, Norway, May 1990. – Bristol: Adam Hilder, IOP Publishing Ltd., 1991. P. 117-127.
79. Boudreaux, D. S., Chance, R. R., Wolf, J. F., et al. Theoretical studies on polyaniline. J. Chem. Phys. 1986, 85, 4584-4590.
80. Pekmez, N., Pekmez, K., & Yildiz, A. Electrochemical behavior of polyaniline films in acetonitrile. J. Electroanal. Chem. 1994, 370, 223- 229.
81. Cattarin, S., Doubova, L., Mengoli, G., & Zotti, G. Electrosynthesis and properties of ring-substituted polyanilines. Electrochim. Acta 1988, 33, 1077-1084.
82. Geniés, E. M., Łapkowski, M., & Penneau, J. F. Cyclic voltammetry of polyaniline: interpretation of the middle peak. J. Electroanal. Chem. 1988, 249, 97-107.
83. Stilwell, D. E., & Park, S.-M. Electrochemistry of conductive polymers. J. Electrochem. Soc. 1988, 135, 2491-2496; 2497-2502.
84. Koval'chuk, E. P., Whittingham, S., Skolozdra, O. M., et al. Copolymers of aniline and nitroanilines. Part I. Mechanism of aniline oxidation polycondensation. Mater. Chem. Phys. 2001, 69, 154-162.
85. Pilar, F. L. Elementary quantum chemistry (2nd Ed.)., McGraw-Hill Publ.: New York, 1990.
86. Karpas, Z., Berant, Z., & Stimac, R. M. An ion mobility spectrometry/mass spectrometry (IMS/MS) study of the site of protonation in anilines. Struct. Chem. 1990, 1, 201–204.

87. Russo, N., Toscano, M., Grand, A., & Mineva, T. Proton affinity and protonation sites of aniline. Energetic behavior and density functional reactivity indices. J. Phys. Chem. A. 2000, 104, 4017-4021.

88. Koval'chuk, E. P., Reshetnyak, O. V., & Błażejowski, J. Protonation-extraction of hydrogen ions during oxidative condensation of aromatic amines. First Russian-Ukrainian-Polish Conf. on Molecular Interaction. School of Physical Organic Chemistry: Book of Abstracts. Gdańsk, June 10-16, 2001. Sopot, Zakład Poligrafii Fundacji Rozwoju Universytetu Gdańskiego, 2001, pp. 111-112.

89. Ćirić-marjanović, G., trchová, m., & stejskal, j. Theoretical study of the oxidative polymerization of aniline with peroxydisulfate: Tetramer formation. Int. J. Quantum Chem. 2008, 108, 318-333.

90. Hedayatullah, M. Oxidation of primary aromatic amines – review. Bull. Soc. Chim. France. 1972, (7), 2957–2974.

91. Ćirić-Marjanović, G., Trchová, M., & Stejskal, J. MNDO-PM3 Study of the early stages of the chemical oxidative polymerization of aniline. Collect. Czech. Chem. Commun. 2006, 71, 1407-1426.

92. Sapurina, I., & Stejskal, J. The mechanism of the oxidative polymerization of aniline and the formation of supramolecular polyaniline structures. Polym. Int. 2008, 57, 1295–1325.

93. Ćirić-Marjanović, G. Recent advances in polyaniline research: Polymerization mechanisms, structural aspects, properties and applications. Synth. Met. 2013, 177, 1–47.

94. Koval'chuk, E. P., Whittingham, S., Skolozdra, O. M., et al. Polyaniline and copolymer aniline and ortho-nitroaniline. Part II. Physicochemical properties. Mater. Chem. Phys. 2001, 70, 38-48.

95. Koval'chuk, E. P., Stratan, N. V., Reshetnjak, O. V., & Whittingham, M. S. Synthesis and properties of polyanisidines. XIV[th] Int. Symp. on the Reactivity of Solids: Program and Abstracts. 27-31 August 2000. Budapest, Hungary. Szeged, Officina Press, 2000, P. 151.

96. Koval'chuk, E. P., Stratan, N. V., Reshetnyak, O. V., et al. Synthesis and properties of the polyanisidines. Solid State Ionics 2001, 142, 217-224.

97. Nakanishi, K. Infrared absorption spectroscopy, practical. Holden-Day, Inc.: San Francisco, 1963.

98. Roßberg, K., Paasch, G., Dunsch, L., & Ludwig, S. The influence of porosity and the nature of the charge storage capacitance on the impedance behavior of electropolymerized polyaniline films. J. Electroanal. Chem. 1998, 443, 49-62.

99. Chiang, J.-C., & MacDiarmid, A. G. 'Polyaniline': Protonic acid doping of the emeraldine form to the metallic regime. Synth. Met. 1986, 13, 193-205.

100. Kobayashi, T., Yoneyama, H., & Tamura, H. Oxidative degradation pathway of polyaniline film electrodes. J. Electroanal. Chem. 1984, 177, 293-297.

101. Mengoli, G., Munari, M. T., Bianco, P., & Musiani, M. M. Anodic synthesis of polyaniline coatings onto fe sheets. J. Appl. Polym. Sci. 1981, 26, 4247-4257.

102. Macdiarmid, A. G., Mu, S.-L., Somasiri, N. L. D., & Wu, W. Electrochemical characteristics of "polyaniline" cathodes and anodes in aqueous electrolytes. Mol. Cryst. Liq. Cryst. 1985, 121, 187- 190.

103. Tsocheva, O., Zlatkov, T., & Terlemezyan, L. Thermoanalytical studies of polyaniline 'emeraldine base'. J. Therm. Anal. Calorim. 1998, 53, 895–904.

104. Anand, J., Palaniappan, S., & Sathyanarayana, D. N. Spectral, thermal, and electrical properties of poly(o- and m-toluidine)-polystyrene blends prepared by emulsion pathway. J. Polym. Sci. Pol. Chem. 1998, 36, 2291-2299.

105. Gonçalves, D., Matvienko, B., & Bulhões, L. O. S. Electrochromism of poly(o-methoxyaniline) films electrochemically obtained in aqueous medium. J. Electroanal. Chem. 1994, 371, 267-271.

106. Widera, J., Grochala, W., Jackowska, K., Bukowska, J. Electrooxidation of o-methoxyaniline as studied by electrochemical and SERS methods. Synth. Met. 1997, 89, 29-37.

107. Patil, S., Mahajan, J. R., More, M. A., & Patil, P. P. Electrochemical synthesis of poly(o-methoxyaniline) thin films: effect of post treatment. Mater. Chem. Phys. 1999, 58, 31-36.

108. Bernard, M.-C., de Torresi, S. C., & Hugot-Le Goff, A. In situ Raman study of sulfonate-doped polyaniline. Electrochim. Acta 1999, 44, 1989-1997.

109. Naudin, E., Gouérec, P., & Bélanger, D. Electrochemical preparation and characterization in non-aqueous electrolyte of polyaniline electrochemically prepared from an anilinium salt. J. Electroanal. Chem. 1998, 459, 1-7.

110. Yang, C.-H. Electrochemical polymerization of aniline and toluidines on a thermally prepared Pt electrode. J. Electroanal. Chem. 1998, 459, 71-89.

111. MacDiarmid, A. G., & Epstein, A. J. New developments in the synthesis and doping of polyacetylene and polyaniline. In: J. L. Brédas, R. R. Chance (Eds.). Conjugated polymeric materials: opportunities in electronics, optoelectronics, molecular electronics: Proceedings of the NATO Advances Research Workshop on Conjugated Polymeric Materials: Opportunities in Electronics, Optoelectronics, and Molecular Electronics. Mons, Belgium, September 3-8, 1989. Dordrecht, Kluwer Academic Publishers, 1990, pp. 53–63.

112. Maia, D. J., das Neves, S., Alves, O. L., & De Paoli, M.-A. Photoelectrochemical measurements of polyaniline growth in a layered material. Electrochim. Acta. 1999, 44, 1945–1952.

113. Sari, B., & Talu, M. Electrochemical copolymerization of pyrrole and aniline. Synth. Met. 1998, 94, 221–227.

114. Ferreira, V., Cascalheira, A. C., & Abrantes, L. M. Electrochemical copolymerization of luminol with aniline: A new route for the preparation of self-doped polyanilines. Electrochim. Acta 2008, 53, 3803-3811.

115. Ferreira, V., Cascalheira, A. C., & Abrantes, L. M. Electrochemical preparation and characterisation of Poly(Luminol– Aniline) films. Thin Solid Films. 2008, 516, 3996–4001.

116. Zhang, G.-F., & Chen, H.-Y. Studies of polyluminol modified electrode and its application in electrochemiluminescence analysis with flow system. Analyt. Chim. Acta. 2000. 419, 25-31.

117. Chen, S.-M., & Lin, K.-C. The electrocatalytic properties of biological molecules using polymerized luminol film-modified electrodes. J. Electroanal. Chem. 2002, 523, 93–105.

118. Chang, Y.-T., Lin, K.-C., & Chen, S.-M. Preparation, characterization and electro-catalytic properties of poly(luminol) and polyoxometalate hybrid film modified electrodes. Electrochim. Acta 2005, 51, 450-461.

119. Lin, K.-C., & Chen, S.-M. Reversible cyclic voltammetry of the NADH/NAD$^+$ redox system on hybrid poly(luminol)/FAD film modified electrodes. J. Electroanal. Chem. 2006, 589, 52-59.

120. Wang, C. H., Chen, S. M., & Wang, C. M. Co-immobilization of polymeric luminol, iron(II) tris(5-aminophenanthroline) and glucose oxidized at an electrode surface, and its application as a glucose optrode. Analyst 2002, 127, 1507–1511.

121. Sassolas, A., Blum, L. J., & Leca-Bouvier, B. D. Electrogeneration of polyluminol and chemiluminescence for new disposable reagentless optical sensors. Anal. Bioanal. Chem. 2008, 390, 865–871.

122. Mendonça, T. P., Moraes, S. R., & Motheo, A. J. Influence of the synthesis param-eters on the polyluminol properties. Mol. Cryst. Liq. Cryst. 2006, 447, 383–391.

123. Koval'chuk, E. P., Grynchyshyn, I. V., Reshetnyak, O. V., et al. Oxidative conden-sation and chemiluminescence of 5-amino-2,3-dihydro-1,4-phtalazinedione. Euro. Polym. J. 2005, 41, 1315-1325.

124. Malinauskas, A., & Holze, R. An in situ UV-vis spectroelectrochemical investigation of the initial stages in the electrooxidation of selected ring- and nitrogen-alkylsubsti-tuted anilines. Electrochim. Acta. 1999, 44, 2613–2623.

125. Alpert, N. L., Keiser, W. E., & Szymański, H. A. IR; theory and practice of infrared spectroscopy. New York, Plenum Press, 1970, 388 p.

126. De Azevedo, W. M., de Souza, J. M., & de Melo, J. V. Semi-interpenetrating polymer networks based on polyaniline and polyvinyl alcohol-glutaraldehyde. Synth. Met. 1999, 100, 241–248.

127. Schweinsberg, D. P., Bottle, S. E., Otieno-Alego, V., & Notoya, T. A near-infrared FT-Raman (SERS) and electrochemical study of the synergistic effect of 1-[(1′,2′-dicar-boxy)ethyl]-benzotriazole and KI on the dissolution of copper in aerated sulfuric acid. J. Appl. Electrochem. 1997, 27, 161–168.

128. Baibarac, M., Cochet, M., Łapkowski, M., et al. SERS spectra of polyaniline thin films deposited on rough Ag, Au and Cu. Polymer film thickness and roughness parameter dependence of SERS spectra. Synth. Met. 1988, 96, 63–70.

129. Hatchett, D. W., Josowicz, M., & Janata, J. Comparison of chemically and electro-chemically synthesized polyaniline films. J. Electrochem. Soc. 1999. 146, 4535–4538.

130. Wells, A. F. Structural inorganic chemistry (5$^{th}$ ed.). Clarendon Press: Oxford, 1987.

131. Ouyang, C. S., & Wang, C. M. Clay-enhanced electrochemiluminescence and its application in the determination of glucose. J. Electrochem. Soc. 1998, 145, 2654−2659.

132. Ouyang, C. S., & Wang, C. M. Electrochemical characterization of the clay-enhanced luminol ecl reaction. J. Electroanal. Chem. 1999, 474, 82–88.

133. Chen, S.-M., & Lin, K.-C. The electrocatalytic properties of biological molecules using polymerized luminol film-modified electrodes. J. Electroanal. Chem. 2002, 523, 93–105.

134. Jenkins, A. D., Kratochvíl, P., Stepto, R. F. T., & Suter, U. W. Glossary of basic terms in polymer science (IUPAC Recommendations 1996). Pure Appl. Chem. 1996, 68, 2287–2311.

135. Purple Book: IUPAC Compendium of Macromolecular Nomenclature. Oxford, Blackwell Scientific Publications, 1991, P.18.

136. Tai, H., Jiang, Y., Xie, G., Yu, J., Chen, X., & Ying, Z. Influence of polymerization temperature on $NH_3$ response of PANI/TiO$_2$ thin film gas sensor. Sens. Actuat. B. 2008, 129, 319–326.

137. Grigore, L., & Petty, M. C. Polyaniline films deposited by anodic polymerization: Properties and applications to chemical sensing. J. Mater. Sci: Mater Electronics. 2003, 14, 389–392.

138. Lakard, B., Herlem, G., Lakard, S., Antoniou, A., & Fahys, B. Urea potentiometric biosensor based on modified electrodes with urease immobilized on polyethylenimine films. Biosens. Bioelectron. 2004, 19, 1641–1647.

139. Binh, P. T. Electrochemical Polymerization of Aniline by Current Pulse Method in the Presence of m-Aminobenzoic Acid in Chlorhydric Acid Solution. Macromol. Symp. 2007, 249–250, 228–233.

# CHAPTER 14

# STUDY OF HEAT PROTECTION POLYMER MATERIALS CONTAINING SILICON CARBIDE AND PERLITE[1]

V. F. KABLOV, O. M. NOVOPOLTSEVA, V. G. KOCHETKOV,
N. V. KOSTENKO, and V. S. LIFANOV

*Volzhsky Polytechnic Institute, Volgograd State Technical University,
Volgograd, 400131, Russia, E-mail: nikol-kosten@yandes.ru*

## CONTENTS

### ABSTRACT

Products made from elastomer compositions are used in rocket, aviation and marine engineering, including special-purpose structures operating at extreme temperatures. Therefore, of special interest are composite polymer

[1]This work has been sponsored under the project 'Developing modifiers and functional fillers for fire and heat resistant polymer materials' implemented within the framework of the state assignment of the Ministry of Education and Science of Russia.

materials with heat resistance higher than 200°C for products capable of prolonged operation at elevated temperatures. This paper shows how functionally active fillers can be used in creating elastomer compositions. Their influence on fire and heat resistance of rubbers based on general-purpose raw rubber has been considered.

## 14.1 INTRODUCTION

Products made from elastomer compositions are used in rocket, aviation and marine engineering, including special-purpose structures operating at extreme temperatures. Therefore, of special interest are composite polymer materials with heat resistance higher than 200°C for products capable of prolonged operation at elevated temperatures.

In engineering, heat and fire resistance of polymer materials and their products is determined by such characteristics as combustibility, ignition point, self-ignition, burn-out rate, surface flame propagation rate, smoke emission during burning, and toxicity of products formed in combustion and pyrolysis [1]. It should be noted that the above characteristics frequently work against one another, and while one of the properties is being improved, others may deteriorate. When fillers are used to increase fire and heat resistance of polymer materials, the physical and mechanical, dielectric and other operational and process properties are usually somewhat reduced, and the cost of materials is increased. For this reason, enhancing heat and fire resistant properties of polymer materials involves optimizing the whole system of characteristics of the material being created.

A promising way of enhancing heat resistance of such materials is to include intumescent (perlite, vermiculite, thermally expanded graphite) and finely dispersed fillers in the elastomer composition, including finely dispersed silicon carbide [2, 3].

The purpose of this study is to find ways of increasing fire and heat resistance of elastomer materials based on different types of raw rubber by adding fillers.

## 14.2 RESEARCH METHODOLOGY

The research is focused on vulcanizates based on SKMS-30ARKM 15 butadiene-styrene rubber with a sulfuric vulcanizing group (Table 14.1)

[4, 5]. The proposed fillers are perlite and finely dispersed silicon carbide.

When finely dispersed silicon carbide is added to the rubber compound, the optimal combination of rheological parameters is observed in composition 3, i.e., 20 pbw of carbon black and silicon carbide per 100 pbw of rubber (Table 14.2). This carbon black and silicon carbide combination results in an increased vulcanization rate with a simultaneous increase of the induction period. Increased silicon carbide content leads to accelerated vulcanization, which may be due to the catalytic action of silicon carbide.

**TABLE 14.1** Rubber Compound Recipe

| Ingredient | Dosage, pbw per 100 pbw of rubber | | | | | | |
|---|---|---|---|---|---|---|---|
| | Control | C1 | C2 | C3 | П1 | П2 | П3 |
| SKMS-30-ARKM-15 | 100.00 | 100.00 | 100.00 | 100.00 | 100.00 | 100.00 | 100.00 |
| Captax | 1.50 | 1.50 | 1.50 | 1.50 | 1.50 | 1.50 | 1.50 |
| Sulfur | 2.00 | 2.00 | 2.00 | 2.00 | 2.00 | 2.00 | 2.00 |
| Zinc oxide | 5.00 | 5.00 | 5.00 | 5.00 | 5.00 | 5.00 | 5.00 |
| Stearine | 1.00 | 1.00 | 1.00 | 1.00 | 1.00 | 1.00 | 1.00 |
| Carbon black П324 | 40.00 | 30.00 | 25.00 | 20.00 | 40.00 | 40.00 | 40.00 |
| Silicon carbide | - | 10 | 15.00 | 20.00 | - | - | - |
| Perlite | - | - | - | - | 10.00 | 15.00 | 20.00 |

**TABLE 14.2** Vulcanization Characteristics of Rubber Compounds*

| Parameter | Control | C1 | C2 | C3 | П1 | П2 | П3 |
|---|---|---|---|---|---|---|---|
| Minimal torque $(M_{min})$, N·m | 1.23 | 1.46 | 1.25 | 1.32 | 1.30 | 1.23 | 1.95 |
| Maximal torque $(M_{max})$, N·m | 9.24 | 7.39 | 9.20 | 9.41 | 9.17 | 9.86 | 11.78 |
| Starting time of vulcanization $(\tau_S)$, min | 2.83 | 3.31 | 3.31 | 4.02 | 3.54 | 3.54 | 4.02 |
| Optimum vulcanization time $(\tau_{90})$, min | 31.00 | 31.89 | 26.93 | 26.46 | 31.00 | 29.50 | 35.91 |
| Vulcanization rate index $(R_v)$, min$^{-1}$ | 3.55 | 3.59 | 4.23 | 4.32 | 3.51 | 3.85 | 3.14 |

* The temperature of vulcanization is 155°C

When perlite is added, an increased induction period is also observed, but the vulcanization rate is practically unchanged. When carbon black and finely dispersed silicon carbide or perlite are added at the same time, an acceptable level of physical and mechanical properties is achieved (Table 14.3). Thus, silicon carbide and perlite can be used to reduce the cost of rubber.

To estimate heat resistance of the obtained vulcanizates, the temperature on the unheated sample surface when exposed to plasmatron open flame was determined. The observed temperature was 2500°C.

When the control sample is exposed to burner flame, practically no 'coke cap' is formed (Figure 14.1$a$), whereas on the surface of samples containing silicon carbide and perlite (Figure 14.1$b$ and $c$), a dense flame-resistant coke layer is formed, which protects the sample from burning.

Silicon carbide microplates can be seen on the coke surface in Figure 14.2. As silicon carbide is a material, which has high heat resistance and is not easily prone to oxidization, the silicon carbide barrier layer effectively protects the rubber exposed to flame from burning out. The plate-shaped silicon carbide particles create a kind of barrier layer protecting the sample from flame.

TABLE 14.3   Physical and Chemical Properties of Vulcanizates

| Parameter | Control | C1 | C2 | C3 | П1 | П2 | П3 |
|---|---|---|---|---|---|---|---|
| Tensile strength ($f_p$), MPa | 10.2 | 11.1 | 6.3 | 4.5 | 11.1 | 10.8 | 8.8 |
| Elongation at break ($\varepsilon_{OTH}$), % | 560 | 637 | 417 | 310 | 476 | 490 | 453 |
| Relative residual elongation after break ($\theta_{OCT}$), % | 21 | 8 | 8 | 4 | 16 | 13 | 20 |
| Hardness, Shore A | 59 | 45 | 50 | 51 | 60 | 61 | 60 |
| Density, g/cm$^3$ | 1.06 | 1.13 | 1.10 | 1.08 | 1.08 | 1.06 | 1.05 |
| Linear combustion rate, mm/min | 24.45 | 15.50 | 13.90 | 14.30 | 22.86 | 18.38 | 20.64 |
| Warm-up time of the sample surface to 100 °C, sec | 119 | 120 | 130 | 150 | 172 | 189 | 202 |
| Change of parameters after aging (100°C × 72 hrs.), %: $\Delta f_p$ $\Delta\varepsilon$ | −38 −60 | −23 −43 | +16 −32 | +13 −35 | −37 −57 | −35 −48 | −38 −50 |

Vulcanization mode: 155°C × 40 min

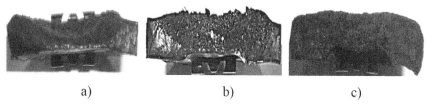

a)                              b)                              c)

**FIGURE 14.1**   Changes in the structure of the sample after exposure to flame: (a) control; (b) silicon carbide; (c) perlite.

a)                                                      b)

**FIGURE 14.2**   Appearance of the coke surface under the action of burner flame: (a) silicon carbide; (b) perlite.

## 14.3   CONCLUSION

Thus, the research conducted has shown that perlite and finely dispersed silicon carbide can be used as fillers enhancing fire resistance in elastomer compositions.

## KEYWORDS

- elastomers
- fillers
- fire resistance
- modifying additives
- rubber

# REFERENCES

1. Kablov, V. F., et al. Teplozashchitnye pokrytiya, soderzhashchie perlit [Heat protection coatings containing perlite]. *In* Mezhdunarodny zhurnal prikladnykh i fundamentalnykh issledovanii [International journal of applied and basic research]. 2012, No. 1. p. 174–175.

2. Kablov, V. F., & Novopoltseva, O. M. Teplozashchitnye pokrytiya na polimernoi osnove, soderzhashchie perlit [Heat protection polymer-based coatings containing perlite]. *In* Sb. tez. nauch.-prakt. konf. mol. uchenykh po napravl.: Khimiya – nauka budushchego. Innovatsii v energosberezhenii i energo-effektivnosti. Inf. tekhnologii – lokomotiv innovats. razvitiya: v ramkakh molodezh. kongressa "Integratsyia innovatsii: region. aspekty" [Collection of abstracts of young scientists' workshop. Area: Chemistry is the science of the future. Innovations in energy saving and energy efficiency. Information technologies are the locomotive force of innovative development: within the frameworks of the youth congress 'The Integration of Innovations: Regional Aspects']. April 19–21, 2012. VPI (branch) of VSTU, Volgograd. p. 25–26.

3. Kablov, V. F., Novopoltseva, O. M., and Kochetkov, V. G. Vliyanie napolnitela perlit na teplostoikost rezin na osnove etilenpropilendienovogo kauchuka [Influence of perlite filler on heat resistance of rubbers based on ethylenepropylenediene rubber]. *In* Sovremennye problemy nauki i obrazovaniya [Current problems of science and education]. 2013, No. 3. p. 56–59.

4. Lifanov, V. S., et al. Issledovanie elastomernykh materialov s mikrodispersnymi otkhodami karbida kremniya [Study of elastomer materials with finely dispersed silicon carbide waste]. *In* Soveremennye problemy nauki i obrazovaniya [Current problems of science and education]. 2013, No. 4. p. 38–42.

5. Zaikov, G. E. Gorenie, destruktsiya i stabilizatsiya polimerov [Polymer combustion, destruction and stabilization]. *In* Zaikov, G.E. (ed.), Nauchnye osnovy tekhnologii [Scientific Basics of Technology]. Saint-Petersburg, 2008.

# CHAPTER 15

# METHOD OF THE INCREASE OF THERMOSTABILITY OF FILLED POLYMERS

N. R. PROKOPCHUK and L. A. LENARTOVICH

*Belarusian State Technological University, Minsk, Sverdlov St., 13-a, 220006, Belarus, Tel. 3275738, E-mail: liliya.popova@mail.ru, prok_nr@mail.ru*

## CONTENTS

### ABSTRACT

The individual contribution of stabilizers and fillers into the increase of thermal stability of a polymer-matrix of composite materials at their joint presence in the composite is estimated in this paper. The increase of thermal stability of polypropylene composites in the process of pre-saturation of the surface of fillers by thermal stabilizers compared to their individual introduction into the polymer is determined. A method of saturation of the surface of fillers by stabilizers is proposed. The mechanism of a prolonged and direct action of stabilizers desorbing from the surface of the filler is explained. It leads to the increase of 10–40% of the thermal stability of stabilized filled polypropylene compositions obtained by the proposed method in comparison with the traditional one. It simultaneously reduces up to 1.25 times the amount of used stabilizers.

## 15.1   INTRODUCTION

Now under existing conditions of the increase of market competition the task of producers of polymeric materials is the constant improvement of the quality of created by them products. Important indicators of comparison are not only physical and mechanical characteristics of the used materials, but also their durability. Created polymer composite materials (PCM) are heterogeneous systems consisting of several components. While creating PCM we use fillers that give to the materials a valuable set of properties, and stabilizers that reduce the negative impact of factors of the environment on the material and protect it in a recycling process. The impact of these components on each property of PCM is sufficiently studied [1–4]. However, the comprehensive researches aimed at studying the individual contribution of stabilizers and fillers in their combined influence on the properties of PCM are not numerous [5–8]. Such studies are highly relevant, as the use of their results is necessary to create a scientific well-grounded approach while selecting components and technology for PCM creation.

   While creating PCM the choice of dosage of stabilizing additives for polymers, containing the fillers is often determined by optimal concentration of stabilizers accepted for the unfilled polymer. Thus, the effect of the filler on the change of properties of PCM because of aging is not taken into consideration. It can lead to the inefficiency of the action of the stabilizer, and to its possible over-expenditure. This may be due to the fact that the necessary concentration of the stabilizer at various filler content may vary compared to the unfilled polymer either in the direction of increasing or decreasing, that depends on the ability of the filler to have a stabilizing or opposite effect on the matrix, and on its ability to adsorb a stabilizer. The study of the combined action of stabilizers and fillers, the creation of new methods of their introduction will allow to increase the resistance of the compositions to various kinds of aging, to exclude cases of overdose of stabilizers, and hence to reduce the cost of the final product.

## 15.2 EXPERIMENTAL PROCEDURE

The composite materials based on PP (Caplene 01030) and containing fillers and stabilizers were the object of the study. Fine marble of Omyacarb 2-UR brand (OMYA RUS, Russia), talc of Fintalk M30 brand (Mondo Minerals, Finland), fiberglass (FG) of "Polotsk Fiberglass" production (Belarus) were used as fillers. Industrially used additives of well-known manufacturers: Irganox 1010, Irganox B561 (BASF, Germany) and Sandostab P-EPQ (Clariant, Switzerland) were applied as stabilizers. The samples were prepared by molding under pressure in the injection molding machine BOY 22A (Dr. Boy, Germany). Thermal aging of the samples was carried out in an oven at 150°C. The concentration of a stabilizer in the composition was up to 1% by weight, filler content – 20% by weight.

In accordance with GOST 11262–80 elongation at break ($\varepsilon_b$, %) before and after the thermal aging (T2020 DC10 SH tensometer, Alpha Technologies, USA) was determined. Ambient air temperature – 20°C, speed of top grip motion – 50 mm/min, number of samples – 10. The calculation of the activation energy of thermal oxidative destruction $E_a$ was carried out by Broido computational method using the TGA curves, filmed on the device TGA/DSCI (Mettler Toledo, Switzerland). We used the method of IRS to prove the adsorption of stabilizers on the filler's surface (Nexus ESP, Thermo Nicolet, USA). The infra-red spectrums were carried out with program Omnic v6.0. Nitrogen adsorption isotherms were obtained by BET method (Brunour, Emmet, Teller) on instrument NOVA 2200. Gaseous nitrogen with operation temperature of 77 K was obtained by evaporation of liquid nitrogen.

## 15.3 RESULTS AND DISCUSSION

To study the influence of mode of introduction of stabilizing additives into the polymer matrix of PCM on the properties of composites they were prepared by traditional process and also with preliminary saturation of filler surface by stabilizers from solutions. The study of the stabilizer adsorption from the polymer melt causes definite difficulties, so the adsorption of stabilizing additives from their solutions was studied. The solvent must meet certain requirements: (i) to dissolve a sufficient amount of investigated stabilizer for the performance of the experiment; (ii) to be optically transparent; (iii) not to be stained when tested substances are dissolved. It was experimentally found that Irganox 1010 and Irganox B561 are highly soluble in acetone. Sandostab P-EPQ is not soluble in acetone but soluble in n-propanol to form a transparent solution. Stabilizer solutions must be optically transparent for the possibility of determining the concentration according to a refractive index (refractometric method). Investigated fillers – marble, talc and chopped fiberglass were used as the adsorbents.

We studied the adsorption interaction at the division border of the phases filler – stabilizer solution. For this purpose adsorption isotherms of stabilizing additives on the fillers were obtained by standard methods [9]. Used fillers were previously washed in organic solvents. The washing of fillers was performed in order to remove possible impurities on their surface and not to skew the results of the experiment. Fillers were placed in a pure solvent and with vigorous stirring were washed three times. The amount of adsorbed material per unit of filler weight at different concentrations of the equilibrium solution was calculated after a preliminary study of the kinetics of adsorption equilibrium at 20°C. According to the results of the experiment the values of limiting adsorption of each stabilizer on various fillers were determined, as well as adsorption isotherms were built.

Adsorption isotherms of stabilizers on fillers are shown in Figure 15.1. The obtained data indicate that the highest value of the maximum absorption is characteristic of the stabilizer Irganox 1010 and varies from 102 to 239 mg/g depending on the filler used.

Irganox 1010 is related to the phenolic type of stabilizers, in the structure of which the polar carbonyl and hydroxyl groups capable of forming hydrogen bonding with the adsorption centers on the filler surface are present. The adsorption of the stabilizer is also possible due to the formation of donor-acceptor complexes, as well as $\pi$-electron – protone complexes [10, 11].

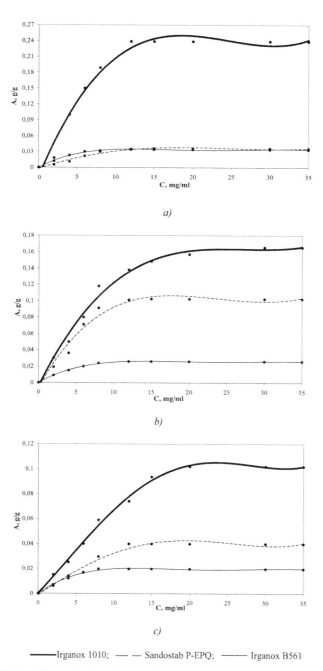

FIGURE 15.1 Adsorption isotherms of different stabilizing additives on fillers: (a) fiberglass; (b) marble; (c) talc.

The Sandostab P-EPQ contains only the benzene rings. In this case the adsorption process is possible only with the use of one mechanism due to the interaction of $\pi$-electrons of the benzene rings with the surface of fillers. For Irganox B561, consisting of phenolic and phosphorus-containing stabilizers, there is the least adsorption on all fillers due to the possibility of competitive adsorption of two components [12].

Therefore, it appears that the interaction of the stabilizer Irganox 1010 with the surface of fillers containing active groups is the strongest. The highest values of the limit adsorption for all stabilizers are observed on fiberglass and marble, the smallest – on talc.

The treatment of the surface of fillers with the solutions of stabilizes through the course of adsorption processes allows to fix stabilizing additives purposefully. The stabilizer retains the ability to desorb from the surface of the filler, it does not lose its function. The presence of adsorbed stabilizes on the filler surface is confirmed by IRS. The spectrum of marble after the treatment and subsequent washing of a stabilizer comprises characteristic absorption strips of Irganox 1010: 3640 cm$^{-1}$, 766 cm$^{-1}$, 1742 cm$^{-1}$. The displacement of the absorption strip in the region of 3000–3600 cm$^{-1}$ indicates the interaction of the surface hydroxyl groups of the filler. The data indicate a strong adsorption interaction of a stabilizing additive and marble. Similar spectra are recorded for all investigated fillers and stabilizers. It has been established that the smallest amount of adsorbed stabilizers is present on the surface of talc.

For the study of adsorption processes on the surface of fillers to modify the surface structure the BET method was used [13].

Figure 15.2 shows the nitrogen sorption curves obtained for marble, before and after its processing by the solution of stabilizer Irganox 1010. Similar curves were obtained for talc and fiberglass.

From the Figure 15.2, it is clear that after the surface treatment by the solution the concentration of the active surface centers is reduced because the pre-adsorption of the stabilizer reducing the adsorption of nitrogen on the filler after the treatment occurs (curve 2). These curves can serve as an indirect confirmation of the adsorption of the stabilizer on the surface of the filler and correlate with the data of adsorption isotherms from solutions of stabilizers.

To study the effect of a new method of the introduction of previously saturated fillers by stabilizers on the properties of the compositions the

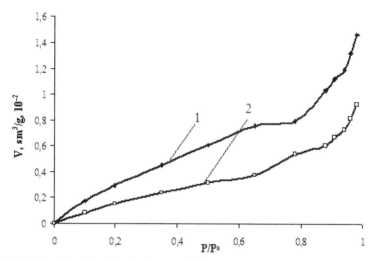

**FIGURE 15.2**   The adsorption isotherms of nitrogen on marble. (a) Adsorption isotherm of the original marble; and (b) adsorption isotherm of marble after saturation of its stabilizer.

change of deformation and strength characteristics of the composition before and after aging was evaluated. The most sensitive indicator of the fall of properties with aging is the change of values of relative elongation at break [14], so the effectiveness of the action of stabilizing additives was evaluated according to this indicator. Figure 15.3 shows the values of relative elongation at break after 250 h of aging of the compositions obtained by the traditional way, and after pre-adsorption of stabilizers on the surface of the fillers. For compositions obtained by the traditional way $\varepsilon_p$ for samples with only optimal concentration of stabilizers was shown.

For compositions containing marble and FG the weakening effect of the action of Sandostab stabilizer and filler on the matrix is observed (samples 11, 18), as $\varepsilon_b$ values are lower than those of unstabilized filled PP (samples 8, 15), indicating that the action of the stabilizer is inefficient in filled PP. The use of Irganox 1010 and Irganox B561 at a conventional method of introduction leads to increased resistance to temperature (samples 2, 6, 9, 13, 16, 20), so there is the enhancing effect of the stabilizer and filler. For the compositions produced by the novel process, there is a significant increase in resistance to the thermal oxidation destruction

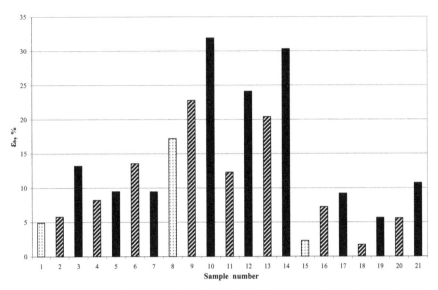

**FIGURE 15.3** The values of relative elongation of the compositions after accelerated aging for 250 h (1 − PP + t; 2 − PP + t + I1010 (1%); 3 − PP + t + I1010 (a); 4 − PP + t + s (0.7%); 5 − PP + t + s (a); 6 − PP +t + Ib561 (0.7%); 7 − PP + t + Ib561 (a); 8 − PP + mr; 9 − PP + mr + I1010 (0.5%); 10 − PP + mr + I1010 (a); 11 − PP + mr + s (0.5%); 12 − PP + mr + s (a); 13 − PP + mr + Ib561 (0.7%); 14 − PP + mr + Ib561 (a); 15 − PP + FG; 16 − PP +FG + I1010 (0.5%);17− PP + FG + I1010 (a); 18 − PP + FG + s (0.7%); 19 − PP + FG + s (a); 20 − PP + FG + Ib561 (0.7%); 21 − PP + FG + Ib561 (a); (a) − preliminary adsorption).

(samples 3, 10, 12, 14, 21). For the composition comprising Irganox B561 such effect is not observed (sample 7). This is due to the negligible adsorption of an additive on the surface of talc that correlates with the adsorption data (Figure 15.1c). Perhaps this amount of a stabilizer is not sufficient for long-term protection of the polymer during the aging time of 250 h. The use of marble with pre-adsorbed stabilizer Irganox 1010 (sample 10) results in a significant (40%) increased thermostability, suggesting the promise of this method.

In order to confirm the results of the increase of the thermal stability of the compositions obtained by a new method an independent method of the energy activation definition of thermo-oxidative destruction was used. The results obtained are shown in the form of examples of non-aged stabilized and non-stabilized PP compositions containing a marble filler (Figure 15.4). When marble and stabilizers − Irganox 1010 and Irganox B561 are jointly introduced the effect of the increase of resistance to thermal aging is

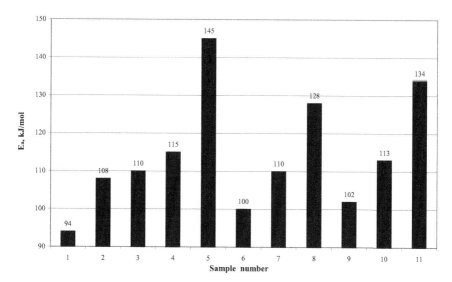

**FIGURE 15.4** The values of the activation energy of thermal oxidative degradation (1 – PP; 2 – PP + I1010 (0.5%); 3 – PP + mr; 4 – PP + mr + I1010 (0.5%); 5 – PP + mr + I1010 (a); 6 – PP + s (0.7%); 7 – PP + mr + s (0.5%); 8 – PP + mr + s (a); 9 – PP + Ib561 (0.4%); 10 – PP + mr + Ib561 (0.7%); 11– PP + mr + Ib561 (a)).

observed (samples 4 and 10). In the case of the use of stabilizer Sandostab P-EPQ there is no increase in the activation energy values (sample 7), the stabilizing effect is not observed, that correlates with the relative elongation at break data (Figure 15.3, sample 11). Thus, the effect of increasing resistance to thermal aging and the opposite one can be observed when stabilizers and fillers are jointly introduced.

The use of pre-treated fillers leads to a significant increase of the activation of the energy of thermal oxidative destruction (samples 5, 8, 11). Of the used stabilizers the greatest stabilizing effect while applying a new method of introduction is observed for Irganox 1010 (sample 5). At the same time there is a thermal stability increase of 26%. It is explained by the high adsorption ability of a given stabilizer due to the presence of polar hydroxyl, carbonyl groups, benzene rings in its structure and a large specific surface of marble, to the presence on its surface of active sites capable to react with the polar groups of the stabilizer. The use of marble, treated with a solution of Irganox 1010, can reduce the required amount of a stabilizer up to 0.4 wt. % (1.25 times less than at a traditional stabilizer introduction) with a significant simultaneous increase of thermal stability. In the case of the individual introduction of components – a filler and a stabilizer the necessary concentration of the stabilizer is 0.5 wt. %.

Thus, one can trace a tendency to the increase of resistance to the durable effect of temperature in the presence of atmospheric oxygen for compositions, obtained using a new method of introduction of PCM components. This may be concerned with the fact that the stabilizers existing on the surface of fillers are more uniformly distributed in the bulk of polymer matrix as the introduced amount of fillers is two orders higher than the amounts of stabilizers, dosed by a conventional manner. As a result, the fields in the polymer matrix with an overdose of the stabilizer, when it begins to function as the initiator of oxidative destruction are eliminated. It is possible to use the less amount of stabilizer as compared to the traditional PCM receiving technology due to its more uniform distribution and direct action in the matrix.

At elevated temperature the stabilizer starts to desorb from the surface of the filler, it neutralizes the action of macroradicals. A gradual stabilizer expenditure takes place, as a result the gradient of stabilizer concentrations appears in the bulk of the polymer and on the filler surface. The stabilizer desorbs from the filler and diffuses in a polymer matrix with a reduced stabilizer content to align the chemical potentials. Since the stabilizer desorbs dosed, the possibility to overdose the stabilizer is eliminated. This fact explains its prolonged and direct action. In addition during the saturation of fillers with stabilizers air is evacuated out of microcracks and pores, and oxygen is evacuated too. In conventional technology it comes together with the filler into a polymer matrix, additionally initiating oxidative processes leading to the destruction of macromolecules.

## 15.4  CONCLUSION

A new method of stabilization of filled polymers by pre-saturation of the surface of fillers with stabilizers is developed. It will provide a significant resistance increase to thermal aging of compositions and a simultaneous significant stabilizer economy due to a more uniform distribution in the matrix, its direct desorption from the surface of the filler and a prolonged action compared with conventional technology of polymer composites production. The totality of data obtained in this work about the individual and joint action of fillers and stabilizers in the compositions of polypropylene can be used in the development of polymer composite materials based on other thermoplastic matrixes.

## KEYWORDS

- a filler
- a polymer composite material
- a stabilizer
- adsorption

## REFERENCES

1. Chatterjee, A., & Deopura, B. L. Thermal stability of polypropylene/carbon nanofiber composite. *J. Appl. Polym. Sci.* 2006, Vol. 100, № 5, pp. 3574–3578.
2. Huang, R., et al. High Density Polyethylene Composites Reinforced with Hybrid Inorganic Fillers: Morphology, Mechanical and Thermal Expansion Performance. *Materials.* 2013, Vol. 6, № 9, pp. 4122–4413.
3. Malik, J., & Sidgi, M. New systems of stabilizers in polyolefin water pipes. *Plast. Massi.* 2006, № 10, pp. 36–39.
4. Kalinchev, E. L., et al. Progressive stabilization technology of polymer products. *Polym. mater.* 2008, № 7, pp. 3–14.
5. Kandare, E., et al. Probing synergism, antagonism, and additive effects in poly(vinyl ester) (PVE) composites with fire retardants. *Polym. Degrad. and Stab.* 2006, Vol. 91, № 6, pp. 1209–1218.
6. Wilen, C. E., Improving Weathering Resistance of Flame-Retarded Polymers. *J. Appl. Polym. Sci.* 2013. Vol. 129, № 6. pp. 925–944.
7. Pena, J. M. et al. Interactions between carbon black and stabilizers in LDPE thermal oxidation. *Polym. Degrad. and Stab.* 2001, Vol. 72, № 1, pp. 163–174.
8. Mashko, T. L., & Kalugin, E. V. Impact of new copper containing supplements on the thermal-oxidative stability of polycaproamide. *Plast. massi.* 2006, № 1, pp. 37–41.
9. Dzgaylis, Ch., et al. Adsorption from solutions on solid surfaces; edited by G. Parfita, K. Rochester. M.: *Mir*, 1986. 488 p.
10. Lenartovich, L. A., et al. Mutual influence of fillers and stabilizers in the polymer composite materials. *Trudi BSTU.* 2011, № 4: Chemistry, Organic Substances Technology and Biotechnology. pp. 98–102.

11. Bulychev, N. A. Study of polymer adsorption on the surface of inorganic pigments by infrared spectroscopy. *Materialovedenie*. 2009, № 10, pp. 2–9.
12. Ivanova, N. I. Adsorption of the surfactant mixture from the aqueous solutions on the surface of calcium carbonate. *Colloid Journal*. 2000, Vol. 62, № 1, pp. 65–69.
13. Greg, C., & Singh, K. Adsorption. Specific surface. Porosity. M.: *Mir*, 1984. 306 p.
14. Babayan, V. G., et al. The accelerated test methods of chemical additives in polymer materials. Methods for the preliminary assessment of antioxidants for plastics and elastomers. M.: *NIITEKHIM*, 1975. 31 p.

# PART III

# INSIGHTS IN MODELING AND CHEMICAL PHYSICS

# CHAPTER 16

# THE PARTICLES' DYNAMIC MOTION MODELING ON CENTRIFUGAL SHOCK MILL BLADES

A. M. LIPANOV and D. K. ZHIROV

*Institute of Mechanics of the Ural Branch of the Russian Academy of Sciences (IM UB RAS), Izhevsk, Russia*

## CONTENTS

## 16.1   INTRODUCTION

Every day over a hundred million tons of various materials is processed in the world. Mechanical grinding is the first step in process of obtaining high-strength structural nanomaterials. Greatest share of crushed raw materials is a bulk material, containing a few non-uniform components in the structure. There are minerals, non-metallic minerals, slag, grain, waste industry, agriculture, etc.

Multi-component materials are characterized by mechanical parameters: hardness, friability, bulk density, chemical composition. If we consider a multicomponent material as a feedstock for further processing, then, as a rule, it's solid particles with air in the voids. The distance between particles in a material during its processing varies continuously. As a result, the bulk density of particulate material is not constant value. Static or dynamic compressive load to the material allows to significantly reduce the distance between solid particles. The dynamic load, unlike static allows to compress material more [1].

Grinding materials can be produced in various ways, the most common: crush breakage, impact, abrasion, cutting or a combination of these methods.

The analysis of the static and dynamic loading research of different materials shows that the most effective milling method for majority of heterogeneous multi-component materials is a free kick [2–7].

The destruction methods and their combinations you can see on Figure 16.1. Irregularity of failure is determined by the difference in the physical-mechanical and chemical properties of each particle. Each particle is individual. Table 16.1 shows the most practical methods for the destruction of materials with different properties.

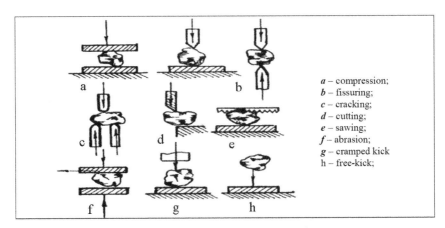

**FIGURE 16.1**    The methods of a materials mechanical grinding.

**TABLE 16.1**  The Methods of the Materials Destruction

| Material properties | Effect methods | | | | | |
|---|---|---|---|---|---|---|
| | Compression | Kick | Abrasion | Crammed kick | Sawing | Cutting |
| Solid, shear | + | + | – | + | – | – |
| Solid, brittle | + | + | – | + | – | – |
| Solid, tough | + | + | – | – | – | – |
| Half-hard | + | + | – | + | O | – |
| Resilient, soft | – | – | + | – | + | + |
| Fibrous | O | – | + | + | + | + |
| Sensitive to heat | – | O | – | O | + | + |

*Note*: + – usable; O – conditionally fit; (–) – unusable.

The research results of P. Guillot, A.R. Demidov, L.A. Glebov, V.A. Denisov, S. Zolotarev [2, 8–11] and other showed that grinding of the most multicomponent structurally inhomogeneous materials using a free pin method allows to obtain a of high quality end product, while expending minimal power. In centrifugal shock mills material milling occurs due to kinetic energy of particles emitted from the rotor and hits the fender deck (plate).

The general factor of grinding in multi-step mill is the impact velocity. Speed increasing enhances the probability of failure. Impact velocity depends on a number of factors: length spreader blades, the shape of blades, rotor speed. A rotor blade of a centrifugal mill may have different shape: straight or curved, arranged at a different inclination to the circumferential velocity.

The main detail of a centrifugal mill is a rotor. It is consisted of a axle, a disk or a discs with acceleration blades. The second disc is located above blades.

The rotor acceleration creates centripetal forces effecting on particles. Particles are driven relatively on an acceleration blade by these forces. In result, particle driving is consisted of rotating driving together with accelerating blade and motion particle relatively on an acceleration blade [12].

The circumferential speed is determined by the formula:

$$v_{okp} = \omega\rho = \frac{\pi\rho n}{30} \tag{1}$$

there $v_{\text{oKp}}$ – the circumferential velocity of the particle, m/s; $\omega$ – an angular rotor speed, rad/s; $\rho$ – radius from the rotational center of the rotor to the center of gravity of the particle, m; $n$ – rotor speed, rpm.

To determine the location of the blades in the acceleration disc you need to know the ratio (Figure 16.2).

$$\frac{0B}{0A} = \frac{r_n}{R} = \sin \angle 0AB = \sin \varphi \tag{2}$$

there $AC$ – accelerating blade; $r_n$ – the perpendicular from the blade rotation axis O to acceleration blade AC; $R$ – the radius from the blade rotation axis O to the end of blade.

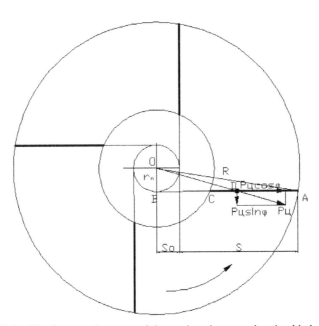

**FIGURE 16.2** The forces acting on particle moving along acceleration blade.

If $r_n = 0$ and $\varphi = 0$ then it's radial blades, else blade is angled.

There are disc with 4 blades in Figure 16.2. The angle ODB determines the blade location on an acceleration disc. (angle ODB= $\varphi$')

$$\frac{0B}{0D} = \sin \angle 0DB = \sin \varphi' \tag{3}$$

During the particle movement along an acceleration disc the angle ODB changes, and it becomes equal $\varphi$ at the end of the blade.

Ru-Bragik was draw relationship of the particles launching speed then particles moves on a radial blades without friction, air drag and rn approximately equals zero.

$$v = \omega R\sqrt{2} = \frac{\pi \rho n}{30}\sqrt{2} \tag{4}$$

there $v$ – the particle escape velocity, m/s; $\omega$ – an angular rotor speed, rad/s; $R$ – the radius of the acceleration disc, m; $n$ – rotor speed, rpm.

## 16.2  THE EQUATIONS OF THE PARTICLES MOTION ALONG BLADES

Let's consider the process of particles motion along the blades located on a horizontal rotor.

The particle is affected by the centripetal force Pc resulted from a vertical and horizontal components. The vertical component $P_c \sin\varphi$ creates *load of particles on the accelerations blade.*

The *horizontal component $P_c \cos\varphi$* motivates particles movement along the *blade.*

The differential equation of particle movement along the acceleration blade has a following view.

$$m\frac{d^2S}{dt^2} = P_u \cos\varphi' - P_u \sin\varphi' - P_k \cdot f \tag{5}$$

there $dS$ – a particle movement along the acceleration blade during $dt$; $P_u = m\omega^2\rho$ – centripetal force; $f$ – the friction coefficient between particle and the blade; $P_k = 2mf\omega\dfrac{dS}{dt}$ – the Coriolis force.

The equation can be converted as:

$$m\frac{d^2S}{dt^2} = m\omega^2\rho\cos\varphi' - fm\omega^2\rho\sin\varphi' - 2mf\omega\frac{dS}{dt} \tag{6}$$

That's why $\cos\varphi' = \dfrac{S_0 + S}{\rho}$; $\sin\varphi' = \dfrac{r_n}{\rho}$ Eq. (6) becomes:

$$\frac{d^2S}{dt^2} = \omega^2\rho\frac{S_0 + S}{\rho} - f\omega^2\rho\frac{r_n}{\rho} - 2f\omega\frac{dS}{dt}$$

$$\frac{d^2S}{dt^2} = \omega^2(S_0 + S) - f\omega^2 r_n - 2f\omega\frac{dS}{dt}$$

$$\frac{d^2S}{dt^2} + 2f\omega\frac{dS}{dt} - \omega^2 S = \omega^2(S_0 - r_n f) \tag{7}$$

Let $S_0$ be Z. The dependence equation of the movement particles distance is obtained by using the wxMAXIMA program.

$$S(t) = e^{-ft\omega}\left\{\frac{\sinh(\sqrt{f^2+1}\cdot t\omega)(2f\omega(2Z - fr))}{2\sqrt{f^2+1}\cdot\omega} + \cosh(\sqrt{f^2+1}\cdot t\omega)(2Z - fr)\right\} - Z + fr$$

$$S(t) = e^{-ft\omega}\left\{\frac{\sin h\left(\sqrt{f^2+1}\cdot t\omega\right)\left[2\left(4f\omega Z - 2f^2 r\omega\right)\right]\left[-2f\omega(2Z - fr)\right]}{2\sqrt{f^2+1}\cdot\omega}\right\} + \left[\cos h\left(\sqrt{f^2+1}t\omega\right)(2Z - fr)\right] - Z + fr \tag{8}$$

$$v(t) = e^{-ft\omega} \left\{ \frac{\cosh(\sqrt{f^2+1} \cdot t\omega)(2f\omega(2Z - fr))}{2} + +\sqrt{f^2+1} \cdot \omega \sinh(\sqrt{f^2+1} \cdot t\omega)(2Z - fr) \right\}$$

$$-f\omega e^{-ft\omega} \left\{ \frac{\sinh(\sqrt{f^2+1} \cdot t\omega)(2f\omega(2Z - fr))}{2\sqrt{f^2+1} \cdot \omega} + \cosh(\sqrt{f^2+1} \cdot t\omega)(2Z - fr) \right\} \quad (9)$$

## 16.3 CALCULATION EXAMPLE

The program for calculation of particle speed depending on the acceleration blade parameters on Turbo Pascal. The following parameters can be changed in this program: a blade length, a blade inclination angle, a coefficient of friction, and an angular rotor speed.

The time for a particle to pass a distance S (the blade length) is determined. Obtained value is substituted into the Eq. (2) then the relative particle speed in moment of particle leaving from the blade end is determined. The liner particle launching speed is calculated as the vector sum of the relative particle speed and the blade end speed (10) (Table 16.2).

Let's consider Figure 16.1:

AOB=$\varphi$
AOC=90
BOF=180 – 90 – $\varphi$
FOD=180 – BOF

$$v_{summ} = \sqrt{v^2_{line} + v^2_{relate} + 2v_{line} \cdot v_{relate} \cdot \cos\beta} \quad (10)$$

$\beta$= 180 – (180 – 90 – $\varphi$) = 180 – 180 + 90 + $\varphi$ = 90 + $\varphi$

Let's initial speed equals zero, $S$ = 0.125 м; $Z$ = $S_0$ = 0.05 м; the blade length $L_{blade}$ = S – $S_0$ = 0.075 m; the rotor speed n = 0–12,000 rpm; the coefficient of friction equals 0.5; $r_n$ = 0.02 m.

**TABLE 16.2**   Particle Launching Speed

| Rn, m | Coefficient of friction | | | | | | |
|---|---|---|---|---|---|---|---|
| | **0.1** | **0.2** | **0.3** | **0.4** | **0.5** | **0.6** | **0.7** |
| 0 | 115.617 | 111.477 | 107.849 | 104.405 | 101.399 | 98.618 | 96.216 |
| 0.01 | 110.316 | 106.185 | 102.54 | 99.307 | 96.275 | 93.647 | 91.376 |
| 0.02 | 104.887 | 100.784 | 97.147 | 93.932 | 91.109 | 88.658 | 86.559 |
| 0.03 | 98.62 | 94.73 | 91.271 | 88.388 | 85.72 | 83.553 | 81.796 |
| 0.04 | 92.031 | 88.463 | 85.29 | 82.667 | 80.501 | 78.676 | 77.303 |
| 0.05 | 85.246 | 82.11 | 79.339 | 77.1 | 75.336 | 74.054 | 73.192 |
| 0.06 | 78.177 | 75.652 | 73.459 | 71.781 | 70.635 | 69.96 | 69.672 |
| 0.07 | 71.167 | 69.287 | 67.881 | 66.961 | 66.563 | 66.6 | 67.009 |
| 0.08 | 64.648 | 63.561 | 63.051 | 63.041 | 63.497 | 64.319 | 65.453 |
| 0.09 | 59.063 | 59.132 | 59.668 | 60.61 | 61.876 | 63.517 | 65.332 |

| Rn, m | Coefficient of friction | | | | | | |
|---|---|---|---|---|---|---|---|
| | **0.8** | **0.9** | **1** | **1.2** | **1.4** | **1.6** | **1.8** | **2** |
| 0 | 94.044 | 92.191 | 90.609 | 88.02 | 86.067 | 84.627 | 83.537 | 82.703 |
| 0.01 | 89.424 | 87.758 | 86.296 | 84.055 | 82.496 | 81.393 | 80.604 | 80.027 |
| 0.02 | 84.786 | 83.357 | 82.169 | 80.445 | 79.316 | 78.605 | 78.149 | 77.864 |
| 0.03 | 80.385 | 79.235 | 78.374 | 77.234 | 76.63 | 76.359 | 76.276 | 76.305 |
| 0.04 | 76.258 | 75.537 | 75.046 | 74.588 | 74.567 | 74.771 | 75.079 | 75.434 |
| 0.05 | 72.646 | 72.382 | 72.33 | 72.643 | 73.241 | 73.931 | 74.636 | 75.311 |
| 0.06 | 69.705 | 69.978 | 70.412 | 71.538 | 72.751 | 73.931 | | |
| 0.07 | 67.668 | 68.519 | 69.457 | 71.382 | 73.198 | | | |
| 0.08 | 66.782 | 68.199 | 69.639 | 72.294 | | | | |
| 0.09 | 67.281 | 69.231 | 71.086 | | | | | |

## 16.4 CONCLUSION

The mathematical model describing of a particles dynamic motion on a centrifugal shock mill blades is developed. Equations allow to determine an angle of jump, a speed of particles from centrifugal mill blades depend on a rotor speed, a blade angle, a frictional coefficient.

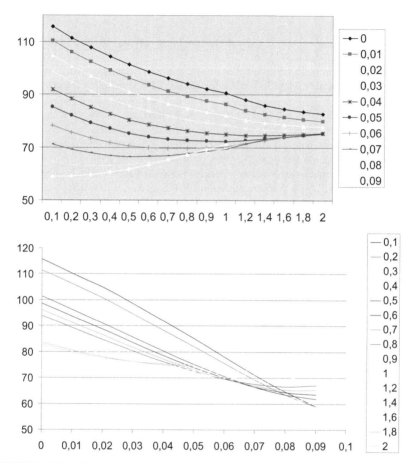

**FIGURE 16.3** The dependence of the particle emission speed from coefficient of friction at different of the blades inclination angle.

## KEYWORDS

- angle of jump
- blade
- centrifugal shock mill
- particles
- shock

## REFERENCES

1. Pugachev, A. V. Control and automation of processing of bulk material. M.: Energoatomizdat, 1989, 152 p.
2. Gio R. The problem of grinding materials and its development. G.S. Hodakova. France. G.G. Luntz. Moscow: Publishing House of Literature on Construction, 1964, 342 p.
3. Gutyer, E. M. By crushing the bulk theory. Proceedings of Moscow. agricultural Academy. Timiryazev. 1961, №.4, pp. 163–166.
4. G. Rumpf. Basic physical problems during grinding. European meeting on grinding. Frankfurt 1962. LA Swallow. M., 1966, pp. 444–472.
5. Reis. Verfahenstechnische und technologie Problemen bei der Zerkleinerung weicher bis mittelharter Stoffe. Aufbereitttechnik. 1964, B. 5, №4, S. 166–178.
6. Riley R. V. Theory and practice of crushing and grinding. Chemical and Process Engineering, 1965, V.46, № 4, p. 189.
7. Walker W. H., Levis W. K., McAdams W. H., & Gilliand E. K. Size reduction in Principles of Chemical Engineering. New York, 1937, 178 p.
8. Glebov, L. A. Intensification of grinding of raw materials in the production of animal feed: PhD Thesis. M. Mosk tehnol. Institute of Food Industry, 1990, 450 p.
9. Denisov, V. A. Improving the efficiency of the grinding process of feed grain components: synopsis. PhD Thesis. M., 1992, 32 p.
10. Zolotarev, S. V. Shock – Centrifugal shredders feed grains (basic theory and calculation). Barnaul HIPD "Altai" 2001, 200 p.
11. Tovarov, V. V., & Oskalenko, G. N. Study of emission of particles from the vane rotor centrifugal grinding machines. Trudy Giprotsement, XXIU, M.: Gosstroiizdat, 1962, pp. 64–90.

# CHAPTER 17

# CONDENSATION-LINGUAL CAPTIVITY ASSOCIATED OIL GAS IN THIN-WALLED PIPES ON MULTIPLE SPEEDS

A. N. BLYABLYAS and M. A. KOREPANOV

*Institute of Mechanics of the Ural Branch of the Russian Academy of Sciences (IM UB RAS), Izhevsk, Russia*

## CONTENTS

### 17.1   INTRODUCTION

Wave film found in many industrial processes and plants, called film devices. The occurrence of waves on the film alters the hydrodynamics of the interaction of the film with a central gas core, since the interfacial friction is added to form an interface resistance. One example is the separation of vapor when the result is a film of fluid.

Compared to the free flow of films, the problem of motion under the influence of an external airflow is more complex and is more profound practical significance.

Chemical composition condensable gas clearly has a significant influence on the picture of the phase transition, respectively, and the flow of condensate by the external airflow will depend on the structure of the original gas.

The research work is aimed at increasing the efficiency of existing technologies condensing gas mixtures.

There is an urgent need to optimize condensing technology for the needs of society, the practice of the economy.

Study of mass transfer in the film flow of condensate will help you find the most effective flow regimes that increase the efficiency of mass transfer devices.

The study heat and mass transfer in the liquid film is one of the major problems of fluid mechanics. The large surface contact at low specific flow rates the liquid film makes a very effective means of interphase heat and mass transfer. Additional intensification of the transfer of gas into the liquid is due to the wave formation. As the experiments [1], the wave modes increase mass transfer between 100 and 400%.

On the other hand the problem of the flow of the film under the influence of an external airflow is very important. Modern aircraft are operated in different climatic conditions. So starts missile carriers in Russia are carried out as from the northern Plesetsk launch site and from the Baikonur Cosmodrome. Accordingly, the environment affects the structure of the rocket. When using cryogenic fuels, the temperature of the outer wall of the carrier rocket, generally below ambient temperature. If the wall temperature is less than or equal to the saturation temperature, the condensation of atmospheric moisture. As a result, in the case of a missile or its elements appear liquid film [2].

Observations show that even at relatively low Reynolds number (Re > 50) appear on the film surface wave.

In terms of wave hydrodynamics surface of the film can be seen as a rough wall, which substantially affects the characteristics of the flow. The presence of the film wave increases the aerodynamic resistance of the body.

## 17.2   OBJECTIVES OF RESEARCH

Study instability wave current liquid hydrocarbon film in an external airflow. Theoretical study of the effect of wave modes, cooling intensity wall velocity and temperature on mass transfer performance in a thin film of condensate hydrocarbon liquid flowing down a vertical surface when the wave modes.

To achieve this goal the following tasks:

- Modeling of cooling the mixture of hydrocarbon gases through the thin wall of the heat exchanger.
- Simulation process of film condensation on the inner wall of the pipe.
- Study of the effect of speed and direction of gas flow especially at flow wave along the inner wall of a vertical tube section.
- Figuring out the basic mechanisms of mass transfer.
- Analysis of the dependence of mass transfer on the flow regime and the search for optimal speed, temperature and wave modes.

## 17.3   FORMATION OF THE PROBLEMS: DIRECTIONS AND METHODS OF SOLUTION

Today many industries use a myriad of facilities used for chemical transformations in the "gas-liquid."

Business Overview-known research suggests that in most cases, structural diagram of a condensing gas phase is chosen arbitrarily. The length of the material tube, the design of the heat exchanger, as well as the type and the cooling temperature are selected and generally very rough, basically using the table data of the last century. The intensity of the cooling rate of the gas phase in the process of condensation has also a huge impact. All these "standardized" assumption, of course, increase the weight and size parameters of the gas phase condensing unit.

Often, it sometimes happens that the parameters do not match those of the condenser of the cooling gas, which usually entails not complete condensation of mass transfer, i.e., no resource management, reducing the efficiency of the condensing unit and the installation in general.

Modeling of processes will significantly reduce the costs of experimental studies on the Heat and Mass Transfer, and in some cases even replace them with theoretical predictions.

For the simulation of heat transfer processes, the condensation heat and mass transfer is considered mass transfer unit, a workstation which is a vertical impermeable tube. With impurity – a single-phase multi-component s hydrocarbon mixture and air is supplied from the dispensing device on top of the working area with definiteness rate, pressure and temperature. Outside the pipe wall is cooled rapidly circulating liquid. By transferring heat from the hydrocarbon mixture to the coolant through the thin wall portion of the hydrocarbon phase, reaching the saturation pressure, is precipitated on the inner wall of the pipe, thereby forming a thin film of condensate on the surface (Figure 17.1).

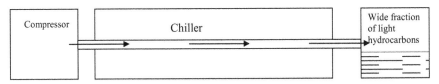

**FIGURE 17.1**   Condensation of the gas mixture in a wide fraction of light hydrocarbons.

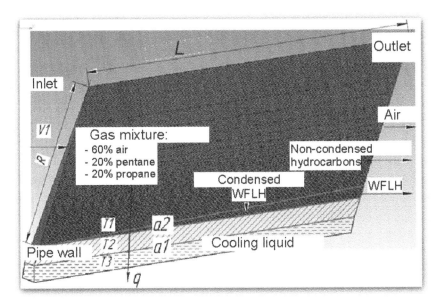

**FIGURE 17.2**   Formulation of a two-dimensional problem.

Because in the process of condensing all the steam does not change the physical state, but only part of it, so the environment is considered a two-phase, multi-component.

The film is a great condensation heat transfer thermal resistance of the phase transition from the condensation surface to the wall. The heat flux through the film determined by Newton's law-Richman [4].

The initial and boundary conditions: the estimated area of the tube length 1000 mm, the inner diameter of the tube 8 mm, the wall thickness of the tube 1 mm, material stainless steel tube (for coefficient conductivity 17.5 W/m•K), time modeling of convective heat and condensation 10 s, the time step of 0.005 s, accounting for the forces of gravity – 9.81 m/s² (similar to the direction of movement of gas in the tube), accounting for the surface tension of the hydrocarbon in the film depending on the wall temperature ($\sigma$ = 50 mN/m at T = 293 K, $\sigma$ = 21 mN/m at T = 373 K), the thermal conductivity of the hydrocarbon film – 0.12 W/m K, the rate of the hydrocarbon mixture in the range of 1–10 m/s, starting temperature of the gas medium in the pipe inlet 373 K, composition of the mixture: 20% propane, pentane 20%, air 60% (75% nitrogen, 23% oxygen), coolant temperature in the range 274–290 K, the coefficient of heat transfer to the cooling liquid = 2500 W/m²K, starting wall temperature equal to the temperature of the cooling liquid, system pressure 1 MPa, the model of turbulence: k – e (This model is the most versatile for a wide range of Reynolds numbers).

## 17.4   THE DYNAMICS OF THE SURFACE WAVES: PECULIARITIES OF HYDROCARBON FILMS

In previous studies [5], the author of meters and it was the behavior of the liquid film in the condensation of vapor. The hydrocarbon film due to the peculiarities of chemical structure and physical properties different from water at various flow regimes behaves differently.

The dynamics of surface waves can identify several specific areas: on the wall of a thin liquid film of random oscillations occur linear waves with a spatial period, then the amplitude of the waves increases.

Natural waves arise as a result of random fluctuations caused by airflow at a constant speed, due to the instability of the plane flow.

Figure 17.3 shows flow regimes of water and hydrocarbon film at the same speed at the inlet of the mixture.

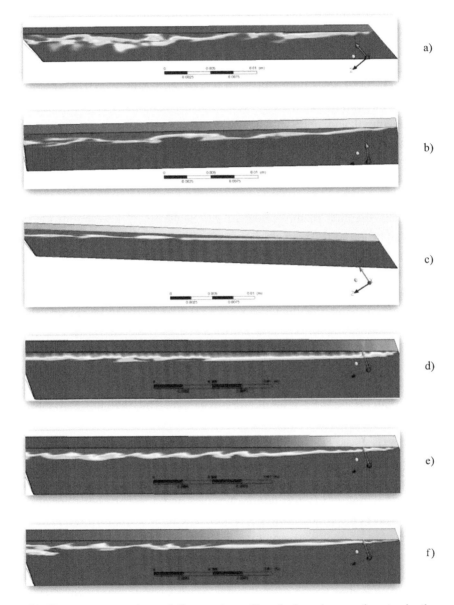

**FIGURE 17.3** Comparison of flow regimes films hydrocarbons and water in the condensation depends on the rate of vapor. (a) Hydrocarbons of 5 m/s; (b) hydrocarbons, 10 m/s; (c) hydrocarbons, 20 m/s; (d) water, 5 m/s; (e) water, 10 m/s; (f) water, 20 m/s.

Undoubtedly, in the condensation of hydrocarbon fractions wavelength, its amplitude and structure appear differently than in the condensation of vapor. No "sawtooth" waves, the wavelength increased by (170–330%).

Warming the tube wall increases, ceteris paribus, in effect at the heat of the phase transition of hydrocarbons because of which increases the amount of condensate. The thickness of the hydrocarbon film is noticeably smaller than water film, under similar conditions, it is associated with lower values of surface tension and viscosity hydrocarbon film. The average speed of the hydrocarbon film more.

With increasing Reynolds number of the film surface covered by more waves of frequency and amplitude, but the average thickness of the film decreases with increasing Re. The rate on the crest of the waves of single-solitons, on average, three times the average speed of the film.

## 17.5   CONCLUSION

With increasing Reynolds number of the film decreases wavelength with respect to its thickness. This means that there is a critical number Re, which will be achieved at the highest wavelength for each flow regime. The exact definition of each mode can be isolated in a separate task.

It builds a picture that describes the movement of the hydrocarbon film on the wall of the mass transfer apparatus, model calculations conducted in CAD ANSYS. The calculation results are in good agreement with experimental data (for falling film of water on a smooth vertical wall), that allows to speak about the adequacy of the model.

## KEYWORDS

- **condensation**
- **flow regime**
- **fluid dynamics**
- **oil associated gas**
- **wave motion of the film**

# REFERENCES

1. Nakoryakov, V. E., Pokusaev, B. G., & Schreiber, I. R. Wave dynamics of vapor-liquid media. M.: Energoatomizdat, 1990, 248 pp. (in Russian).
2. Alekseenko, S. V., Nakoryakov, V. E., & Pokusaev, B. G. The impact of waves on the transport processes. Characteristic for liquid films. Nauka, Novosibirsk, 1992, pp. 191–207 (in Russian).
3. Kapitsa, P. L. Characteristic for the thin layer of viscous liquid: 3 hours. Part 1 Free flow. Journal of Experimental and Theoretical Physics. 1948, T. 18. Vol. 1, pp. 1–28 (in Russian).
4. Kapitsa, P. L., & Kapitsa, S. P. The wave flow of thin layers of viscous fluid: 3 hours. Part III. Experimental study of wave flow regime. Journal of Experimental and Theoretical Physics. 1949, T. 19, Vol. 2, pp. 105–120.
5. Blyablyas, A. N., & Korepanov, M. A. Modeling the hydro-process in the condensation of vapor in the pipe. Vestnik IzhGTU. 2014, № 1 (61), pp. 43–45. (in Russian).

## CHAPTER 18

# STUDY OF LAMINAR DIFFUSION COMBUSTION MODELS

A. A. SHAKLEIN

*Institute of Mechanics of the Ural Branch of the Russian Academy of Sciences (IM UB RAS), Izhevsk, Russia*

## CONTENTS

### 18.1   INTRODUCTION

Substantial influence of burning processes on engineering systems such as internal combustion engines, turbines, reactors, and rocket engines establishes strong requirements for theoretical description of combustion. Special attention must be focused on pollutants formation.

Combustion and pollutants formation have been comprehensively studied over last decades (e.g., [1–10]). Methane and propane diffusion combustion experimental research have been carried out [1]. Combustion chamber has square cross section, and fuel is supplied from porous cylinder. The results consist of fuel, oxidizer, CO and $CO_2$ concentrations. Similar experiments are carried out in Ref. [2]. Numerical and experimental study of the partial premixed methane air mixture burning in cylindrical combustion chamber is described in Ref. [3]. The results contain temperature and $N_2$, $O_2$, $CH_4$, $CO_2$, $H_2O$, CO, $H_2$, OH and NO concentration distributions. Coflow premixed combustion of the methane oxygen mixture is studied in Ref. [4]. Numerical results and experimental data are presented for temperature and $CH_4$, $CO_2$, $H_2O$ and CO concentrations. Coflow diffusion combustion of methane and air mixture is investigated in [5]. There are wide set of experimental data (temperature, $CH_4$, $O_2$, $H_2O$, $CO_2$, CO, $H_2$, $C_2H_2$, $C_2H_4$, $CH_2O$, $CH_2CO$, $N_2$, $C_3H_3$, $C_3H_4$, $C_6H_6$, etc.). Numerical and experimental study of gravity, feeding velocity and fuel delusion influence on coflow diffusion combustion of methane and air is shown in Ref. [6]. Temperature fields and flame size are presented. Quenching effects of methane-air flame in microgravity are investigated [7]. Temperature, fuel and oxidizer concentrations profiles are presented. Photos of flume are shown. In the next work [8] authors evaluated fuel (methane) and oxidizer (air) influence on combustion process. The main species, CO and NO concentrations are shown. NO level is evaluated in study [9], dedicated to investigation of temperature and fuel to oxidizer mixture fraction influence on combustion. Concentration of the CH radicals formed in methane air mixture flame is studied in Ref. [10].

Thoroughly studied reaction mechanisms of some fuels allow to obtain detailed description of large amount of processes and wide set of chemical reacting species.

One approach in the scope of all combustion models has special features – the Flamelet model. It based on mixture fraction of fuel and oxidizer. The Flamelet model is considerably refined to date in comparison with first versions. It allows to model diffusion and partial remixed combustion of laminar and turbulent reacting gas flow.

The current work is dedicated to study the numerical approaches to model laminar diffusion combustion.

## 18.2  NUMERICAL METHOD

The set of equations describing a laminar reacting flow of a multicomponent gas is as follows

$$\frac{\partial \rho}{\partial t} + \frac{\partial \rho u_j}{\partial x_j} = 0 \tag{1}$$

$$\rho \frac{\partial u_i}{\partial t} + \rho u_j \frac{\partial u_i}{\partial x_j} = -\frac{\partial p}{\partial x_i} + \frac{\partial}{\partial x_j} \mu \frac{\partial u_i}{\partial x_j} + (\rho_a - \rho) g_i \tag{2}$$

$$\rho C \frac{\partial T}{\partial t} + \rho C u_j \frac{\partial T}{\partial x_j} = \frac{\partial p}{\partial t} + \frac{\partial}{\partial x_j} \lambda \frac{\partial T}{\partial x_j} + Q_c - Q_r \tag{3}$$

$$\rho \frac{\partial Y_i}{\partial t} + \rho u_j \frac{\partial Y_i}{\partial x_j} = \frac{\partial}{\partial x_j} \rho D \frac{\partial Y_i}{\partial x_j} + \omega_i \tag{4}$$

$$p = \rho R T \tag{5}$$

Radiative energy transfer effects on the combustion process are taken into account by the fourth term at the right side of the Eq. (3). Radiation heat loss evaluation is known to be very important even for free flame configurations (i.e., without walls irradiation) study, because of strong dependency of reactions on temperature, which can lead to overprediction of species concentration. The radiative heat loss is described by the following equation [3]

$$Q_r = 4\sigma \left( T^4 - T_a^4 \right) \sum P_i \kappa_i \tag{6}$$

where $5.67 \cdot 10^{-8}$ W/(m$^2 \cdot$K$^4$) is Stefan-Boltzmann constant, $T_a$ is the ambient temperature, $P_i$ is the partial pressure of the i$^{th}$ specie, $\kappa_i$ is the averaged over whole spectrum absorption coefficient of the i$^{th}$ specie.

The summation in the Eq. (6) is conducted over two components only (i.e., $H_2O$ and $CO_2$), since it is the main radiative heat transfer source in the combustion products. The absorption coefficient dependency upon temperature is taken into account by the fifth order polynomial

$$\kappa_i = a_0 + a_1 \left(10^3 / T\right) + a_2 \left(10^3 / T\right)^2$$
$$+ a_3 \left(10^3 / T\right)^3 + a_4 \left(10^3 / T\right)^4 + a_5 \left(10^3 / T\right)^5 \tag{7}$$

with coefficients being approximated based on the averaged over whole spectrum absorption coefficient [12] (Table 18.1).

**TABLE 18.1**    Polynomial Coefficients for Absorption Coefficient Calculation

| Species | $a_0$ | $a_1$ | $a_2$ | $a_3$ | $a_4$ | $a_5$ |
|---------|-------|-------|-------|-------|-------|-------|
| $CO_2$ | 0.3263608 | −1.412508 | 2.6311964 | −1.7518103 | 0.4935058 | −0.0501506 |
| $H_2O$ | −0.1201951 | 0.6279217 | −1.0526880 | 0.8538732 | −0.2955810 | 0.0366692 |

## 18.3   COMBUSTION MODELING

The most crucial attention must be focused on modeling of the source terms located in the energy (3) and species transport (4) equations. In general case this sources can be evaluated in the following way [13]. The $i^{th}$ specie rate of creation/destruction is as follows

$$\omega_i = W_i \sum_{k=1}^{r} v_{ik} w_k \tag{8}$$

where $r$ is the total number of reactions, $v_{ik} = v''_{ik} - v'_{ik}$, $w_k$ is the $k^{th}$ reaction rate expressed in the following way

$$w_k = k_{fk} \prod_{j=1}^{n} \left(\frac{\rho Y_j}{W_j}\right)^{v'_{jk}} - k_{bk} \prod_{j=1}^{n} \left(\frac{\rho Y_j}{W_j}\right)^{v''_{jk}} \tag{9}$$

where $k_{fk}$ and $k_{bk}$ are forward and backward reaction rate coefficients, $v'_{ik}$ and $v''_{ik}$ are stoichiometric coefficients of $k^{th}$ forward and backward reaction.

The forward reaction rate coefficient is expressed as [14]:

$$k_{fk} = A_{fk} T^{\beta_k} \exp\left(-\frac{E_k}{R_0 T}\right) \tag{10}$$

The energy source term results from chemical reactions occurred during combustion process is as follows

$$Q_c = \sum_{i=1}^{n} h_i \omega_i \qquad (11)$$

An mechanism is the main parameter, i.e., the number of reactions taken into consideration and the number of species participating in the process. One of the most detailed and available mechanism for methane oxidation is the Gri-Mech 3.0 [15] based on 325 reactions and 53 species including $NO_x$ formation. A weak side of the approach described is high computational costs requirement. Time elapsed on solving the Eq. (9) written for the whole reaction set can be much higher than time spent on solving the gas dynamics equations.

It seems quite useful to utilize simplified mechanisms consisted of small amount of species (e.g., [16]). Generally these species contain fuel, oxidizer, inert specie, combustion products and pollutants (e.g., CO, $NO_x$, $SO_x$). If only one reaction is used to model hydrocarbon combustion, the resulted equation is expressed as

$$v'_F \, C_m H_n + v'_{O_2} \, O_2 \rightarrow v''_{CO_2} \, CO_2 + v''_{H_2O} \, H_2O \qquad (12)$$

where $m$ and $n$ are the numbers of carbon and hydrogen atoms in fuel specie. The equation (12) composed for methane oxidation is as follows

$$CH_4 + 2O_2 \rightarrow CO_2 + 2H_2O \qquad (13)$$

More detailed mechanisms have to be incorporated for accurate pollutant prediction. Carbon monoxide concentration for methane combustion can be calculated as [16]

$$CH_4 + 1.5O_2 \rightarrow CO + 2H_2O \qquad (14)$$

$$CO + 0.5O_2 \rightarrow CO_2 \qquad (15)$$

Nitrogen monoxide concentration can be estimated from [17]

$$N_2 + O_2 \rightarrow 2\,NO \qquad (16)$$

Reaction rate coefficients in the Eq. (8) and power coefficients in the Eq. (9) are shown in Table 18.2.

**TABLE 18.2**   Reaction Rate Coefficients

| Reaction | $A_{fk}$ | $\beta_k$ | $E_k$, kJ/mol | Power |  |  |  |
|---|---|---|---|---|---|---|---|
|  |  |  |  | Specie | Value | Specie | Value |
| (13) | $5.2 \times 10^{16}$ | 0 | 124 | $CH_4$ | 1.0 | $O_2$ | 1.0 |
| (14) | $1.5 \times 10^7$ | 0 | 125 | $CH_4$ | −0.3 | $O_2$ | 1.3 |
| (15) | $3.98 \times 10^{14}$ | 0 | 167 | $CO$ | 1.0 | $O_2$ | 0.25 |
| (16) | $2.26 \times 10^{15}$ | −0.5 | 577 | $N_2$ | 1.0 | $O_2$ | 0.5 |

Other approach (Flamalet [13]) consists of coordinate transformation for the Eqs. (3) and (4) to new space Z (mixture fraction) computed in the following way

$$Z = \frac{Z_C / (mW_C) + Z_H / (nW_H) + 2\left(Y_{O_2}^{in} - Z_O\right) / \left(v'_{O_2} W_{O_2}\right)}{Z_C^{in} / (mW_C) + Z_H^{in} / (nW_H) + 2 Y_{O_2}^{in} / \left(v'_{O_2} W_{O_2}\right)} \qquad (17)$$

where $Z_i$ is an element mass fraction, $v'_{O_2}$ is the oxygen molecule molar stoichiometric coefficient used in complete fuel oxidizing reaction equation. It worth to mention, the expression (17) is written for hydrocarbon fuel and oxygen oxidizer.

Following relationships after coordinate transformation

$$T = T(Z, \chi), \quad Y_i = Y_i (Z, \chi) \qquad (18)$$

are computed as preprocessor step (i.e., before the main solution procedure) by modeling counterflow diffusion combustion with the Eqs. (1)–(5) and tabulated. The equations can be reduced to one-dimensional space significantly lowering computational costs [13].

The scalar dissipation rate allowing to take into account non equilibrium effects for laminar flow regime is expressed in the following way

$$\chi = 2D \frac{\partial Z}{\partial x_j} \tag{19}$$

The Z transport equation can be expressed as

$$\rho \frac{\partial Z}{\partial t} + \rho u_j \frac{\partial Z}{\partial x_j} = \frac{\partial}{\partial x_j} \frac{\mu}{Sc} \frac{\partial Z}{\partial x_j} \tag{20}$$

Temperature and species concentration are interpolated based on Eq. (16) if needed.


## 18.4  SOLUTION PROCEDURE

The finite volume method is used for integrating the set of equations shown. OpenFOAM software [18] is extended to solve the equations. Convective and diffusive fluxes are computed with second order of approximation. First order of approximation is used for time stepping because of steady-state nature of the process simulated. Pressure and velocity fields are coupled by the PISO method.

A solution algorithm has distinctive features based on a combustion model chosen.

Chemical kinetics mechanism:

- The momentum equation (2) is solved.
- Chemical kinetics is evaluated (i.e., reaction rates, chemical source terms).
- The species transport equations (4) are solved.
- The energy equation (3) is solved.
- Physical properties of mixture are corrected, absorption coefficients are updated.
- Pressure and velocity fields are corrected in order to satisfy continuity equation (1).

Flamelet approach:

- $T = T(Z,\chi)$, $Y_i = Y_i(Z,\chi)$ relations are tabulated in the following way. Diffusion combustion is modeled in Cantera software [3], with the scalar dissipation parameter being varied in the range from equilibrium regime to quenching and radiation heat transfer being taken into account. Mixture fraction is computed based on Eq. (17). Tabulation is carried out as preprocessor step (i.e., before the main computation).
- The momentum equation (2) is solved.
- The mixture fraction equation (20) is solved.
- Physical parameters of mixture are corrected based on species mass fraction and temperature field taken from the tables computed.
- Pressure and velocity fields are corrected in order to satisfy continuity equation (1).

## 18.5   RESULTS AND DISCUSSION

The model is verified on diffusion laminar combustion burners with coflow and counterflow fuel and oxidizer supply. First flame configuration is studied in Ref. [4]. The device used in work is based on cylindrical burner and coflow diffusion combustion regime. The case parameters are shown in Table 18.3.

The 2d axisymmetrical simplified configuration of the burner is used, with cylindrical geometry being modeled by wedge-type computational mesh. For simplicity the boundary condition set shown below is consisted of all studied models description. Thermodynamic parameters are computed based on Gri-Mech 3.0 data [15].

TABLE 18.3   Physical and Geometrical Parameters

| Parameter | $V_F$, cm/s | $V_O$, cm/s | $T_F$, K | $T_O$, K | Oxidizer, volume fraction | Fuel, volume fraction | h, cm | $R_F$, cm | $R_O$, cm |
|---|---|---|---|---|---|---|---|---|---|
| Value | 4.5 | 9.88 | 300 | 300 | 21% $O_2$. 79% $N_2$ | 100% $CH_4$ | 10 | 0.635 | 2.54 |

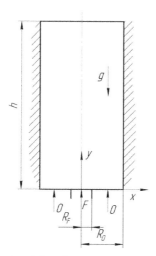

**FIGURE 18.1**    The computational domain.

The boundary condition set for the computational domain (Figure 18.1) is formulated as follows:

$$x = 0 : \frac{\partial \varphi}{\partial x} = 0, \varphi = \{v, T, Y_i, Z\}, u = 0;$$

$$x = R_O : \frac{\partial \varphi}{\partial x} = 0, \varphi = \{T, Y_i, Z\}, u = 0, v = 0;$$

$$y = 0, x < R_F : u = 0, v = V_F, T = T_F, Y_F = 1, \quad Y_O = 0, Z = 1;$$

$$y = 0, R_O < x < R_F : u = 0, v = V_O, T = T_O, Y_F = 0, \quad Y_O = 1, Z = 0;$$

$$y = h : \frac{\partial \varphi}{\partial y} = 0, \varphi = \{u, v, T, Y_i, Z\}$$

where the subscript $O$ stands for oxidizer, and $F$ for fuel.

3D computational mesh with one sell over small angle is $\alpha_{\text{sect}}$ built to model radial sector, because mathematical model is formulated in the Cartesian coordinate system. The both surfaces are set to symmetry boundary condition type as follows

$$\alpha = 0, \alpha = \alpha_{\text{sect}} : \quad u_n = 0, \frac{\partial \varphi}{\partial n} = 0, \varphi = \{u_\tau, T, Y_i, Z\}$$

Grid convergence study is carried out for different finite-volume non-uniform meshes: $15 \times 50 \times 1$ (1), $30 \times 100 \times 1$ (2) and $60 \times 200 \times 1$ (3). Substantial quantitative and qualitative differences are observed in results obtained based on first and second meshes. However, difference between results obtained based on second and third mesh is very small. Therefore, the main simulation is carried out on second mesh $30 \times 100 \times 1$ (Figure 18.2).

The results are shown at Figure 18.3. Distinction between simulated and experimental results of the methane mole fraction distribution over the radius at 1.2 cm above the burner head can be explained by boundary condition formulation. Velocity profile at the fuel and oxidizer inlet boundary conditions has constant distribution over the surface. However, velocity profile at the burner head in the experiment have parabolic type because of wall friction.

Flamelet model (curve 4) shows two times lower carbon dioxide level in the flume zone in comparison with the experimental data (Figure 18.3, b). The simplified mechanism (curve 2) predicts good agreement with the experiment. The coupled gas dynamics and chemical kinetics model based on the Gri-Mech 3.0 mechanism (curve 3) gives the worst prediction of the $CO_2$ concentration between the approaches studied.

Nevertheless, carbon monoxide distribution picture is quite different (Figure 18.3, c). The Flamelet model (curve 4) allows to obtain carbon monoxide concentration being in good agreement with the experimental data. However, the usage of coupled models with the both simplified and detailed mechanisms doesn't guarantee accurate CO prediction.

Despite the results shown above the temperature profiles computed based on different models used have quantitative and qualitative similarity in comparison with the experimental results (Figure 18.3, d). However, the model based on the Gri-Mech 3.0 mechanism (curve 3) gives slightly lower temperature values.

Considering the results obtained the following conclusion can be stated. Solving the coupled gas dynamics and chemical kinetics equations with the detailed mechanism (Gri-Mech 3.0) requires thorough analysis and testing. Perhaps, time step should be lowered to resolve stiff equations of chemical kinetics. However, the approach proposed is bounded by high computational costs. THEREFORE, this article is focused on the Flamelet model. The results of the simulation carried out based on the simplified mechanism show good agreement with the experimental data for carbon dioxide and temperature distributions.

**FIGURE 18.2**   The computational mesh.

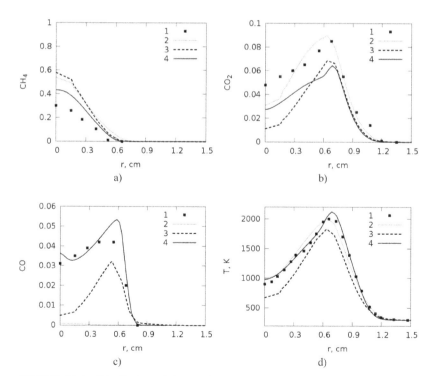

**FIGURE 18.3**   Distribution of the species mole fraction over the line 1.2 cm above the burner head, with r being measured from the symmetry axis (1 – experiment [4], 2 – simplified mechanism, 3 – Gri-Mech 3.0, 4 – Flamelet).

Second flame configuration simulated is studied in Ref. [9]. The case parameters are shown in Table 18.4.

The boundary condition set for the computational domain (Figure 18.4) is formulated as follows

$$x = 0: \quad \frac{\partial \varphi}{\partial x} = 0, \varphi = \{v, T, Y_i, Z\}, u = 0;$$

$$x = R: \quad \frac{\partial \varphi}{\partial x} = 0, \varphi = \{u, v, T, Y_i, Z\};$$

$$y = 0: \quad u = 0, v = V_F, T = T_F, Y_F = 1, \quad Y_O = 0, Z = 1;$$

$$y = h: \quad u = 0, v = -V_O, T = T_O, Y_F = 0, \quad Y_O = 1, Z = 0;$$

3D computational mesh with one sell over small angle $\alpha_{sect}$ is built to model wedge domain, because mathematical model is formulated in the Cartesian coordinate system. The both surfaces are set to symmetry boundary condition type as follows

**TABLE 18.4**  Physical and Geometrical Parameters

| Parameter | $V_F$, cm/s | $V_O$, cm/s | $T_F$, K | $T_O$, K | Fuel, volume fraction | Oxidizer, volume fraction | h, cm | R, cm |
|---|---|---|---|---|---|---|---|---|
| Value | 10.68 | 22.3 | 531 | 717 | 50.7% $CH_4$ 49.3% $N_2$ | 23.9% $O_2$. 76.1% He | 2.9 | 1.45 |

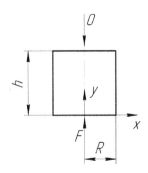

**FIGURE 18.4**  The computational domain.

$$\alpha = 0, \, \alpha = \alpha_{\text{sect}} : \quad u_n = 0, \, \frac{\partial \varphi}{\partial n} = 0, \, \varphi = \left\{ u_\tau, T, Y_i, Z \right\}$$

Grid convergence study is carried out for different finite-volume non-uniform meshes: $15 \times 30 \times 1$ (1), $30 \times 60 \times 1$ (2) and $60 \times 120 \times 1$ (3). Substantial quantitative and qualitative differences are observed in results obtained based on first and second meshes. However, difference between results obtained based on second and third mesh is very small. Therefore, the main simulation is carried out on second mesh $30 \times 60 \times 1$ (Figure 18.5).

The results are shown at Figure 18.6. The NO mole fraction and temperature profiles are compared with the experimental data. Temperature values are much higher for coupled gas dynamics and chemical kinetics

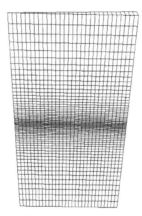

**FIGURE 18.5** The computational mesh.

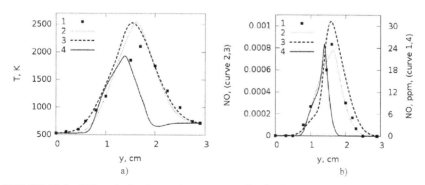

**FIGURE 18.6** NO mole fraction and temperature distribution over the symmetry axis (y) (1 – experiment [9], 2 – simplified mechanism, 3 – Gri-Mech 3.0, 4 – Flamelet).

model (curve 2 and 3, Figure 18.6, a) in comparison with the experimental results despite the fact that the radiative heat transfer is taken into account in the model. In result, computed NO values are two times higher than the experimental data.

Third flame configuration simulated is studied in Ref. [8]. The case parameters are shown in Table 18.5.

The computational domain and boundary conditions set are the same as in second flame configuration.

Grid convergence study is carried out for different finite-volume uniform meshes: $15 \times 30 \times 1$ (1), $30 \times 60 \times 1$ (2) and $60 \times 60 \times 1$ (3). Substantial quantitative and qualitative differences are observed in results obtained based on first and second meshes. However, difference between results obtained based on second and third mesh is very small. Therefore, the main simulation is carried out on second mesh $30 \times 60 \times 1$ (Figure 18.7).

**TABLE 18.5**    Physical and Geometrical Parameters

| Parameter | $V_F$, cm/s | $V_O$, cm/s | $T_F$, K | $T_O$, K | Fuel, volume fraction | Oxidizer, volume fraction | h, cm | R, cm |
|---|---|---|---|---|---|---|---|---|
| Value | 70 | 70 | 300 | 300 | 100% $CH_4$ | 21% $O_2$, 79% $N_2$ | 1.5 | 1.0 |

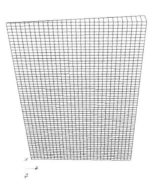

**FIGURE 18.7**    The computational mesh.

The results are shown at Figures 18.8 and 18.9. Methane mole fraction distribution over the symmetry axis y (Figure 18.8, a) obtained by all models is in good agreement with the experimental data. The $CO_2$ profile (Figure 18.8, b) computed based on the simplified model (curve 2) distinctly differs from other results.

Reaction rate coefficients of the simplified mechanism, obviously, have to be thoroughly tested for every flame configuration, since the model is failed to predict NO and CO concentrations (the error is more than ten times).

The results obtained by solving the coupled gas dynamics and chemical kinetics equations with detailed mechanism (Figure 18.8, curve 3) are in good agreement with the experimental data except CO concentration. Nevertheless, the good agreement of NO concentration with the experimental results (Figure 18.8, d) is seems to be just a side effect of low temperature values (Figure 18.9).

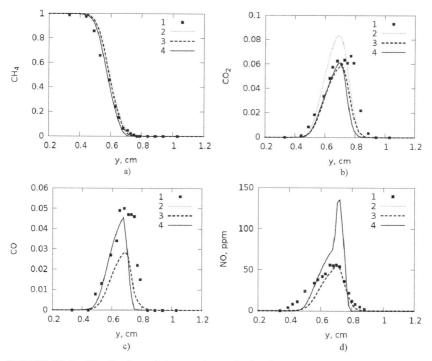

**FIGURE 18.8**  Distribution of the species mole fraction over the symmetry axis (y) (1 – experiment [8], 2 – simplified mechanism, 3 – Gri-Mech 3.0, 4 – Flamelet).

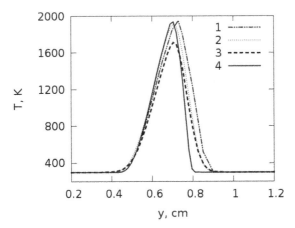

**FIGURE 18.9**    The temperature distribution over the symmetry axis (y) (1 – simulation [8], 2 – simplified mechanism, 3 – Gri-Mech 3.0, 4 – Flamelet).

The Flamelet-model used in combustion simulation allows one to obtain results, which are in good agreement with experimental data. (Figure 18.8, curve 4). However, the maximum value of the NO mole fraction distribution is two times higher than the measured value.

## 18.6   CONCLUSION

In the study presented different combustion models have been evaluated and tested, with results being compared with the experimental data. The simplified mechanism is able to show fast performance, but its reaction rate coefficients have to be adjusted for every simulation. Therefore, necessity of searching for suited experimental data is the main difficulty for simplified mechanism usage. On the other hand, solution of the coupled fluid dynamics and chemical kinetics equations based on a detailed mechanism is very sensitive to algorithm parameters (e.g., time stepping, chemical kinetics equations set solution method), which leads to high computational costs. Diffusion combustion simulation based on Flamelet model described allows to achieve compromise between time spent on computations, accuracy and information obtained.

The future work will be focused on Flamelet model extension to accounting for non-adiabatic combustion except the radiative heat transfer already presented in model (e.g., fuel supply with different temperature being the main feature of coupled combustion of solid fuel) and premixed or partially premixed turbulent combustion.

## KEYWORDS

- carbon monoxide
- Flamelet approach
- laminar diffusion combustion
- methane combustion
- nitrogen monoxide

## REFERENCES

1. Tsuji, H., & Yamaoka, I. The structure of counterflow diffusion flames in the forward stagnation region of a porous cylinder. Proc. Combust. Inst., 1969, V. 12, pp. 997–1005.
2. Tsuji, H. Counterflow diffusion flames. Prog. Energy Combust. Sci., 1982, V. 8, pp. 93–119.
3. Barlow, R. S., Karpetis, A. N., Frank, J. H., & Chen, J. Y. Scalar profiles and NO formation in laminar opposed-flow partially premixed methane/air flames. Combustion and flame, 2001, V. 127, pp. 2102–2118.
4. Mitchell, R. R., Sarofim, A. F., & Clomburg, L. A. Experimental and numerical investigation of confined laminar diffusion flames. Combustion and flame. 1980, V. 37, pp. 227–244.
5. Cuoci, A. Experimental and detailed kinetics modeling study of PAH formation in laminar co-flow methane diffusion flames. Proc. Combust. Inst., 2013, V. 34, pp. 1811–1818.
6. Cao, S. A computational and experimental study of coflow laminar methane/air diffusion flames: Effects of fuel dilution, inlet velocity, and gravity. Proc. Combust. Inst., 2015, V. 35, pp. 897–903.
7. Hamins, A. The structure and extinction of low strain rate non-premixed flames by agent in microgravity. NISTIR 7445, 2007, 177 p.
8. Lim, J., Gore, J., & Viskanta, R. A study of the effects of air preheat on the structure of methane/air counterflow diffusion flames. Combustion and flame. 2000, V. 121, pp. 262–274.
9. Mungekar, H., & Atreya, A. NO formation in counterflow partial premixed flames. Combustion and flame. 2007, V. 148, pp. 148–157.

10. Gibaud, C., Snyder, J. A., Sick, V., & Lindstedt, R. P. Laser-induces fluorescence measurements and modeling of absolute CH concentrations in strained laminar methane/air diffusion flames. Proc. Combust. Inst., 2005, V. 30, pp. 455–463.

11. Schmitt, P., Schuermans, B., Geigle, K. P., & Poinsot, T. Effects of radiation, wall heat loss and effusion cooling on flame stabilization and pollutant prediction in LES of gas turbine combustion. A numerical study on NOx formation in laminar counter-flow CH4/air triple flames. Combustion and Flame, 2005, V. 143, pp. 282–298.

12. Modest, M. F. Radiative heat transfer. Academic Press. 2003, 822 p.

13. Peters, N. Turbulent combustion. Cambridge university press, 2000, 320 p.

14. Poinsot, T., & Veynante, D. Theoretical and numerical combustion. Edwards, 2005, 538 p.

15. Smith, P. G. Optimized mechanism for natural gas combustion Gri-Mech 3.0 (Web: http://www.me.berkeley.edu/gri_mech/).

16. Westbrook, C. K., & Dryer, F. L. Chemical kinetic modeling of hydrocarbon combustion. Prog. Energy Combust. Sci. 1984, V. 10, pp. 1–57.

17. Warnatz, J., Maas, U., & Dibble, R. Goreniye. Fizicheskiye i himicheskiye aspecti, modelirovaniye, eksperimenti, obrazovaniye zagryaznyayushih veshestv. M.: Fizmatlib, 2006, 352 p.

18. Weller, H. G., Tabor, G., Jasak, H., & Fureby, C. A tensorial approach to computational continuum mechanics using object-oriented techniques. Computers in physics. 1998, V. 12, N. 6, pp. 620–631.

19. Goodwin, D. G., Moffat, H. K., & Speth, R. L. Cantera: An Object-oriented Software Toolkit for Chemical Kinetics, Thermodynamics, and Transport Processes. 2015. Web: http://www.cantera.org.

# CHAPTER 19

# MATHEMATICAL MODELING OF FLOW WITH HOMOGENEOUS CONDENSATION

M. A. KOREPANOV[1] and S. A. GRUZD[2]

[1]Institute of Mechanics of the Ural Branch of the Russian Academy of Sciences (IM UB RAS), Izhevsk, Russia

[2]Kalashnikov Izhevsk State Technical University, Izhevsk, Russia

## CONTENTS

### 19.1   INTRODUCTION

In our time, the increase in capacity of computer technology leads to the fact that mathematical modeling is becoming one of the main tools in the study of objects whose behavior previously studied only experimentally.

An analytical solution of some problems of thermodynamics and physics is not always possible, and numerical methods in this act as the only way out to determine the thermodynamic parameters of the system (pressure, temperature, density, energy, etc.), described by differential equations of conservation.

One such phenomenon, which has always attracted the attention of researchers is condensation. Such concepts as undercooling and over-heating are an integral part of the process of transition of substance from one physical state to another. Similar phenomena are common in life, but they are still poorly studied, because of their transience. Modeling of processes of phase transition is of fundamental importance in atmospheric physics, astronomy, aerodynamics, nuclear and nanomaterials physics.

The smallest group of molecules called clusters appears in the system when parameters are close to saturation, and have a significant influence on the thermodynamic parameters of the system as a whole. Therefore, in building a model the phenomenon of condensation, it is important to consider their presence, especially in calculating flows in the nozzle and the jets of jet engines, the formation of aerosols, in the calculation of steam turbines and other thermodynamic processes.

## 19.2 EQUILIBRIUM CONSTANT OF SMALL AGGLOMERATES

In classical theory of homogeneous nucleation in accordance with the ideas of Mayer [1] and Frenkel [2] considered that in real gas, along with monoparticles present agglomerates, which form by supercooling supercritical clusters, giving rise to the formation of a new phase.

Monoparticles can form a double particles via reaction I+I=II, triple particles (trimer) – III = I + II, etc. It is assumed that the agglomerates behave as gas particles resulting mixture of monoparticles and agglomerates satisfies the equation of state of an ideal gas.

Thus the basic question remains determining the concentration or partial pressure of agglomerates. In the view of the process of the formation of agglomerates as a chemical reaction and the assumption of local chemical equilibrium the partial pressures of the agglomerates can be found from the condition of equality of chemical potentials of the agglomerates and monogaz:

$$\varphi_g = g \cdot \varphi_1 \tag{1}$$

where $\varphi = I^0 - T\left(S^0 - R_0 \ln p\right)$ In this case, the equilibrium constant of the formation of the agglomerate is equal to

$$K(g,T) = \frac{p_1^g}{p_g} \tag{2}$$

The main problem in this case is to determine the chemical potential $\varphi_g$ agglomerates. In the works of Mayer, Frenkel, Vukalovich and Novikov [1, 2] proposed to calculate the partition function of the agglomerates, which is a very time consuming task, in view of the large number of different configurations and their agglomerates.

At the same time classical theory of homogeneous nucleation uses next equation for calculating the chemical potential of the agglomerates [3] in approximation of macroscopic drops:

$$\varphi_g = 4\pi r^2 \cdot \sigma(T) + g \cdot \left(I^0 - TS^0\right) \tag{3}$$

where $I^0$ and $S^0$ – enthalpy and entropy of condensed phase, $\sigma$ – surface energy, $r$ – the cluster (agglomerate) radius. In determining the surface energy is generally believed that at the interface, the molecules do not have enough for only one bond. This is not true for near-critical clusters, whose size is not large, and the number of missing bonds is greater than one. It makes adjustments to the value of the surface free energy, for example in Ref. [4] is proposed to use the Tolman's length for correction of the surface energy values. However, in current work to determine surface free energy of the gas-cluster interface an equation, based on earlier results [5], are proposed:

$$\sigma(T) = \left(\frac{\Delta z \cdot \mu}{z} \frac{\Delta H_{vap}(T) - pv}{N_A} - \frac{1}{k} \cdot \frac{q^2}{\pi \varepsilon_0 \cdot (2r_l)^3}\right) \cdot \frac{1}{\pi \cdot r_l^2} \tag{4}$$

$\Delta H_{vap}$ – latent heat of vaporization, $z$ – bulk coordination number of liquid, $\Delta z$ – number of missing bonds for surface particle (Figure 19.1), $r_l$ – equimolar particle radius, $q$ – dipole moment of molecules (Eq. (2) can be used for substances with polar molecules, for example, water), $k$ – coefficient taking into account the interaction of the dipoles. It is shown [2] that for most liquids in the limit of macroscopic drops $k = 3$, however, for small clusters $k \rightarrow 1$. It should be noted that the cluster radius $r$ and equimolar particle radius $r_l$ are related $r/r_l \sim g^{1/3}$, therefore after substituting (2) into (1) the chemical potential of the cluster is a function of the number of particles in it.

In the works of Zhukhovitskii [6] have shown that for small clusters to determine the equilibrium constant can be used Einstein solid model. In accordance to it the equilibrium constant of the reaction of cluster formation can be found as

$$K(g,T) = \frac{1}{\lambda^{3(g-1)} Z_C(g,T)} \left( \frac{k_B T}{101325} \right)^{g-1} \qquad (5)$$

**TABLE 19.1**  Number of Contacts per Particle in Argon-Like Cluster

| Number of particles in the cluster, $g$ | Average number of contacts per particle, $k(g)$ |
| --- | --- |
| 2 | 1/2 or 2/2 = 1 |
| 3 | 3/3 = 1 |
| 4 | 6/4 = 1.5 |
| 5 | 9/5 = 1.8 |
| 6 | 12/6 = 2 |
| 7 | 16/7 ≈ 2.286 |
| 13 (1st coordination sphere) | 42/13 ≈ 3.231 |
| 55 (2nd coordination sphere) | 224/55 ≈ 4.073 |
| 147 (3rd coordination sphere) | 666/147 ≈ 4.531 |
| 309 (4th coordination sphere) | 1500/309 ≈ 4.854 |
| 561 (5th coordination sphere) | 2870/561 ≈ 5.112 |
| 923 (6th coordination sphere) | 4884/923 ≈ 5.291 |
| 1415, 2057, 2869, 3871 (etc.) | – |
| ∞ (argon-like liquid) | 6 |

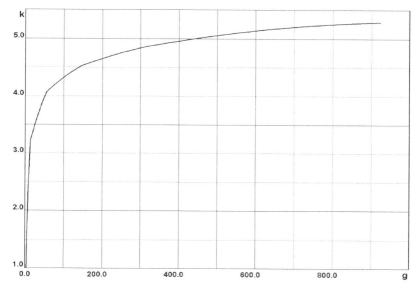

**FIGURE 19.1**    The average number of the particles bonds in argon-like cluster ($z = 6$) depending on its size.

where $\lambda = \sqrt{\dfrac{2\pi \cdot \hbar^2}{M\,k_B T}}$ – thermal wavelength, $\hbar = h/2\pi$ - the reduced Planck constant, $M$ – molecular mass, $Z_C(g,\,T)$ – partition function of the agglomerates of size $g$. The multiplier $k_B T/101325$ serves to bring the equilibrium constants of to one type – to the partial pressures.

The Einstein solid model considered that each particle is an oscillator having 3 degrees of freedom and interaction on several links with neighboring particles. To determine the partition function in general ($g \geq 3$) uses the following expression:

$$Z_C\left(g,T\right)=\frac{V}{\lambda^3}C_r\left(g\right)\left(\frac{a}{\lambda}\right)^3 C_v\left(g\right)\left(\frac{k_B T}{\hbar\omega_0}\right)^{3g-6}\exp\left(\frac{g\cdot k(g)\cdot D_0}{k_B T}\right) \qquad (6)$$

where $D_0$ – the depth of the potential well, $a$ – intermolecular distance corresponding to $D_0$, $r_0$ – the radius of the potential well, $V = 4\pi a^3/3$ – molecular volume, $\omega_0 = \left(2/r_0\right)\sqrt{D_0/M}$ – the oscillation frequency of the dimer, $k(g)$ – average number of the particles bonds in cluster, $C_r(g)$ and

$C_v(g)$ – form factors determined according to the close-packed structure of Einstein solid model $C_r(3) = C_r(4) = 2\pi^2/3$, $C_v(3) = 4/3\sqrt{2}/3$, $C_v(4) = \sqrt{2}$.

The interaction of molecules is described by the short-range potential [3, 4]:

$$u(r) = \begin{cases} +\infty & r < a - r_0 \\ \left(M\omega_0^2/4\right)(r-a)^2 - D_0, & a - r_0 \le r \le a + r_0 \\ 0 & r > a + r_0 \end{cases} \quad (7)$$

By determining the short-range potential (5) considered that the length of the parameters $a$ and $r_0$ satisfy $a/r_0 \gg 1$ (Figure 19.2).

The parameters of the short-range potential $a$ and $D_0$ can be taken from the Lennard-Jones potential: $a \approx \sigma$, $D_0 \approx \varepsilon$. At the same time, these parameters can be determined from the thermophysical properties of the material:

$$a = \sqrt[3]{\frac{3\mu}{4\pi \cdot \rho \cdot N_A}}, \quad D_0 = \frac{\mu \cdot \Delta H_{vap}}{z \cdot N_A} \quad (8)$$

From the known equilibrium constant of formation of agglomerates can be obtained quasi-equilibrium cluster size distribution using the method of calculation of equilibrium composition, for example [7]. However, use of this technique is only possible in subcritical zone clusters [8], since when supercritical clusters disturbed the equilibrium distribution.

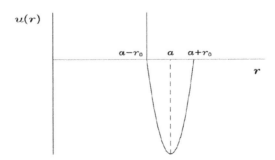

**FIGURE 19.2**    Model of short-range potential [6].

Thus, following the ideas of Frenkel [2] and Zhukhovitskii [6], it is proposed to determine the partial pressure of the agglomerates used in the case of small clusters (no more than 13 particles) Einstein solid model and in the case of larger clusters – droplet model using chemical potential (3) and the expression for surface free energy (4). In the range where there is a transition from the small to critical clusters size equilibrium constant defined by the formula, in which its value is smallest. Figure 19.3 shows the values of the equilibrium constant for the small droplets and small agglomerates of $H_2O$.

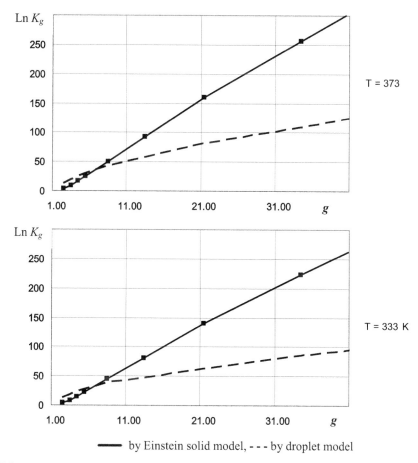

— by Einstein solid model, - - - by droplet model

**FIGURE 19.3**   Equilibrium constant of water clusters by size.

The graphs show that, depending on the temperature of the transition from the small agglomerates to small droplet passes at about the same size of the cluster: by g ≤ 5 the equilibrium constant by Einstein solid model is smaller, but by g > 8 – by droplet model.

## 19.3 AGGLOMERATES OF PARTICLES NEAR THE SATURATION LINE

Using the proposed method of calculation of the equilibrium constants of reactions of formation of agglomerates of particles, it is possible to determine their partial pressure and the assumption that a mixture of various agglomerates and monogas satisfies the equation of state of an ideal gas, we can determine the properties of the mixture, such as the total pressure and molar mass:

$$P = \sum_{g=1}^{n} P_g \tag{9}$$

$$\mu = \frac{1}{P}\sum_{g=1}^{n} \mu_g P_g = \frac{\mu_1}{P}\sum_{g=1}^{n} g P_g \tag{10}$$

Most interesting is the account of agglomerates near the saturation line, when the vapor has a significant amount of agglomerates of particles, resulting in a difference in properties of a real gas from ideal gas representation.

By calculation the pressure and molar mass of saturated water vapor (Figures 19.4 and 19.5) [8] short-range potential parameters determined by the expression (8) wherein $D_0/k_B$ = 1086.9 K, $a$ = 1.953 Å. By calculation of the saturated vapor pressure of carbon dioxide (Figure 19.6) and benzene (Figure 19.7) the short-range potential parameters determined also by the expression (6) wherein for carbon dioxide $D_0/k_B$ = 306.1 K, $a$ = 2.455 Å, and for benzene – $D_0/k_B$ = 616.9 K, $a$ = 3.348 Å. By calculation the pressure of water vapor and carbon dioxide used parameter $r_0$ = 0.049$a$ and for benzene $r_0$ = 0.015$a$, that satisfies $a/r_0 \gg 1$ [6].

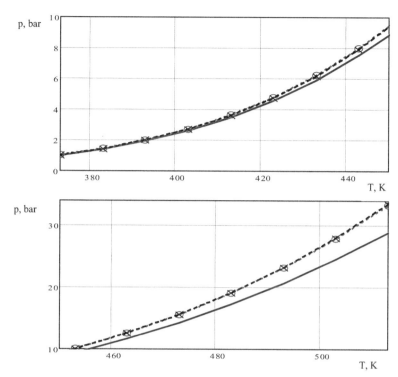

**FIGURE 19.4** Saturated vapor pressure of water (theoretical value (chemical equilibrium, Eq. 9) – solid line, experimental – x, calculated taking into account the small agglomerates – o).

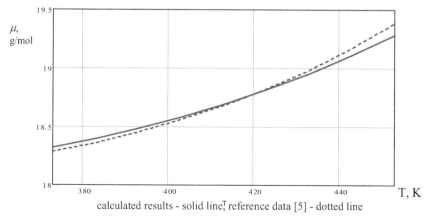

**FIGURE 19.5** The molar mass of the saturated water vapor (calculated results – solid line, reference data [5] – dotted line).

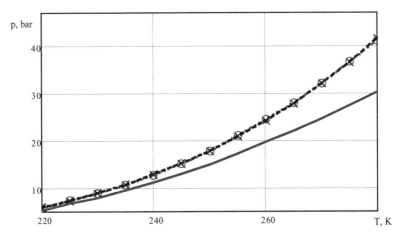

**FIGURE 19.6** Saturated vapor pressure of carbon dioxide (theoretical value (chemical equilibrium, Eq. 9) – solid line, experimental – x, calculated taking into account the small agglomerates – o).

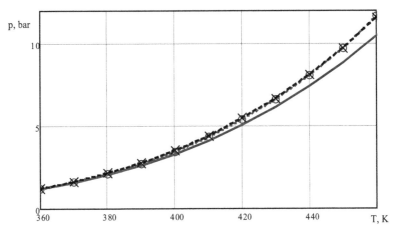

**FIGURE 19.7** Saturated vapor pressure of benzene (theoretical value (chemical equilibrium, Eq. 9) – solid line, experimental – x, calculated taking into account the small agglomerates – o).

The theoretical value of the saturated vapor pressure $p_s$ is calculated from the equilibrium condition of the chemical potentials of the gas and condensed phase:

$$\ln p_s = \frac{S_g^0 - S_z^0}{R_0} - \frac{I_g^0 - I_z^0}{R_0 T} \tag{11}$$

where $I^0$, $S^0$ – standard enthalpy and entropy, $R_0$ – universal gas constant, z – condensed phase, g – gas.

Calculations based on the proposed model are in good agreement with the experimental data in a wide range of temperatures. Accuracy of calculations for given temperature was not more than ±1% for water and benzene and ±2% for carbon dioxide.

Figure 19.8 shows the values of partial pressures of various agglomerates in the saturated vapor of water depending on the temperature. We see that with increasing size of the agglomerates their partial pressures decrease rapidly. Increasing the partial pressure of the agglomerates with rising temperature associated with general increase in the saturated vapor pressure

The results show that for the determination of the saturated vapor pressure and molar mass is sufficient to take into account the size of the agglomerates of particles up to 8, as partial pressures of the agglomerates with the size quite quickly fall.

## 19.4  NUCLEATION

As the temperature of vapor decreases below the saturation they go into a state of supersaturation and at some point in them having nuclei of a new phase.

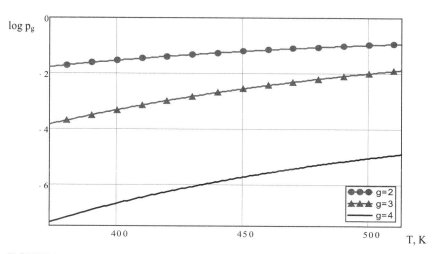

**FIGURE 19.8**   Partial pressure of small agglomerates depending on the temperature.

The reason for the transition from one phase to another can be illustrated by a graph of the chemical potential of each of the phases as a function of temperature at a constant pressure. At the temperature T, above the equilibrium $T_s$ (Figure 19.9), the first phase (liquid, l) has large chemical potential than the second (gas, g), and at a temperature lower $T_s$, on the contrary, the first phase has a lower chemical potential, i.e., at a temperature above $T_s$, the liquid will pass into the gas, and vice versa.

At the same time, the chemical potential of the liquid in this case corresponds to the case when a molecule (particle) liquid is surrounded with the same molecule, i.e., when the influence of surface effects can be neglected. However, by homogeneous nucleation of a new phase surface effect cannot be neglected. Accounting for surface effect leads to an increase of the chemical potential.

In addition, the formula (3) generally illustrates a graph of the Gibbs energy change during the formation of the cluster (nucleus condensed phase) on the cluster size – its radius or number of particles in it (Figure 19.10). In classical nucleation theory [3] noted that the by small size of the clusters the greatest contribution to the sum (3) makes the term responsible for surface energy $4\pi R_g^2 \cdot \sigma$, and thus, on a plot there is the area in which $\Delta\varphi > 0$. This area has a maximum, which is in the classical nucleation theory mapped to "critical" cluster having a size $r_c$ and consisting of $g_c$ particles. This cluster

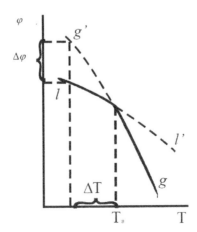

**FIGURE 19.9**   Dependence the chemical potential on temperature at $p=const$.

is considered to be capable of further growth, as its increase is accompanied by a decrease in $\Delta\varphi$.

In general, knowing $\Delta\varphi$ and equilibrium constants of reactions of formation of agglomerates (clusters) can get their quasi-equilibrium distribution of clusters, using the method of calculation of the equilibrium composition [7]. However, using this method it is possible only for the clusters in the subcritical zone [9]. Since at occurrence of supercritical clusters the equilibrium distribution is disturbed and it is replaced with a stationary distribution (Figure 19.11) [9].

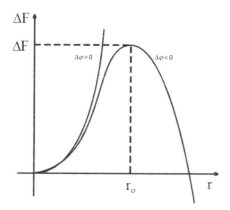

**FIGURE 19.10**    Gibbs energy change $\Delta F$ during the formation of the cluster.

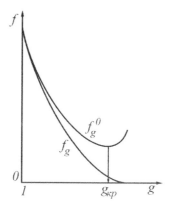

**FIGURE 19.11**    The cluster size distribution: the equilibrium $f^0$ and stationary $f$ [9].

To determine the rate of transition from cluster size greater $g_{cr}$ from the gas phase to the condensed necessary to consider a system of equations that describes the kinetics of the formation of a new phase. This system was first recorded by Zeldovich [10]:

$$\frac{d\tilde{n}_g}{dt} = \tilde{n}_{g-1}v^+_{g-1} + \tilde{n}_{g+1}v^-_{g+1} - \tilde{n}_g(v^+_g + v^-_g) \qquad (12)$$

where $\tilde{n}_g$ – the concentration of clusters of size $g$, $v^+_g, v^-_g$ – the rate of condensation and evaporation, respectively. When writing the Eq. (12) the transformation of the cluster is considered as a chain of randomly alternating acts of condensation and evaporation of single particles. Thus a simultaneous change in two or more particle is considered unlikely and not included [10]. Thus, the concentration of clusters g size may increase due to the condensation of particles in the cluster size of g–1 (the first term) or by evaporation of a particle with a cluster size of g + 1 (second term), and reduced by switching the cluster size g into clusters sizes g–1 and g + 1 during evaporation or condensation of particles respectively.

The main problem is the point that in the direction of increasing $\Delta\varphi$ in accordance with the basic provisions of the thermodynamic process cannot go. Thus, increasing the subcritical cluster on one particle makes it less stable, which should result in a next time to reduce it to one particle.

In this regard it is proposed to use the idea of Smoluchowski [11], which in the theory of rapid coagulation considered the formation of agglomerates as follows: first single primary particles collide and form double particles; then double face similar and form a quaternary, etc., i.e., formed by a number of agglomerates consisting of 2, 4, 8, 16, etc. particles. The increase of agglomerates is primarily due to the collision of particles of similar size. Another example of such a series is the Fibonacci series, wherein each subsequent row number is formed by adding the two previous series of numbers: 1, 2, 3, 5, 8, 13, 21, 34, 55, etc. This process can be described as follows:

$$g_n = g_{n-1} + g_{n-2}$$

Figure 19.12 [12] shows the dependence of the partial pressures of the agglomerates in supercooled water vapor. In the case of occurrence of supercritical clusters is necessary to check the possibility of condensation, and if the rate of formation of clusters or their concentration is negligible, to exclude from the calculation of the maximum cluster of quasi-equilibrium composition and to continue without him.

Coagulation rate per unit volume per unit of time, suggesting that the formation of near-critical clusters is due to the collision of two sufficiently large variety of agglomerates can be written as:

$$I_g = k n_{g-1} n_{g-2} \tag{13}$$

where $n$ – the concentration of particles, $k$ – coagulation rate constant, which characterizes the probability of convergence and equal to:

$$k = 8\pi R D \tag{14}$$

where $R = R_{g-1} + R_{g-2}$ – the cluster radius, $D = D_{g-1} + D_{g-2}$ – the diffusion coefficient, which is the sum of the diffusion of two colliding clusters form another larger. To determine the diffusion coefficient can be used for the known dependence of the binary diffusion coefficient [13].

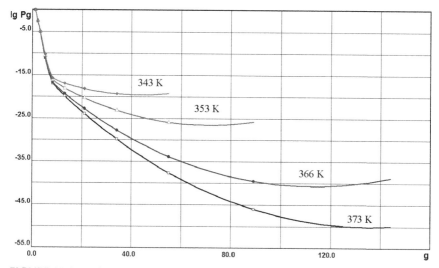

**FIGURE 19.12**    The partial pressure of agglomerates in the water vapor during supercooling.

## 19.5  FLOW WITH CONDENSATION IN SUPERSONIC NOZZLES

Simulation of the vapor flow in a supersonic nozzle was made in one-dimensional formulation in the variables density – temperature and based on a system of equations from [14]. In developing the mathematical model has adopted the following assumptions:

- until a sufficient number of supercritical cluster not appear, their growth is neglected, and after their formation the nucleation process will be terminated and considered only the growth of supercritical nuclei. This assumption is generally accepted [15];
- used model of single-speed single-temperature flow due to the small size of the nascent phase, whereby the heat of condensation is related to the density of the entire stream (equal to the initial density of the gas phase) and the heat capacity of the medium is assumed constant and equal to the gas and condensed phase.

The system of equations for determining the flow parameters before particle growth, i.e., until a sufficient amount of supercritical cluster contains the following equation:

- for the number (concentration) of supercritical clusters

$$\frac{dN}{dt} = I_g \tag{15}$$

- for the mass of supercritical clusters

$$\frac{dm}{dt} = 0 \tag{16}$$

- for the density of the gas phase

$$\frac{d\rho}{dt} = -\frac{M^2}{M^2 - 1} \cdot \frac{\rho}{F(x)} \cdot \frac{dF}{dx} \cdot V - I_g \cdot g_{cr} \cdot m_1 \tag{17}$$

- for the temperature

$$\frac{dT}{dt} = -\frac{(k-1)M^2}{M^2 - 1} \cdot \frac{T}{F(x)} \cdot \frac{dF}{dx} \cdot V + I_g \cdot g_{cr} \cdot m_1 \cdot \frac{\Delta H_{cond}}{\rho_0 C_V} \cdot \frac{k_{sv}}{z} \tag{18}$$

- for the flow velocity

$$\frac{dV}{dt} = \frac{1}{M^2 - 1} \cdot \frac{V}{F(x)} \cdot \frac{dF}{dx} \cdot V \qquad (19)$$

- the current position of the nozzle

$$\frac{dx}{dt} = V \qquad (20)$$

where $m$ – the mass of the cluster, which is after the formation of a sufficient amount of supercritical clusters and at the beginning of their growth can be assumed to be equal weight of the supercritical cluster $m_0 = m_1 \cdot g_{cr}$, or with the gradual accumulation of supercritical clusters of different sizes possible, for example, on rapid cooling (supersonic flow in the nozzle) – $m_0 = (\rho_0 - \rho) / N$, $m_1 = \dfrac{\mu_1}{1000 \cdot N_A}$ – weight monoparticles of gas, kg, $\mu_1$ – molar mass of monogas, g/mol. Calculations show that in the flows in the nozzles condensation rate increases very rapidly, so that most of the initial nucleation of a new phase is critical clusters corresponding thermodynamic conditions at the time of transition to the calculation of their growth, i.e., mass supercritical clusters obtained from different conditions are practically the same.

The equation for the density is considered the mass removal in the formation of clusters of supercritical that negligible almost to the dew point because of the very low values of the rate of condensation $I_g$. In the equation for the temperature was added a term allowing heat during the formation of clusters of supercritical, thus due to the fact that the small clusters (g <300), the number of bonds between the particles are much smaller than their number in the fluid is considered in terms of the factor $k_{sv}/z$ [5].

In the second step calculates the particle growth of a new phase, with the formation of new clusters supercritical terminated primarily because of lower supersaturation. The system of equations is as follows:

- for the number (concentration) of supercritical clusters

$$\frac{dN}{dt} = 0 \qquad (21)$$

- for the mass of supercritical clusters (growth in a free molecular regime [10])

$$\frac{dm}{dt} = 4\pi \cdot r_{cr}^2 \frac{p_1 - p_s}{\sqrt{2\pi \frac{R_0}{\mu_1} T}}, \quad r_{cr} = r_1 \cdot \sqrt[3]{\frac{3}{4\pi} \cdot \frac{m}{m_1}} \tag{22}$$

- for the density of the gas phase

$$\frac{d\rho}{dt} = -\frac{M^2}{M^2 - 1} \cdot \frac{\rho}{F(x)} \cdot \frac{dF}{dx} \cdot V - N \cdot \frac{dm}{dt} \tag{23}$$

- for the temperature

$$\frac{dT}{dt} = -\frac{(k-1)M^2}{M^2 - 1} \cdot \frac{T}{F(x)} \cdot \frac{dF}{dx} \cdot V + N \frac{dm}{dt} \cdot \frac{\Delta H_{cond}}{\rho_0 C_V} \tag{24}$$

The equations for the velocity (19) and the current position of the nozzle (20) remain unchanged.

In all equations $F(x)$ – the cross sectional area of the supersonic nozzle (the conical nozzle is a linear function of $x$), $\frac{dF}{dx}$ – derivative, for ease of calculation determined analytically.

At each step time integration of well-known thermodynamic parameters of state $\rho$, $T$ the composition of the vapor is calculated taking into account the subcritical clusters (supercritical thus excluded from the calculations).

One of the most important problems is to determine the number of supercritical clusters, after which the can begin examining their growth. For this was made a simulation of the cooling process of water vapor from $T_{st} = 373$ K, which corresponds to the boiling point. The system of equations describing the condensation process in this case is reduced, it is removed from the equations and terms relating to flow by the nozzle, and the equation for the temperature term is added for the removal of energy in the form of $-Q/C_V/\rho_0$.

Figure 19.13 shows a plot of temperature versus time for the homogeneous condensation of water vapor $H_2O$. The threshold value of the concentration of supercritical clusters was assumed to be $10^{12}$, $10^{13}$, $10^{14}$ $M^{-3}$.

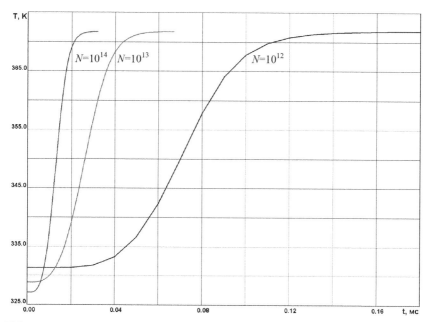

**FIGURE 19.13**   Temperature of water vapor by condensation.

Calorification of condensation after drops rise leads to an increase in temperature. At this temperature curve has two inflection points. Initially, this is due to the fact that the critical nucleation size is very small, whereby the mass flow of steam condensing on their surface area is limited. At the end of the process – this is due to a decrease in supersaturation to 1, so that the mass flow of condensing steam drops to 0. Nevertheless, it is important that the whole process takes about 0.1 ms for $N = 10^{12}$ m$^{-3}$ and about 0.02 ms $N = 10^{14}$ m$^{-3}$.

Figure 19.14 shows a similar temperature versus time for a homogeneous condensation of the vapor of carbon dioxide $CO_2$. The threshold value of the concentration of supercritical clusters was assumed to be $10^{10}$, $10^{12}$, $10^{14}$ m$^{-3}$.

Figures 19.15 and 19.16 shows the distribution of temperature and pressure along the length of the nozzle according to accepted sufficient supercritical clusters $N = 10^{10} \div 10^{14}$ m$^{-3}$, the radius of the nozzle throat 0.01 m, half-angle of the cone 5.71°, parameters of vapor are given corresponding to saturation point of T = 393 K, $\rho = 1,12$ kg/m$^3$.

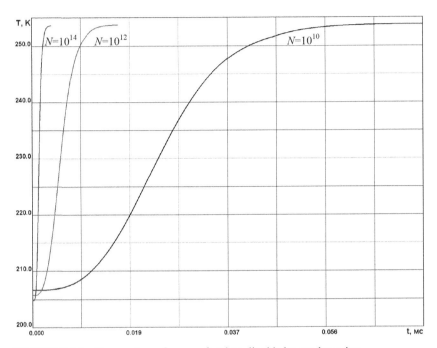

**FIGURE 19.14**    Temperature of vapor of carbon dioxide by condensation.

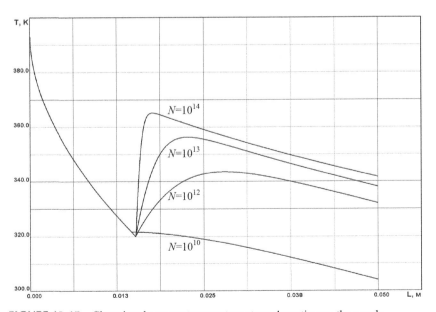

**FIGURE 19.15**    Changing the vapor temperature at condensation on the nozzle.

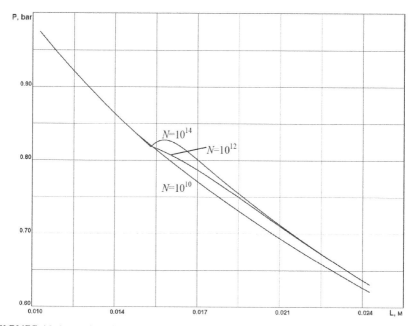

**FIGURE 19.16**   Changing the vapor pressure at condensation on the nozzle.

Figure 19.17 shows the experimental data on the condensation of water vapor in the nozzles [16]. We see good qualitative agreement of the results for $N=10^{12} \div 10^{14}$ m$^{-3}$.

The results of calculations show that the threshold number of clusters of supercritical $N = 10^{10}$ their number is insufficient, so that the maximum temperature in the graph is weakly expressed, and the pressure curve inflection caused by condensation, virtually invisible, which disagrees with the known experimental data.

Figure 19.18 shows the change in temperature, the degree of supersaturation and the size of supercritical the cluster over time for supersonic flow of nozzle at $N = 10^{14}$ m$^{-3}$.

Additional explanations to graphs listed in Table 19.2. The results show that as the temperature decreases the vapor supersaturation increases, and with it decreases the size of the supercritical cluster. By decreasing the size of the supercritical cluster increases their rate of formation, which is associated with an increase in the rate of diffusion, as well as the fact that the concentration of the smaller clusters is much higher. Especially noticeable

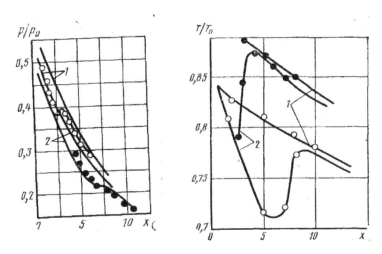

**FIGURE 19.17** Experimental data on the distribution of the relative pressure and temperature on the relative length of the conical nozzle [16] ($r = 1$ cm: $\circ - p_0 = 4{,}9 \cdot 10^5$ Pa, $T_0 = 442$ K, $T_S = 415$ K; $\bullet - p_0 = 2 \cdot 10^5$ Pa, $T_0 = 445$ K, $T_S = 370$ K; 1 – equilibrium flow; 2 – nonequilibrium flow).

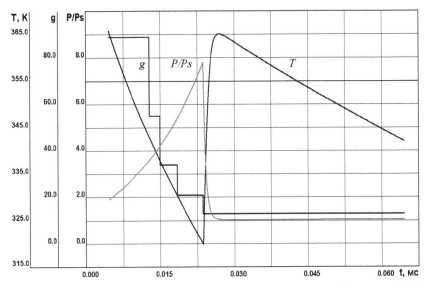

**FIGURE 19.18** Temperature, degree of supersaturation and the size of the supercritical cluster over time for supersonic flow of nozzle.

**TABLE 19.2**   The Rate of Formation and the Concentration of Supercritical Clusters at the Nozzle

| Number of particles in the cluster, $g$ | The rate of formation of supercritical clusters, $I_g$, 1/(m³·s) | The concentration of supercritical clusters, $N$, 1/m³ | The degree of supersaturation, $P/P_s$ |
|---|---|---|---|
| 89 | $1.60 \times 10^{-7}$ | $1.60 \times 10^{-16}$ | 1.84 |
|  | $1.05 \times 10^{9}$ | $2.60 \times 10^{2}$ | 3.60 |
| 55 | $4.86 \times 10^{9}$ | $2.64 \times 10^{2}$ | 3.60 |
|  | $6.50 \times 10^{11}$ | $2.90 \times 10^{5}$ | 4.21 |
| 34 | $2.72 \times 10^{12}$ | $2.93 \times 10^{5}$ | 4.21 |
|  | $1.54 \times 10^{14}$ | $1.30 \times 10^{8}$ | 5.40 |
| 21 | $5.54 \times 10^{14}$ | $1.31 \times 10^{8}$ | 5.40 |
|  | $9.38 \times 10^{15}$ | $1.62 \times 10^{10}$ | 7.78 |
| 13 | $1.92 \times 10^{21}$ | $1.93 \times 10^{12}$ | 7.78 |
|  | $1.95 \times 10^{21}$ | $1.01 \times 10^{14}$ | 7.81 |

*Note*: for one cluster size the top row of values corresponds to the beginning stair in the graph (Figure 19.18), and the second – the end of the stair.

difference in the rate of formation of supercritical clusters (5–6 orders of magnitude) in the transition from a particle size of 21 to 13 at almost the same supersaturation.

From the analysis of the size of the supercritical cluster graphics and data of Table 19.2 it is shown that even if you define the size of the critical cluster of up to one particle that does not lead to a significant change in the condensation beginning, i.e., the moment of growth of supercritical nuclei.

At calculations the condensation in the nozzle with different thresholds concentration of supercritical clusters following values were obtained for size of formed by condensation drops: $r_{cl} = 3.16$ μm at $N = 10^{12}$ m⁻³, $r_{cl} = 1.55$ μm at $N = 10^{13}$ m⁻³ and $r_{cl} = 0.72$ μm at $N = 10^{14}$ m⁻³. From the obtained results it is evident that the change in radius of the droplet is related to the change in their number – when changing N is ten times the droplet radius varies about $\sqrt[3]{10} = 2,15$ times.

Figure 19.19 shows graphs of temperature on the reduced length ($L/R_{cr}$) for nozzles of different sizes – with radii of the nozzle throat

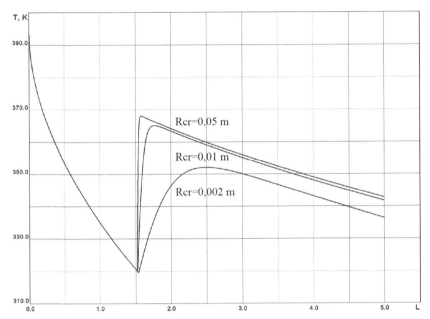

**FIGURE 19.19**    Temperature at condensation of water vapor in the nozzles of different sizes.

$R_{cr}$ = 0.002; 0.01; 0.05 m and a threshold number of supercritical clusters $N$ = $10^{14}$ m$^{-3}$. It is known that the geometric similarity nozzles, gas-dynamic parameters of in them are the same [14], but this does not apply to the flows with the physical and chemical transformations, as in this case a very important role is played by the total process time. The graphs show that for very small nozzles $R_{cr}$ = 0.002 m and a length of the supersonic nozzle $L$ = 0,01 m process of condensation is more elongated by the relative length of the nozzle, and, obviously, with even smaller nozzles condensation process may not be completed in the nozzle.

In the literature [3, 15] as a criterion for the start of a new phase growth process is proposed to use the rate of formation of supercritical clusters, however, from our point of view, the most appropriate is the use the concentration of supercritical clusters per unit volume, because when rapid processes, such as supersonic flows in nozzles, the residence time may not be enough to accumulate the required number of nuclei of a new phase. For supersonic nozzles, despite their geometric and gas-dynamic similarity, the physical size will play a significant value.

Furthermore, in Ref. [15] on the basis of analysis of the results of numerical studies of the condensation of water vapor at flow in a supersonic nozzle concludes adequacy determining the condensation point temperature to within ±5°. The our data (Figure 19.15 and Table 19.2) shows that the change in the threshold concentration of supercritical clusters from $10^{10}$ to $10^{14}$ leads to a change in temperature of the beginning of condensation of about 2 degrees. Thus, the use of a threshold concentration of supercritical clusters allows to determine the temperature of the beginning of condensation with sufficient accuracy.

Figure 19.20 shows the change of temperature during the condensation of the carbon dioxide vapor in the nozzles with different half-angle of the cone: 14 and 16.7 degrees. If the angle is greater than the temperature decreases faster, respectively, condensation begins earlier.

In Ref. [16] it was carried out a generalization of large number of experimental data of different authors. Figure 19.21 shows a comparison

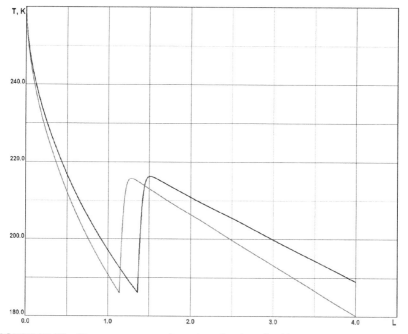

**FIGURE 19.20**    Temperature at condensation of carbon dioxide vapor in the nozzles with different half-angle of the cone.

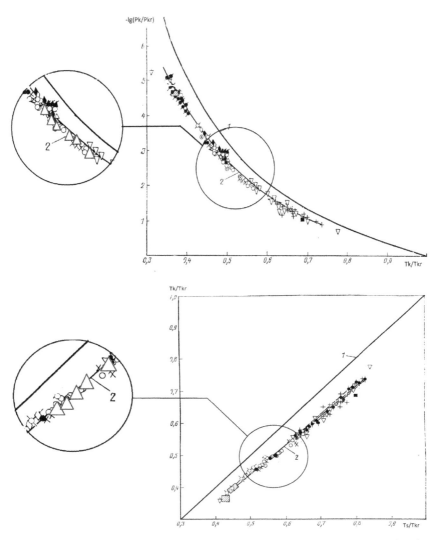

**FIGURE 19.21** Generalized line of Wilson (1 – saturated vapor line, 2 – approximation of experiments on water vapor and △ our results).

of our results for the water vapor with the experimental data [16]. On the graph of the results of temperature and pressure, at which the condensation of saturated vapor begins, reduced to a dimensionless form a relatively the critical parameters of a substance.

## 19.6 CONCLUSION

The proposed model for calculating the saturated vapor pressure taking into account the existence of small agglomerates gives values close to the experimental data.

The formation of supercritical clusters occurs through the merger of subcritical clusters. As criterion of condensation beginning is proposed to use not condensation rate, but the concentration of supercritical clusters. According to the comparison of the results with experimental data for water and carbon dioxide that number ranges: $N=10^{12} \div 10^{14}$ m$^{-3}$.

The numerical study of the condensation process in moving flow with the formation of clusters showed the adequacy of the proposed mathematical model.

## KEYWORDS

- cluster
- coagulation
- homogeneous condensation
- mathematical model
- nuclei
- numerical simulation
- saturated vapor
- small agglomerates
- supercooling
- supersonic nozzle

## REFERENCES

1. Vukalovich, M. P., & Novikov, I. I. Uravnenie sostoyania realnykh gasov. Moscow, 1948. 340 p. (in Russian).
2. Frenkel, Ya. I. Kineticheskaya teoriya zhidkostey. Leningrad: Nauka. 592 p. (in Russian).
3. Anisimov, M. P. Nucleation: Theory and Experiment. Russian Chemical Reviews. 2003, Vol. 72. № 7, pp. 591–628.

4. Zhukhovitskii, D. I. Surface Tension of the Vapor-Liquid Interface With Finite Curvature. Colloid Journal, 2003, Vol.65, № 4, pp. 440–453.
5. Korepanov, M. A. Raschet koeffizienta poverkhnostnogo natyazhenia. Vestnik IzhGTU, 2006, №1, pp. 6–9 (in Russian).
6. Zhukhovitskii, D. I. Hot clusters in supersaturated vapor. Progress in Physics of Clusters, eds. G. N. Chuev, V. D. Lakhno, and, A. P. Nefedov, World Scientific Publ., Singapore, 1998, pp. 71–101.
7. Termodynamicheskie i teplofizicheskie svoistva produktov egorania. Moscow: VINITI, 1971, Vol. 1. (in Russian).
8. Korepanov, M. A., & Gruzd' S. A. The Calculation of the Saturated Vapor Pressure Including Small Agglomerates. Khimicheskaya fizika i mesoskopiya, 2013, Vol. 15, №2, pp. 223–230. (in Russian).
9. Skripov, V. P., & Koverda, V. P. Spontannaya kristallizacia pereokhlazhdennekh zhidkostey. Moscow: Nauka, 1984, 232 p. (in Russian).
10. Zeldovich, Ya. B. Izbrannye trudy. Khimicheskaya fizika i gidrodinamika. Moscow, 1984. 374 p. (in Russian).
11. Smoluchowski coagulation equation. https://en.wikipedia.org/wiki/Smoluchowski_coagulation_equation.
12. Korepanov, M. A., Gruzd' S. A. Modeling Homogeneous Condensation Considering the Quasiequilibrium Concentration of Small Agglomerates. Khimicheskaya fizika i mesoskopiya, 2014, Vol. 16. № 1, pp. 63–67. (in Russian).
13. Reid, R. C., Prausnitz, J. M., & Sherwood, T. K. The properties of gases and liquids. 1977, 688 p.
14. Barilovich, V. A. Osnovy termogazodinamiki dvukhfaznykh potokov i ikh chislennoe reshenie. S.-Petersburg, 2009. 425 p. (in Russian).
15. Gidaspov, V. Yu., Pirumov, U. G., Ivanov, I. E., Severina, N. S. Modeli obrazovaniya nanochastiz v potokakh gaza. Moscow, 2011, 214 p. (in Russian).
16. Gorbunov, V. N., Pirumov, U. G., Ryzhov Yu.A. Neravnovesnaya kondensazia v vysokoskorostnykh potokakh gaza. Moscow, 1984, 200 p. (in Russian).

# CHAPTER 20

# SIMULATION OF PHASE TRANSFORMATIONS IN THE FE-CR ALLOYS AT HIGH TEMPERATURES

N. V. GONCHAROVA and T. M. MAKHNEVA

*Institute of Mechanics of the Ural Branch of the Russian Academy of Sciences (IM UB RAS), Izhevsk, Russia*

## CONTENTS

### 20.1 INTRODUCTION

Nitrogen alloying of iron-chromium steels is a promising trend with a view to the formation of nitrous austenite in their structures, which is much more stable than carbonic austenite. Numerous investigations have been conducted and technologies have been developed, wherein ammonia, nitrogen and other nitrogen-containing gases and mixtures are used as a nitrifying medium. Despite the unquestionable scientific and practical interest to the problem of choosing ecological and economical gas medium for nitrification, the information on investigations of the solubility of air nitrogen is absent in the Russian and foreign literature. Usually, the influence of the air on steels at high-temperature holding is considered only with a view to the investigation the surface oxidizing process. However, it should be taken into account that the main component of the air is nitrogen; therefore it is practical to consider the air not only as an oxidizing medium but also as a nitrifying medium.

The objective of the present paper is the investigation of the phase and chemical transformations in the Fe-Cr alloys in the air at high temperatures.

## 20.2 EXPERIMENTAL RESEARCHES

30 μm thick H15 alloy foils (chemical composition: Fe-85%, Cr-14.94%, N-0.03%, C-0.028%) with a mass $m_s = 1 \times 10^{-4}$ kg ($\Delta m_s = \pm 0.1 \times 10^{-6}$ kg) were placed into sealed quartz vessels of a certain size. The error in the measuring of the vessel volume was $\Delta V = \pm 0.1 \times 10^{-6}$ m$^3$. The pressure in the vessels was controlled with a pressure gauge with an accuracy $\Delta P = \pm 0.1 \times 10^{-3}$ MPa. The heat treatment of the specimens was conducted at the temperature 1000°C ($\Delta T = \pm 5$°C) for 1 h; after that they were water-quenched.

The air content ($m_g$) in the vessels was varied by the change of the ratio of the gaseous phase mass to the solid mass ($m_g/m_s$): *variant I* – the air pressure ($P_a$) in the range of 0.001–0.1 MPa at the constant vessel volume ($V = 22.5 \times 10^{-6}$ m$^3$); *variant II* – the volume of the vessels ($V_p$ was varied in the range of $22.5 \times 10^{-6}$–$2.25 \times 10^{-6}$ m$^3$ at the constant air pressure (0.1 MPa).

The construction of thermodynamic models was performed using the method [1]. The correctness of the models constructed was assessed using the following methods: the X-ray phase analysis (XPA), Mössbauer effect study, X-ray photoelectron spectroscopy (XPS), and reducing fusion in the gas-carrier medium.

## 20.3 RESULTS AND DISCUSSION

The experimental data on the influence of the air pressure 0.001–0.1 MPa on the phase composition of the H15 ferrite alloy at the temperature 1000°C are given in the Table 20.1. It is shown that after the heat treatment

**TABLE 20.1** INFLUENCE of the Air Content on the Phase Composition of the H15 Alloy at Heat Treatment

| Variant | Initial conditions | | | Phase composition of quenched speci mens (XPA, MES, XPS) | | | |
|---------|---------|---------|---------|---------|---------|---------|---------|
| | $m_g/m_s$ | P, MПa | $V\times10^{-6}M^3$ | | | | |
| | 0.003 | < 0.001 | 22.5 | α-Fe(Cr) | – | – | – |
| I | 0.06 | 0.02 | 22.5 | α-Fe(Cr) | γ-Fe(Cr) | $Cr_2O_3$ | – |
| | 0.18–0.3 | 0.06–0.1 | 22.5 | α-Fe(Cr) | – | $Cr_2O_3$ | FeO |
| II | 0.03–0,018 | 0.1 | 2.25–9 | α-Fe(Cr) | – | $Cr_2O_3$ | $Fe_xCr_{3-x}O_4$ |

of the specimens (variant I) the phase composition does not change at the air pressure smaller than 0.001 MPa; however, in the air pressure range of 0.001–0.1, oxide phases are formed on the surface of the specimens. At the air pressure 0.02 MPa, Mössbauer spectra of the quenched specimens show a central paramagnetic line, which indicates that retained austenite (~30%) is present. The obtained $\gamma_{retained}$ is stable down to the temperature $-70°C$; heating and holding at 650°C for 27 h do not lead to the complete decomposition of this phase ($\gamma_{retained}$ ~5%) [2].

For the investigation of the reasons for the appearance and stabilization of the γ-phase in the H15 ferrite alloy in the air at the temperature 1000°C, a method for investigating phase transformations in iron-chromium alloys has been developed, which comprises the construction of a thermodynamic model, the calculation of the equilibrium composition, the estimation of the nitrogen solubility, the conformation of the calculation results with the experiment (Figure 20.1) [3].

A closed heterogeneous system "alloy H15 – air" has been considered as a model system; it has the following composition (mass%): 85%Fe and 15%Cr in the solid phase, and 76%$N_2$ and 24%$O_2$ in the gaseous phase. Three variants of the model are considered. In the first variant, the formation of solid solutions is not taken into account; in the second variant, the possibility of the solid solution formation (1 – Fe-Cr and 2 – FeO, $Fe_2O_3$, $Fe_3O_4$, $Cr_2O_3$, $CrO_3$) is assumed; in contrast to the second variant, in the third variant the possibility of the formation of the chromium nitrides CrN and $Cr_2N$ is precluded. The results of the calculations of the second and third variants of the model are shown in Figure 20.2a-c.

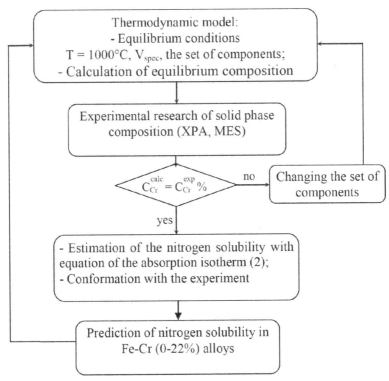

**FIGURE 20.1**   The algorithm of the thermodynamic simulation of the "Fe-Cr (alloy) – air" system.

The simulation results were compared with the XPA data (the presence of oxides and chromium nitrides was determined qualitatively) and MES data (the chromium content in the α-solid solution was determined quantitatively). The chromium content in the quenched specimens was determined by the value of the medium superfine field on the nucleus $\overline{H}$ (kE) from the equation [2, 3]:

$$C_{Cr}\% = 114.81 - 0.35H \qquad (1)$$

It is shown that the more exact description of the model is given by the third variant (Figure 20.2b,c), from which it follows that at the temperature 1000°C at any air content, oxygen completely interacts with the alloy forming oxides; in the gaseous phase only nitrogen is present. The value of the nitrogen pressure (Figure 20.2c) and the content of iron and

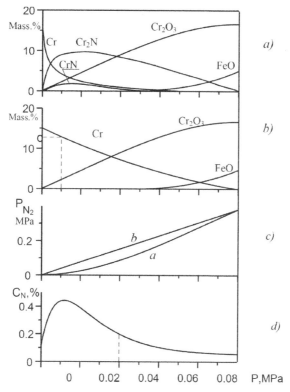

**FIGURE 20.2** Dependences of the equilibrium compositions of the solid (*a, b*) and gas phases (*c*) and nitrogen solubility (*d*) on the pressure of the air in the system "alloy H15 – air" at the temperature 1000°C (variant I): (*a*) a model with the formation of solid solutions, (*b*) a model without the CrN and Cr₂N formation.

chromium oxides (Figure 20.2*b*) are determined by the pressure of the air. Chromium nitrides are not formed in the alloy, and the interaction of the gaseous nitrogen with the Fe-Cr solid solution is only the stage of the diffuse saturation of the g-solid solution.

The nitrogen amount in the g-phase has been calculated with the use of the equation of the absorption isotherm [4]:

$$[C_N\%] = K_N \times \sqrt{;P}_{N2} \times 10 \times (eN^{Cr\cdot} C_{Cr}\%)$$

where $P_{N2}$ is the nitrogen pressure in the reaction volume at 1000 °C (Figure 20.2*c*), $C_{Cr}\%$ is the chromium content in the Fe-Cr solid solution

(Figure 20.2*b*); $K_N$ is Henry constant (0.025); $eN^{Cr}$ is the interaction parameter of the first order ($eN^{Cr} = -0.114$ at T = 1000°C).

The obtained dependence of the nitrogen solubility in the γ-solid solution at 1000°C on the pressure of the air, $C_N\% = f(P)$, is shown in Figure 20.2*d*. It is seen that the dependence has an extreme character and the largest values are reached in the region of the air pressures from 0. 001 to 0.04 MPa. Consequently, at the air pressure 0.02 MPa, at heat treatment the dissolution of nitrogen takes place in the H15 alloy, which leads to the stabilization of the high-temperature γ-phase.

As is seen from the thermodynamic calculation results (Figure 20.2*d*), in the studied system "alloy H15 – air" the nitrogen absorption by the alloy is possible in the wide air pressure range of 0.001 – 0.04 MPa. The experimental data on the nitrogen content in the quenched specimens qualitatively correspond to the calculation dependence obtained (Figure 20.3, variant I). Thus, the model under consideration describes adequately the composition of the studied system "alloy H15 – air" at the temperature 1000°C.

Different contents of nitrogen and chromium in the quenched specimens explains the extreme character of the obtained dependence of the retained austenite amount on the air pressure $\gamma_{retained} = f(P)$ (Figure 20.4*a*).

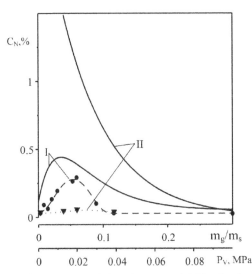

**FIGURE 20.3** The dependences of the nitrogen solubility in the alloy H15 on the content of the air: I and II are variants (lines are thermodynamic calculation, dots are the experimental data).

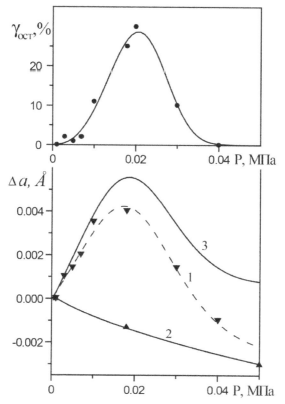

**FIGURE 20.4**    The dependences of the retained austenite amount (a) and the change of the crystal lattice spacing $\alpha$-Fe(Cr) (b) of the quenched alloy H15 specimens on the air pressure: $\Delta a = f(C_{Cr}\%, C_N\%)$ is line 1, $\Delta a = f(C_{Cr}\%)$ is line 2, and $\Delta a = f(C_N\%)$ is line 3.

The largest content $\gamma_{retained}$ ~30% corresponds to the following composition: $Cr - 9.4\%N$; $N - 0.29\%$.

The variation of the chemical composition of the studied alloy H15, which is the result of the oxidation of chromium and the dissolution of nitrogen during heat treatment, is accompanied by the change of the crystal lattice spacing of the $\alpha$-solid solution $\Delta a = f(C_{Cr}\%, C_N\%)$ (Figure 20.4b, line 1). A decrease in the chromium content in the alloy leads to the decrease of the lattice spacing $\Delta a = f(C_{Cr}\%)$ (line 2), and the dissolution of nitrogen is accompanied by the lattice spacing increase $\Delta a = f(C_N\%)$ (line 3) (the nitrogen contribution is the difference in lines 1 and 2). Thus, the increase of the crystal lattice spacing and the stabilization of the $\gamma_{retained}$

in the quenched alloy H15 specimens are obtained in the conditions of the largest nitrogen solubility (Figure 20.2*d*) [5].

Using the algorithm developed (Figure 20.1), the simulation of the nitrogen solubility in binary alloys with the chromium content 0–22 mass% was conducted depending on the air pressure. As a result, a three-dimensional diagram of the nitrogen solubility in the solid solution γ-Fe-Cr at 1000°C has been constructed, in which solubility is the function of the air pressure and the chromium content in the alloy $C_N$ = f (P, $C_{Cr}$%) (Figure 20.5). It is shown that in the alloys containing less than 5 mass% of chromium, with the air pressure decreasing the nitrogen solubility decreases and at the larger chromium contents the nitrogen solubility increases and has the largest values at the pressure of the air in the range of 0.001–0.04 MPa [6].

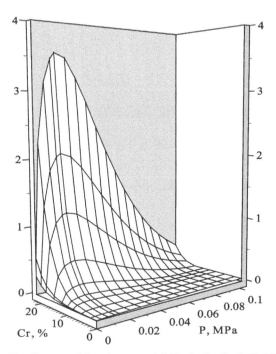

**FIGURE 20.5**  The diagram of the nitrogen solubility in the Fe-Cr(0–22%) alloys in the air at the temperature 1000°C.

With the use of the algorithm developed (Figure 20.1), the air nitrogen solubility in the alloy H15 at 1000°C has been calculated for variant II (Figure 20.3, line II). The dependence obtained is changed exponentially and has higher values than those in variant I (Figure 20.3, line I). However, despite the thermodynamic probability of the nitrogen absorption in variant II, an increase in the nitrogen content in the alloy bulk has not been found experimentally (Figure 20.3, dashed line II). In addition, the retained austenite in the quenched specimens is absent and the crystal lattice spacing of the $\alpha$-solid solution decreases.

The XPS investigations show that the main phases of the oxide layer on the alloy surface in variant II are chromium oxide $Cr_2O_3$ and iron-chromium spinel with variable content $Fe_xCr_{3-x}O_4$ (see the Table 20.1). At the air content $m_g/m_s = 0.04$ on the boundary 'alloy – oxide', the spinel $Fe_xCr_{3-x}O_4$ is only formed, which concentrates up to 17 at.% of nitrogen, i.e., it is a barrier layer for the nitrogen absorption into the alloy bulk. Thus, the nitrogen solubility takes place if chromium oxide $Cr_2O_3$ is formed on the surface (variant I), and nitrogen dissolving does not occur if the spinel phase $Fe_xCr_{3-x}O_4$ is formed together with chromium oxide $Cr_2O_3$ (variant II) [7].

## 20.4  CONCLUSIONS

1.  In the present paper a method for investigating phase transformations in the Fe-Cr alloys at high temperatures is offered, which comprises the construction of a thermodynamic model, the calculation of the equilibrium composition of the model system, the determination of the nitrogen solubility, and the conformation of the calculation results and the experimental data.

2.  With the use of the above method, the system "alloy H15 – air" has been investigated at the temperature 1000°C. It is shown that in addition to the oxidation of the alloy at the air pressures from 0.001 to 0.04 MPa the dissolution of nitrogen takes place without nitride formation. Nitrogen dissolving is accompanied by an increase in the $\alpha$-Fe crystal lattice spacing and the stabilization of the high-temperature $\gamma$-phase; the largest amount of the $\gamma_{retained}$ (30%) is observed at the pressure 0.02 MPa when the content of chromium reaches 9.4% and that of nitrogen is 0.29%.

3. It has been found that the formation of iron-chromium spinel ($Fe_xCr_{3-x}O_4$) on the alloy surface prevents the nitrogen absorption into the H15 alloy bulk.
4. The diagrams of the nitrogen solubility in the $\gamma$-solid solution of the iron alloys containing 0–22% nitrogen were calculated at 1000°C. It is shown the dependence of the nitrogen solubility on the air pressure for the alloys with the chromium content more than 5%Cr has an extreme character; the largest values for the nitrogen solubility are observed at the air pressures in the range of 0.001–0.04 MPa.

## KEYWORDS

- air
- austenite
- closed space
- high temperatures
- iron-chromium alloys
- lattice spacing
- nitrogen
- oxidation absorption
- phase transformations
- pressure
- solubility
- spinel
- thermodynamic model

## REFERENCES

1. Vatolin, N. L., Moiseev, G. K., & Trusov, B. G. Thermodynamic simulation in high-temperature inorganic systems. –M.: Metallurgia, 1994, 352 p.
2. Goncharova, N. V., Makhneva, T. M., Yelsukov, Ye. P., & Voronina, E. V. The influence of gaseous medium on the phase transformations in the Fe-Cr ferrite alloy at heat treatment. FHOM. 1997, №4, pp. 112–117.
3. Goncharova, N. V. Simulation of the phase transformations in the Fe-Cr alloys at high temperatures/Avtoref. diss. kand. fiz-mat.nauk. Izhevsk: UdGU, 2000, 20 p.

4. Morozov, A. N. Hydrogen and nitrogen in steel. M: Metalurgia, 1968, 284 p.
5. Goncharova, N. V., Makhneva, T. M., Yelsukov, Ye. P., Voronina, E. V., & Titor-ova, D. V. Retained austenite in the Fe-Cr ferrite alloy. FMM. 1998, Vol. 86, №6, pp. 53–58.
6. Goncharova, N. V., & Makhneva, T. M. Thermodynamic determination of the nitro-gen solubility in Fe-Cr alloys. FHOM. 2000, №1, pp. 81–85.
7. Goncharova, N. V., Makhneva, T. M., Yelsukov, Ye. P., Voronina, E. V., Kanunnikova, O. M., & Titorova, D. V. Phase transformations in the chromium-containing ferrite alloy at heat treatment in closed space in the air. Perspektivnye materially. 2000, №4, pp. 83–90.

# CHAPTER 21

# THE MOBILITY OF TECHNOGENIC ARSENIC IN THE SOILS OF UDMURTIA UNDER THE INFLUENCE OF ATMOSPHERIC PRECIPITATION

M. A. SHUMILOVA, V. G. PETROV, O. S. NABOKOVA, and A. YU. KARPOVA

*Institute of Mechanics of the Ural Branch of the Russian Academy of Sciences (IM UB RAS), Izhevsk, Russia*

## CONTENTS

### 21.1   INTRODUCTION

In accordance with the implementation of the International Convention on the prohibition of the development, production, stockpiling and use of chemical weapons and on their destruction in the Udmurt Republic in the city of Kambarka in the period from 2006 to 2009 work was carried out for the destruction of chemical weapon lewisite, and to 2012 – with the reaction mass after it destruction. Since the arsenic oxides and inorganic compounds in accordance with GOST 17.4.1.02–83 "Protection of nature. Soil" (GOST – State standard of the Russian Federation) belong to the first class of danger, the obvious need to determine the presence of contaminated land specific substances and products of their degradation, and to establish a system of regular observations of the status of these territories, to exclude the possibility of negative impact of pollutants on human health and the environment. The aim of the present work is the improvement of the approach to the organization of ecological monitoring of natural environments with consideration of the peculiarities of the chemical behavior of technogenic compounds of arsenic and its influence on precipitation.

## 21.2  MATERIALS AND METHODS

The sampling, preservation, storage and transportation of soil specimens for ecological and analytical studies were carried out in accordance with GOST 17.4.3.01-83 and GOST 28168-89. Chemical analyzes of soil samples in the area of the facility for destruction of chemical weapons in Kambarka that are implemented with the aim environmental monitoring in the sanitary protection zone (SPZ) and in the area of protective measures (APM), was made in the Central ecological and analytical laboratory (CEAL) of Regional Centre of the State environmental control and monitoring in the Udmurt Republic, on the techniques developed by the Federal state institution "State research Institute of industrial ecology" (Saratov). Analysis of soil samples for arsenic was performed on x-ray fluorescence SPECTROSCAN "MACS GFIE (S)" and "MACS GFIE (P)."

The maintenance of arsenic and other heavy metals in soil extracts field experiment was determined by atomic absorption method with a spectrophotometer "Shimadzu- AA7000" with electrothermal atomizer according to certified methods M-02-902-125-2005 [1]. All reagents used for the spectral analysis, had the qualification of "high purity."

Precipitation data taken from meteorological observations in 2009 and 2010 for the city of Sarapul – nearest meteorological station for city of Kambarka [2,3].

## 21.3 RESULTS AND DISCUSSION

Technology decomposition of lewisite on the CWD object was low-temperature alkaline hydrolysis, the essence of which is described by the following chemical equation:

$$C_2H_2AsCl_3 + 6NaOH \rightarrow C_2H_2 + 3NaCl + Na_3AsO_3 + 3H_2O \quad (1)$$

The reaction mass decomposition of lewisite were evaporated to the dry state and stored, and then transported in the Saratov region for further processing. Formed after the hydrolysis of the light-yellow solution consisted of 13.88% of sodium arsenite, 12.69% sodium chloride, 0.06% of sodium hydroxide and 73.22% water. It is well known that sodium arsenite is one of the most easily soluble arsenic compounds (28.3 g/100 g of water [4]), with a fairly high degree of toxicity ($LD_{50} - 41$ mg/kg [5]). Based on the high solubility of sodium arsenite, it can be assumed that the releases to the environment resulting from operation of the facility will not always lead to the accumulation of this element in the soil.

For discussion of this hypothesis let us consider the obtained earlier experimental data. The study in the laboratory parameters of the mobility of some HM in some types of regional soil as a pollutant allowed to formulate a conclusion about the desire arsenite ion to delocalization of pollution unlike other metals [6]. In particular, the value of the half-life ($T_{0.5}$) of sodium arsenite under the influence of the characteristic of the region of precipitation is a few days and is close to the values obtained for sand, whereas in the case of other oxides of HM this parameter in similar terms has values of the order of several tens or hundreds of years. In Table 21.1 shows the calculated values of mobility of HM oxides and sodium arsenite for sand.

**TABLE 21.1** The Parameters of the Mobility of a Number of Cat-ions of HM (10 MAC) in the Sand

| Contaminant | The order process | The rate constant $k$, $s^{-1}$ | The half-life, years ($\alpha = 0.5$) |
|---|---|---|---|
| CuO | $n \sim 1$ | $1.238 \times 10^{-9}$ | 32 |
| $Cr_2O_3$ | $n \sim 1$ | $7.791 \times 10^{-10}$ | 51 |
| $NaAsO_2$ | $n \sim 2$ | $4.449 \times 10^{-7}$ | 1.19 |

The degree of separation of substances from contaminated surface soil from rainfall, calculated by the formula:

$$\int_0^\alpha \frac{d\alpha}{(1-\alpha)^n} = \kappa_{_H} T_\alpha S \sum_{i=0}^{m} \frac{H_{_\mathrm{i}}}{\omega_i} \qquad (2)$$

where $\alpha$ is the amount of soil contaminants in in parts from of the initial maintenance; k- is the observed rate constant of discharge of the pollutant from the soil; $T\alpha$ is the time required for the drainage of substances from contaminated soil to the degree $\alpha$, in years; S – area of the soil cover, which was provided to the technogenic impact; Hi – amount of individual types of precipitation as rain (light rain, rain, heavy rain), in mm; $\omega_i$ is the rate of passage of water through contaminated soil, ml/s, m – the number of types of precipitation as rain [7].

As follows from the Eq. (2), at the same time precipitation the degree of separation of a substance from the soil layer increases with the increase of precipitation in the same area, reducing its content in the soil layer.

To clarify the legality of the findings about the reduction in the content of the arsenite ion in the soil to the volume of precipitation was field-tested in the mobility of the pollutant. Several types of soils, common in our region was contaminated with sodium arsenate in the amount of 10 MAC (maximum allowable concentration) for arsenic and placed in natural conditions. The results obtained the degree of excretion of sodium arsenite under the influence of precipitation as rain in the summer-autumn period and in the form of meltwater in the spring-winter period are presented in Table 21.2.

As follows from the Table 21.2. data and is confirmed in the literature [8], the mobility of arsenite ions under the action of rainfall increases considerably with the decrease in the content of humus substances in the topsoil.

In accordance with a General characteristic of landscape-climatic and sanitary conditions of the Kambarka district terrain is characterized by flat terrain with a predominance of light loamy and sandy soils with well-developed hydrographic and gully network. As is known, the MAC of arsenic in the soil corresponds to the level of 2.0 mg/kg. However, for podzolic soils, which are dominant in the study area, background levels of

**TABLE 21.2**   Agrochemical Characteristics of the Soils and the Degree of Allocation of NaAsO$_2$ of Them in field Experiments During 1 Year of Observation

| No | The soil type | The maintenance | | | | pH$_{KCl}$ | The degree of allocation, % | |
|---|---|---|---|---|---|---|---|---|
| | | Humus, % | Mn mg/kg | Fe mg/kg | Al mg/kg | | The summer-autumn period | Winter-spring period |
| 1 | Calcareous | 3.73 | 2.10 | 150.56 | 6753.20 | 7.46 | 0.280 | – |
| 2 | Dark-gray forest | 5.96 | 2.22 | 131.98 | 4175.80 | 6.34 | 0.003 | – |
| 3 | Sod-podzolic | 2.55 | 2.89 | 149.27 | 4620.40 | 7.45 | 1.042 | 1. 439 |
| 4 | Light gray forest | 3.95 | 2.27 | 127.37 | 5464.5 | 7.38 | 0.037 | 0.300 |
| 5 | Sand | 0.00 | – | – | – | – | – | 9.910 |

arsenic in the region of Kambarka (including the surrounding area within a radius of 5 km) is taken to be 3.0 mg/kg [9].

Table 21.3 contains data on the maintenance in some sampling points of the soil series of HM in the SPZ and in APM of the CWD object resulting from monitoring of CEAL, and precipitation during the study period. The data show that the content of ions of copper and nickel in the surface layer has slight variations in magnitude and almost not associated with the amount of precipitation during the observation period. In some cases, recorded a sharp increase in the content of ions of nickel and copper (point No. 68, dated 19.05.10. and point No. 9 dated 15.04.09, etc.) in the soil, and the amount of precipitation also tended to increase (Figure 21.1*b* and *c*). In those periods when there was a decrease in precipitation, the number of ions of copper and Nickel in the soil also decreased (e.g., point No. 9 of 30.06.09 and point No. 68, dated 11.08.09), which is reflected in the graph (Figure 21.1*b* and *c*). Given the observed trends, we can assume that the HM cat-ions sufficiently strongly absorbed by the soil and all the recorded fluctuations of the concentrations in soil are associated largely with technogenic impacts.

For the arsenite ion installed another variant according to the number of adsorbed ion on the volume of precipitation. In all sampling points, the

**TABLE 21.3** The Maintenance of Heavy Metals in Soil and the Amount of Precipitation

| No. of locations | Date sampling | The maintenance, mg/kg | | | The amount of precipitation, mm |
|---|---|---|---|---|---|
| | | As | Cu | Ni | |
| 2 | 16.04.2009 | 7.2 | 35.0 | 30.0 | 130.2 |
| | 30.06.2009 | 9.9 | 37.0 | 33.0 | 59.5 |
| | 26.04.2010 | 8.6 | 33.0 | 29.0 | 286.2 |
| 9 | 15.04.2009 | 2.0 | 42.0 | 40.0 | 130 |
| | 30.06.2009 | 11.0 | 11.0 | 28.0 | 59 |
| | 26.04.2010 | 9.2 | 37.0 | 33.0 | 286 |
| 68 | 01.07.2009 | 8.1 | 21.0 | 16.7 | 91 |
| | 11.08.2009 | 13.8 | 23.0 | 16.1 | 52 |
| | 07.10.2009 | 9.8 | 23.0 | 17.5 | 52 |
| | 19.05.2010 | 9.6 | 36.0 | 33.0 | 138 |

concentration of arsenic in soil increases with decreasing rainfall and vice versa, with increasing rainfall amount adsorbed by soil arsenic decreases (Table 21.3). Comparison of the arsenic in the soil and precipitation during the observation period are presented graphically in Figure 21.1. The observed patterns are in full compliance with the assumption of weak absorption of the soil absorbing complex arsenite ions, which due to the high solubility of salt is relatively easily washed out of the soil, particularly when sufficiently abundant precipitation. Therefore, a slight increase of arsenic content in soil was recorded in the dry period, with the increase in precipitation pollutant content again returns to some average value, typical for the specific point of sampling. Possibly can talk about low ability is particularly light podzolic and sandy loam soils to accumulation of arsenic in the form of arsenite; this conclusion is in agreement with the available literature data [10]. Thus, the obtained data can indicate technogenic contamination of soil with arsenic and legitimacy of conclusions made on the basis of the made laboratory and field experiments. The influence of precipitation on the degree of absorption of arsenite ion soil must be considered in the organization of ecological monitoring of environment in the area of technological impact, and also at carrying out of rehabilitation of the contaminated territories.

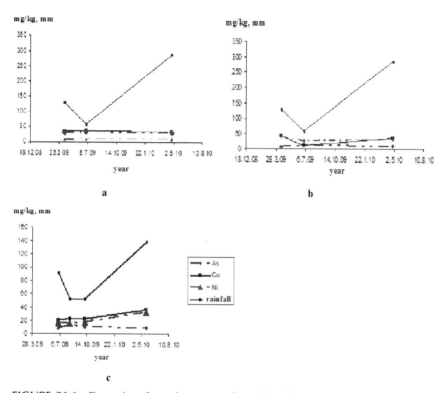

**FIGURE 21.1** Dynamics of arsenic compounds and HM in the upper soil horizon in the background of the atmospheric precipitation in sampling points: (a) No 2, (b) No. 9, (c) No. 68.

## 21.4 CONCLUSION

Data obtained in laboratory experiments and a field experiment on typical regional soils show a relatively high mobility of arsenite ion under the influence of precipitation. Comparison of arsenic content in sampling points in the area of object the CWD in Kambarka the precipitation shows a commitment to reducing its content in the soil with increasing amounts of precipitation, in contrast to ambiguous changes in the content of some of the HM, which confirms the results of the experiments. It is shown that in the lungs podzolic and sandy loam soils under the influence of precipitation as rain rather easily and quickly proceed desorption processes arsenite ion, so this effect must be taken into account in the organization of ecological monitoring of environment in the area of operation of industrial facilities with man-made impacts and also at carrying out of rehabilitation of the contaminated territories, in particular, as a result of work on the CWD.

## KEYWORDS

- arsenite ion
- heavy metal ions
- lewisite
- soil pollution
- the impact of atmospheric precipitation

## REFERENCES

1. Method for quantitative chemical analysis. Determination of As, Cd, Co, Cr, Cu, Hg, Mn, Ni, Pb, Sb, Sn, Zn (acid-soluble forms) in soils and sediments by atomic absorption method. M-02-902-125-2005. S.-Pb, 2005.
2. On the state of the natural environment of the Udmurt Republic in 2009: State report. Izhevsk: Publishing house of ISTU, 2010. 288 p.
3. On the state of the natural environment of the Udmurt Republic in 2010: State report. Izhevsk: Publishing house of ISTU, 2011. 238 p.
4. Lurie, Y. Handbook of analytical chemistry. M.: Chemistry, 1989, 448 p.
5. Harmful chemical substances. Inorganic compounds V – VIII groups. L.: Chemistry, 1989, 592 p.
6. Petrov, V. G., Shumilova, M. A., Nabokova, O. S. & Lebedeva, M. G. Improvement of methods of control of technogenesis products for online tracking of the destruction of chemical weapons. Theoretical and Applied Ecology. 2012, No. 4, pp. 63–66.
7. Petrov, V. G., Shumilova, M. A., & Nabokova, O. S. The parameters of the mobility of contaminants by sodium arsenite for soils of the Kambarka district. Chemical Physics and Mesoscopy. 2013, T. 15. No. 3, pp. 465–470.
8. Babushkina, S. V., Puzanov, A. V., Elchaninov, O. A., & Gorbachev, I. V. Arsenic in the soils of technogenic landscapes of Altai. Polzunov Bulletin. 2005, No. 4, pp. 153–156.
9. Scotich, P. E., Zheltobrjuhov, V. F., & Klocek, V. V. Ecological and hygienic aspects of the problem of chemical weapons destruction. – Volgograd: Publishing House of the Volga, 2004, 236 p.
10. Popova L. F. Assessment of heavy metal pollution typical of soils Arkhangelsk. Fundamental Research. 2014, No. 8, pp. 849–853.

# CHAPTER 22

# SYNTHESIS OF DIAMOND PRECURSOR MW CVD METHOD FOR FORMING DIAMOND FILMS ON TUNGSTEN CARBIDE MATERIALS

D. S. VOKHMYANIN

*Perm National Research Polytechnic University, Perm, Russia*

## CONTENTS

### 22.1  INTRODUCTION

The unique combination of physical properties of diamond is attracting attention to this material in terms of its practical use. Great interest to the diamond material originated after the development of methods for chemical vapor deposition of carbon from the gas phase (CVD), which allowed to obtain thin films. Such films are attractive due to their properties for use in fields ranging from electronics to mechanical processing of various composite materials [1]. In the latter case the diamond film is deposited on the carbide materials. Despite numerous studies, there are certain aspects that remain unclear to date. These aspects are: the absence of the necessary adhesion to the solid alloy and low density of nucleation of the diamond phase. To solve the problem of poor adhesion of the diamond film using various embodiments of cemented carbide and etching subcoating [2]. To solve the problem of low-density diamond nucleation phase is used in the treatment of precursors, which includes: applying a thin polymer films [3], which is synthesizing a complex task, and the use of detonation diamonds that do not always produce predictable results, which is characterized by their production [4].

For controlling the properties and the reproducibility of diamond coatings, diamond suggested to synthesize a precursor with a specific crystallographic orientation. As the substrate material, the precursor for the synthesis of diamond has been chosen metallic nickel. As is known, nickel has a face-centered cubic cell, which is different in the parameters of the diamond cell to 1.2%. In addition, it is used as an additive in the catalytically active diamond synthesis by HP-HT (high pressure and high temperatures method), and thus is a suitable material for the growth of heteroepitaxial diamond precursor. The literature indicates that it was possible to obtain diamond crystals using HF CVD method: the concentration of methane of not more than 0.8% of the total volume of the gaseous mixture to form the intermediate layers type Ni-CH, owing to the solubility of carbon in the material [5].

The purpose of the study is a precursor in the synthesis of diamond and nickel substrates to investigate the possibility of its use for the formation of diamond films on tungsten carbide materials.

## 22.2   EXPERIMENTAL RESEARCHES

As we noted above, as a starting material for producing diamond precursor using spark plasma sintering of the substrate were manufactured from nickel. Before synthesis, the substrates were subjected to polishing for surface smoothing after sintering, as well as washing in isopropyl alcohol. Synthesis of diamond precursor was carried out at the facility AX5200S-ECR Seki Technotron (Japan), by the pressure in the chamber 25 Torr, the concentration of methane in the gas mixture ranged from 1 to 2%, the processing time for 2 hours. Separating the precursor to the plate after the synthesis was carried out ultrasound treatment in an environment of isopropyl alcohol.

Before the deposition of diamond coatings, carbide inserts subjected alternately etched in $H_2SO_4$, Caro's acid, and $H_2O_2$, for 10, 40, and 15 minutes, respectively. The next step after the etching treatment in isopropyl alcohol with the addition of precursor. Processing was carried out in an ultrasonic bath Branson 1510 for 40 minutes. Thereafter carbide inserts were placed into the reactor chamber. The deposition of diamond films on tungsten carbide materials were carried out in a mixture of methane-hydrogen by the pressure in the chamber 25 Torr, the methane concentration of 1%, the processing time of 4 hours.

Differential thermal analysis (DTA) was performed on diamond powders derivatograph Q-1500D system Paulic-Paulic-Erdey to a temperature of 900°C at a heating rate of 5°C/min. Based on the experimental data cleaning mode of diamond powders from graphite impurities formed during synthesis of powders was developed.

The phase composition of the coating precursor and examined by Raman scattering (RS) of light. Raman spectra were obtained on a multifunction spectrometer Raman "SENTERRA" (Bruker) at a wavelength of 532 nm emitting laser. The microstructure of the materials was studied by scanning electron microscopy (SEM) on the analytical field emission scanning electron microscope ULTRA 55 (Carl Zeiss, Germany).

## 22.3   RESULTS AND DISCUSSION

The morphology of the synthesized precursor was investigated using scanning electron microscopy and shown in Figure 22.1. The precursor obtained at a concentration of 1% CH4, one can distinguish two types of particles differing in size and morphology. Larger particles have a size of about 250 nm, and are surrounded by the small size of the embryos, which have a large variation in the size of 10–80 nm. Large particles are characterized in the literature as "cauliflower", and consist of a plurality of nanometer crystallite united together into one globule. When the concentration of methane in the gas mixture up to 2%, morphology changes on

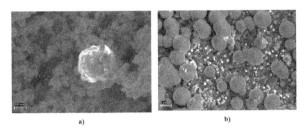

a)                                             b)

**FIGURE 22.1**    SEM images of the synthesized precursor on Ni substrates, at concentrations of 1% (a) and 2% (b) and an increase of 50,000 and 4,500 respectively.

the surface of the formation of large spherical particles reaching 4 micron geometrical parameters. On the surface of the spheres shows no signs of crystalline inclusions, indicating that the resulting amorphous material, the observed fused together portions.

The Raman spectra (Figure 22.2) derived precursors there are several peaks that characterize the different modifications of carbon. The peak in the vicinity of 1334 cm$^{-1}$, associated with the crystalline diamond and diamond phase is confirmation of the obtained precursor. Thus it cannot be identified with increased methane concentration of 2%. The presence in the spectra of the peak 1380 cm$^{-1}$, having the name D-peak means having sp2-bonded carbon. The peak of 1580 cm$^{-1}$ called the G-peak, characterized by ordered graphite sample [6]. The maximum intensity of the D and G peaks is observed at a concentration of 2%. These data suggest that the amorphous spherical particle is graphite, which in the process of synthesis of diamond nucleation suppressing components, due to high catalytic activity of the surface of the substrate.

The degradation of the synthesized precursor has significant differences which depend on the initial concentration of the carbon-containing gas (Figure 22.3). So at a concentration of 1% CH4, it is a two-stage process, which indicates the presence of two phases of carbon precursors with different levels of defects. On the DTA curve is clearly visible exothermic peak in the temperature range 500–735°C, which consists of two components. The first peak is achieved at a temperature of 640°C, with the relative mass loss reaches 50% of the original sample. The next peak is reached at 660°C, and indicates the oxidation of the heavier component in the composition. The opposite situation is observed at a concentration of

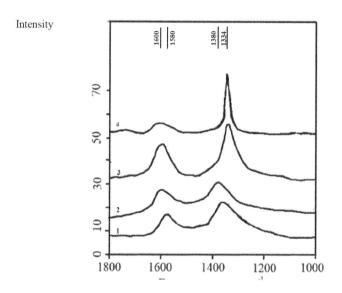

**FIGURE 22.2** Raman spectra of the synthesized precursor: 1 – (2% CH4) and 3 – (1% CH4) and heat treated at 620 0 C: 2 – (2% CH4) and 4 – (1% CH4).

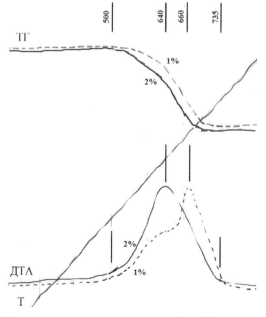

**FIGURE 22.3** Thermograms precursor at concentrations of 1% and 2% methane.

2% CH$_4$. In the above temperature range and holds only one exothermic peak at a temperature of 640°C, while achieving an almost complete loss of mass of the order of 85%.

Analysis of the Raman spectra of precursors annealed in an oven at 620°C in order to remove a defective material showed the restriction line at 1334 cm$^{-1}$, as well as the almost complete absence of graphite lines on 1600 cm$^{-1}$. Thus, after the spent thermal influence, the obtained precursor (1% CH$_4$) contains a fraction of diamond, and the contribution of the line corresponding to the carbon G-line is insignificant. For a concentration of CH$_4$ 2%, annealed precursor spectrum is shifted toward higher peak values up to 1600 cm$^{-1}$ for G-line and 1390 cm$^{-1}$ for the D-line, which is associated with the oxidation of the graphite parts. The presence of the diamond component is not revealed, and after thermal exposure to the precursor.

By results of research of the phase composition obtained precursors for further testing was selected material with a 1% concentration of methane.

Figure 22.4 shows the structure of the resulting film on tungsten carbide wafer. After 4 hours, the synthetic diamond on the surface of the formed film thickness 3.5 microns, which corresponds to the growth rate of 0.8 microns/hour. The film has a tree structure, with the growth of diamond plate faces corresponding one preferential growth of crystallographic orientation.

The Raman spectrum (Figure 22.5) shows a narrow peak near 1332 cm$^{-1}$ with a half-width not more than 9 cm$^{-1}$, which confirms the high quality

a)                                          b)

**FIGURE 22.4**   SEM images of a diamond film grown using the precursor, while increasing fracture 13000 (a), a plan view at magnification of 7000 (b).

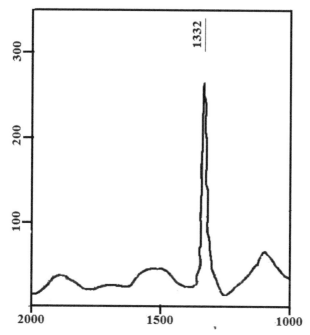

**FIGURE 22.5**    Raman spectrum of film on carbide plate.

and purity of the resulting film. Low intensity peaks near 1450 cm$^{-1}$ and 1150 cm$^{-1}$, is characterized by small defects in the structure.

## 22.4  CONCLUSION

The possibility of the formation of diamond crystals on the surface of the nickel substrate at a concentration of 1% methane. The resultant particles have the shape of "cauliflower" with the geometrical dimensions of 250 nm. At high concentrations there is a formation of spherical graphite particles.

Based on the TG and DTA curves, as well as to analyze the Raman intensities annealed powder in the furnace and the original powder, the temperature of purification precursor synthesized at a concentration of 1% methane in the gas mixture is selected at 620°C.

Solid diamond film obtained using the precursor synthesized in the nickel substrate at a concentration of carbonaceous gas is not more than 1%.

Diamond film has a thickness of 3.5 mm, with the growth of the diamond phase plate according to one of the crystallographic directions.

## KEYWORDS

- carbide materials
- diamond films
- diamond precursor
- nickel

## REFERENCES

1. Vyrovec, I. I., Gricina, V. I., Dudnik, S. F., & Opalev, O. A. Nanokristallicheskie almaznye CVD-plenki: struktura, svojstva i perspektivy primenenija. i dr. FIP PSE, 2010, T.8, vol.1 №1. pp. 4–19.
2. Haubner, R., & Kalss, W. Diamond deposition on hard metal substrates – Comparison of substrate pre-treatments and industrial applications. International Journal of Refractory Metals and Hard Materials. 2010, Vol. 28. № 4, pp. 475–483.
3. Sedov, V. S. Sintez tonkih mikro-i nanokristallicheskih almaznyh plenok v SVCh plazme: avtoreferat dis. na kon-ta fiz-mat. nauk – Moskva 2013. (in Russian).
4. Dolmatov, V. Y. Detonation synthesis ultradispersed diamonds: properties and applications. Russian Chemical Reviews, 2001, 70(7), pp. 607–626.
5. Diamond films. Chemical Vapor Deposition for oriented and heteroepitaxial growth. Koji Kobashi. 2005, pp. 350.
6. Prawer, S., & Nemanich, R. J. Raman spectroscopy of diamond and doped diamond. Philosophical transaction of the royal society of London. A (2004) 362, pp. 2537–2565.

# CHAPTER 23

# THE INFLUENCE OF NON-SPECIFIC BINDING OF ANTIBODIES ON FORMING OF ANALYTICAL RESPONSE OF AMPEROMETRIC IMMUNOSENSOR

I. CHERENKOV, V. POZDNJAKOV, and V. SERGEEV

*Udmurt State University, Izhevsk, Russia*

## CONTENTS

### 23.1  INTRODUCTION

Immunobiosenory – a kind of affinity biosensors, in which as a recognition element used antibodies [1, 2]. Many of the approaches used in the immunosensors design borrowed from immunosorbent assay [1–4].

Advantages of immunosensors analysis is rapidity, the unity of time and place of receipt of the result, as well as the ability to use a variety of physical methods of signal conversion – amperometry, potentiometry, conductivity, gravimetry [1, 2].

The most widely amperometric immunosensors in which the analytical signal is the change of current [1, 2]. The catalytic current is formed by the enzymatic activity of oxidases conjugated antibodies. The most commonly used horseradish peroxidase. For this enzyme a direct electron transfer is shown [5]. However, the effectiveness of this process is not sufficient for use in the immunosensor where peroxidase is often localized at the material, from the viewpoint of electron exchange, the distance from the electrode. The ability to produce analytical signal in such systems is achieved by using electron transfer mediators – substances capable of specifically oxidized peroxidase and subjected electroreduction on the electrode surface, forming a catalytic current [6]. The range of such compounds is quite wide: quinones, phenothiazines, ferrocene and its derivatives, and other biogenic amines. Use of mediators can reduce operating voltage and improves specificity analysis. There is a new research task – harmonization of enzymatic and electrochemical reactions. The problem lies in the need to create optimum conditions for the formation of the analytical signal, taking into account the kinetics of enzymatic and electrochemical reactions, sorption interactions mediator and electrode material, and so on [1, 2].

Another problematic aspect of the design of this type of biosensors consists in the fact that, as a rule, amperometric immunosensor – disposable device (the balance in the "antigen-antibody" is shifted toward the formation of an immune complex – binding constant of $10^6$–$10^{12}$, i.e., the reaction is practically irreversible) [1]. On the basis of this specificity, the layout immunosensor should be as simple and cheap.

Demand for such immunosensors practice beyond doubt. Thus, one object of the clinical diagnosis, the detection of specific antibodies in the blood. Their presence in the blood, and the characteristic dynamics of the titer is an important diagnostic and prognostic indicators [7].

The aim of this work was to develop a prototype immunosensor based planar graphite electrode and study the effect of non-specific binding of antibodies to the formation of the analytical signal.

## 23.2  MATERIALS AND METHODS

### 23.2.1  PREPARATION OF IMMUNOSENSOR

As basis of immunosensor served planar graphite electrodes with a surface area of the working electrode of 0.3 cm$^2$, which was preincubated in ethanol and dried in a dust chamber. Prior to immobilization of the antigen electrode was electrochemically cleaned by repeated potential cycling in a medium of phosphate buffer solution (pH 7.2). Then the electrode was washed with distilled water, dried in a dust chamber and conducted immobilizing the antigen.

To simulate the reaction of "antibody-antigen" on the surface of the electrode used human immunoglobulin IgG, which served as a model antigen (AG) and antispecies antibody, goat to human IgG conjugated with horseradish peroxidase (AT-HRP).

On the surface of the working electrode was applied 2 µL of the antibody solution (0.5 mg/mL). The electrode was incubated in a humid chamber for 18 hours at a temperature of 4°C, then washed with sterile buffered saline.

To block nonspecific binding was used a solution of bovine serum albumin (BSA) (Sigma-Aldrich, USA), which was applied after immobilization of the model antigen.

Before measuring on the surface of the electrode was applied a solution of goat antibodies to human IgG labeled with horseradish peroxidase. Incubated for 30 min at room temperature, then washed with sterile buffered saline and electrochemical measurements were carried out.

### 23.2.2  ELECTROCHEMICAL MEASUREMENTS

Background electrolyte was sterile buffered saline, on the basis of which all working solutions are prepared. The concentrations of the mediator (hydroquinone) and hydrogen peroxide in the cell was 0.1 mmol/L and was constant for all measurements.

For the electrochemical measurements was used potentiostat EcoLab 2A-100. When operating in a cyclic voltamperometry (CV) used potential range from $-100$ mV to $+700$ mV at a potential sweep rate of 100 mV/s at a temperature of 20–25°C and natural saturation of the working solution with oxygen. The measurement accuracy of the current is ± 1 nA. All potentials are given relative to silver chloride electrode.

## 23.3   RESULTS AND DISCUSSION

The biosensor element used in operation, a planar graphite electrode with adsorbed on the surface of the working electrode human immunoglobulins. Signal processing is as follows: at the working electrode surface applied antispecies antibody conjugated with horseradish peroxidase which specifically interact with the immobilized antigen (in this case its role of human antibodies). After incubation the electrode was washed with a buffer solution to remove non-associated with an antibody and add into cell solution of hydrogen peroxide and hydroquinone in equimolar concentrations. Analytical signal is electroreduction current of benzoquinone, which is the product of the enzymatic reaction catalyzed by peroxidase (Figure 23.1).

The object of the first step was to obtain the analytical signal corresponding to the interaction with horseradish peroxidase conjugated antibody to the immobilized antigen on an electrode surface.

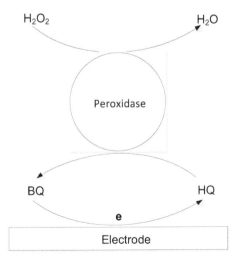

**FIGURE 23.1**   Scheme of catalytic cycle of peroxidase in the electrochemical system in the presence of hydrogen peroxide and hydroquinone (HQ – hydroquinone; BQ – benzoquinone).

When studying a solution of hydroquinone it was found that in the absence of peroxidase at neutral pH, a characteristic curve of a cyclic voltamperometry corresponding redox reactions of the mediator. The presence of peroxidase on the electrode is reflected by a sharp increase in the current restoration of benzoquinone and the reduction current of electrooxidation hydroquinone because oxidation reaction in the presence of peroxidase is carried out mainly due to the enzyme.

Similar results were obtained in the processing of the biosensor antispecies antibodies conjugated to peroxidase.

Obviously, the possible non-specific binding to AT-HRP with electrode surface, which will lead to formation of background signal unrelated to the presence of a model antigen. Factors contributing to the increase of background currents are nonspecific adsorption AT-HRP on graphite and the presence in the solution of the oxidized form of the mediator (Figure 23.2).

Applying a solution of AT-HRP on the electrode not containing a model antigen and BSA wasn't pretreated leads to the formation of signal strength value comparable with the current analytical. Thus the electrode material exhibits a sorption capacity in respect of AT-HRP regardless of the presence of antigen molecules on it. It was shown that protein molecules immobilized on the surface of the carbon electrode materials do not form a continuous monolayer [5]. A substantial portion

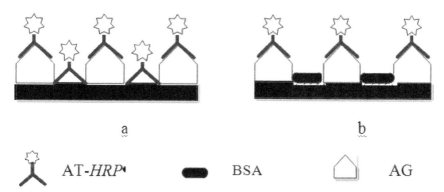

a                                                       b

⚛ AT-*HRP*◄          ▬ BSA          ⬠ AG

**FIGURE 23.2** Scheme of interaction "antigen-antibody" on the electrode surface of immunosensor in the conditions of nonspecific binding (a) and block of free surface of the electrode with BSA (b).

of the electrode remains free and available for nonspecific binding of peroxidase-conjugated antibodies. Therefore, there is a need for processing of the working electrode to prevent nonspecific binding. For these purposes may be used a solution of BSA or other inert protein. In the experiments, we used a BSA concentration 1 and 0.1 mg/mL. Applying to the surface of the electrode BSA solution undermines the quality of nonspecific binding AT-HRP, which was accompanied by a decrease in power recovery current of benzoquinone at the electrodes not containing a model antigen (Figure 23.3, graph 2). Amperage while significantly lower level of the analytical signal.

An interesting fact of the impact of the BSA on the formation of the analytical signal. Figure 23.4 shows the analytical signal amplification with increasing concentrations of BSA on the electrode (group "AG+AB").

Albumin has not expressed an electrochemical activity in the investigated range of potentials – the maximum values of the current in the absence of peroxidase on the electrode close to different concentrations of BSA (Figure 23.4, group "AG"). However, the presence of BSA on

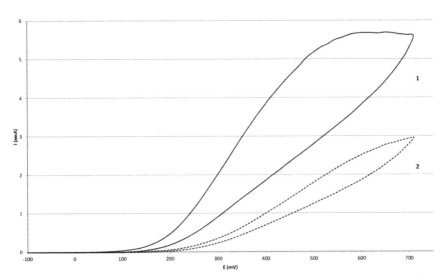

**FIGURE 23.3** The data of cyclic voltamperometry after application of antibodies conjugated to peroxidase for an electrode containing the adsorbed antigen and blocked with BSA free surface (1) and an electrode containing no antigen coated BSA (2).

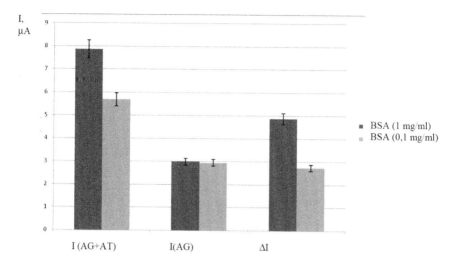

**FIGURE 23.4** Changes to the maximum values of restoration current of benzoquinone using different concentrations of BSA to reduce non-specific binding antibodies.

the electrode leads to a significant increase in analytically significant component of the signal (Figure 23.4, group "ΔI").

As an explanation of this phenomenon may be assumed that the treatment electrode BSA improves of interaction of the mediator to the surface of the electrode. Perhaps benzoquinone – oxidation product of the hydroquinone takes sorption of interaction with albumin, which leads to an increase in restoration current.

## 23.4  CONCLUSION

Thus, the provided scheme of immunosensor allows to obtain analytical signal, significantly different from the value of background currents, and can be used to construct immunosensors to detect antibodies. Modification of the surface of the working electrode with BSA conducted after immobilization of the antigen, can significantly reduce the values of currents corresponding to nonspecific tying of antibodies. BSA can amplify the signal of the biosensor due to the interaction with the mediator (hydroquinone). Such of interaction should be taken into account when creating immunosensors.

## KEYWORDS

* bioelectrochemistry
* bovine serum albumin
* cyclic voltammetry
* immunosensor

## REFERENCES

1. Budnikov, G. K., Evtugin, G. A., & Maistrenko, V. N. Modificirovannie elektrody dlja voltamperometrii d himii, biologii i medicine (Modified electrodes for voltammetry in chemistry, biology and medicine) Moscow, "BINOM. Laboratorija znanyi" Publ., 2010, pp. 134–154; 226–238; 293–308.
2. Banica, F.-G. Chemical Sensors and Biosensors: Fundamentals and Applications, Wiley, 2012, pp. 101–118; 347–367 (Rus. ed. Banica, F.-G. Khimicheskie i biologicheskie sensory osnovy i primenenija. Moscow, Technosfera Publ., 2014, p. 185–210; 571–589.
3. Ordonez, S. S., & Fabregas, E. New antibodies immobilization system into a graphite–polysulfone membrane for amperometric immunosensors. Biosensors and Bioelectronics, 2007, Vol. 22, pp. 965–972.
4. O'Connor, M., Kim, S. N., Killard, A. J., Forster, R. J., & Smyth, M. R. Papadimitrakopoulos, F., Rusling, J. F. Mediated amperometric immunosensing using single walled carbon nanotube forests. Analyst, 2004. Vol. 129, pp. 1176–1180.
5. Bogdanovskaja, V. A. Mnogocomponentnie kataliticheskie sistemy katodnogo vosstanovlenija kisloroda. Avtoreferat diss. dokt. chim. nauk [Multicomponent catalyst systems of cathodic reduction of molecular oxygen. synopsis of dr. chem. sci. diss.]. Moscow, 2011. 52 p.
6. Katz, E., Shipway, A. N., & Willner, I. Mediated electron-transfer between redox-enzymes and electrode supports. Encyclopedia of Electrochemistry, Vol. 9: Bioelectrochemistry, Edited by, G. S. Wilson. Wiley-VCH GmbH. Weinheim. 2002, pp. 559–626.
7. Yanamandra, K., Gruden MA, Casaite, V., Meskys, R., Forsgren, L., et al. Synuclein reactive antibodies as diagnostic biomarkers in blood sera of Parkinson's disease patients. PLoS ONE, 2011, vol.6, no 4, e18513. doi: 10.1371/journal.pone.0018513.

# CHAPTER 24

# MOLECULAR MODELING OF THE 2-(PYRIDIN-2-YL)-1H-BENZIMIDAZOLE INTRAMOLECULAR DYNAMICS

E. V. RAKSHA,[1] A. B. ERESKO,[1] YU. V. BERESTNEVA,[1] A. V. MURATOV,[1] and G. E. ZAIKOV[2]

[1]L.M. Litvinenko Institute of Physical Organic and Coal Chemistry, Donetsk 83114, Ukraine, E-mail: elenaraksha411@gmail.com

[2]Institute of Biochemical Physics, Russian Academy of Sciences, Kosygin Street, 4, Moscow, 117 334, Russian Federation, E-mail: chembio@sky.chph.ras.ru

## CONTENTS

## ABSTRACT

Results of the DFT and MP2 theoretical investigation of 2-pyridin-2-yl-1H-benzimidazole intramolecular dynamics are presented. Structural parameters of 2-pyridin-2-yl-1H-benzimidazole conformers were obtained by these methods; barriers of internal rotation were estimated. GIAO-calculated NMR chemical shifts ($^1$H and $^{13}$C) as obtained at various computational levels are reported for the 2-pyridin-2-yl-1H-benzimidazole conformers. Comparative analysis of experimental and computer NMR spectroscopy results revealed that the GIAO method with B3LYP/6–31G(d,p) level of theory and the PCM approach can be used to estimate the NMR $^1$H and $^{13}$C spectra parameters of the 2-pyridin-2-yl-1H-benzimidazole.

## 24.1 INTRODUCTION

Benzimidazole derivatives are promising leading compounds in the design of substances with antimicrobial, antiviral and anticancer activity [1]. Introduction of pyridine fragment to the benzimidazole structure provides additional coordination center and offers opportunities to create new biomimetic catalytic and sensor systems. 2-Pyridin-2-yl-1H-benzimidazole (PBI) is a very versatile multidonor ligand displaying three potential donor atoms, one sp$^3$- and two sp$^2$-hybridized N-donors, and is particularly interesting in view of its own pharmacological properties as an antibacterial [2] and anti-inflammatory agent [3]. PBI is a key structural element in the design of the allosteric activators of glucokinase [4], PBI and other benzimidazole 2-aryl derivatives show high anticancer activity [5]. PBI complexes with Pd (II) are effective catalysts for the Heck reaction [6], in the case of Co (II) complexes are sensors for the amino acids determination in aqueous media. Neutral and cationic mono- and dinuclear Au(I)/Au(III) complexes derived from PBI show antitumor properties [7]. PBI and its derivatives have interesting photochemical and photophysical properties [9]. This provides PBI using as a model compound for the determination of water content and proton transfer processes investigations in membrane fuel cells [9–12].

The efficiency and selectivity of these systems will depend on the conformational properties of the piridynil as well as benzimidazole fragments. The intramolecular dynamics of the PBI was investigated experimentally by NMR $^1$H and $^{13}$C spectroscopy [13]. The aim of this work is a comprehensive study of the PBI intramolecular dynamics as well as its NMR $^1$H and $^{13}$C spectra by DFT and MP2 methods.

## 24.2  EXPERIMENTAL PART

PBI was obtained as reported elsewhere [14]. $^1$H and $^{13}$C spectra were recorded in DMSO-d$_6$ on 400/100 MHz NMR spectrometer (Bruker Avance II 400) and chemical shifts values ($\delta$) are given in parts per million relative to tetramethylsilane (TMS). Solvent, DMSO-d$_6$ was Sigma-Aldrich reagent and was used without additional purification.

Molecular geometry and electronic structure parameters, thermodynamic characteristics of the 2-pyridin-2-yl-1H-benzimidazole conformers were calculated using the Gaussian 09 [15] software package. Geometric parameters, harmonic vibrational frequencies, and the vibrational contribution to the zero-point vibrational energy were determined after full geometry optimization in the framework of B3LYP/6-31G and B3LYP/6-311G(d,p) density functional calculations as well as MP2/6-31G ones. The optimized geometric parameters were used for total electronic energy calculations by the B3LYP/6-31G, B3LYP/6-311G(d,p), and MP2/6-31G methods. The 6-31G basis set was used in this work because it has a low computational cost. The B3LYP/6-311G(d,p) method was used to elucidate the effect of basis set extension on the results of calculations.

Figure 24.1 presents PBI molecule atom numbering used for geometric and $^1$H and $^{13}$C NMR spectra parameters presenting.

The magnetic shielding tensors ($\chi$, ppm) for $^1$H and $^{13}$C nuclei of the PBI conformers were calculated with the MP2/6–31G(d,p) and MP2/6–31G(d,p)/PCM optimized geometries by standard GIAO (Gauge-Independent Atomic Orbital) approach [16]. The calculated magnetic isotropic shielding tensors, $\chi_i$, were transformed to chemical shifts relative to TMS, $\delta_i$, by $\delta_i = \chi_{ref} - \chi_i$, where both, $\chi_{ref}$ and $\chi_i$, were taken from calculations at the same computational level. The solvent effect was considered

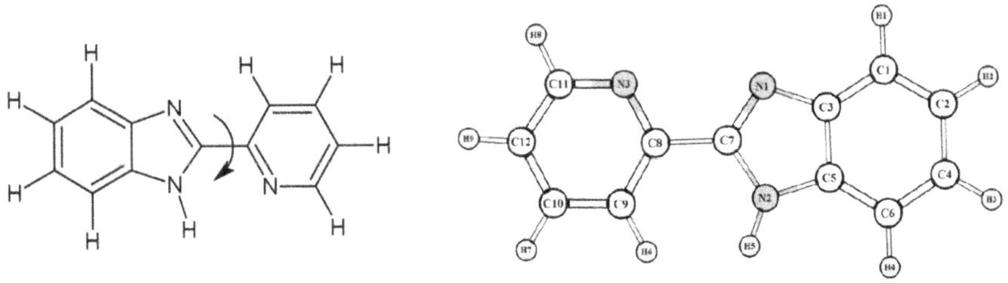

**FIGURE 24.1**    Atom-labeling scheme of PBI molecule.

in the PCM approximation [17, 18]. $\chi$ values for magnetically equivalent nuclei were averaged.

Inspecting the overall agreement between experimental and theoretical spectra RMS errors ($\sigma$) were used to consider the quality of the $^1$H and $^{13}$C nuclei chemical shifts calculations. Correlation coefficients (R) were calculated to estimate the agreement between spectral patterns and trends.

## 24.3  RESULTS AND DISCUSSIONS

### 24.3.1  COMFORMERS AND ROTATIONAL BARRIERS

The potential energy pathway for internal rotation in PBI molecule was estimated by optimizing the molecular geometries with different dihedral angle between the two aromatic planes. The barriers to internal rotation were calculated within the framework of B3LYP/6–31G, B3LYP/6–311G(d,p), MP2/6–31G and MP2/6–311G(d,p) methods. The dependence of the potential energy on the dihedral angle (Figure 24.2) was determined by scanning the dihedral angle N(1)-C(7)-C(8)-C(9) (used as internal rotation coordinate $\Theta$) from 0 to 360° with an increment of 15° and geometry optimization in each step. Rotation was performed sequentially about the C(7)-C(8) bond. At each energy minimum, the molecular geometry was fully optimized. The analysis of vibrational frequency was also performed at the same level of theory, and the calculation results revealed that the PBI conformers have no imaginary frequency.

The internal dynamics curves of the pyridinyl moiety around C(7)-C(8) bond in PBI molecule obtained within B3LYP/6–31G and MP2/6–31G methods are presented on the Figure 24.2. There are three minima on

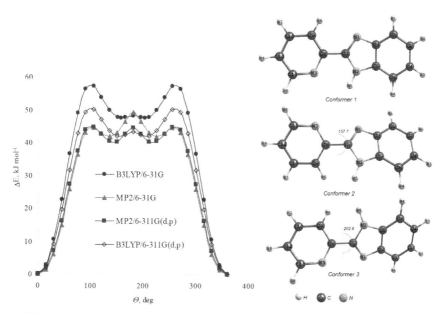

**FIGURE 24.2**  Conformational energy *vs.* dihedral angle plots for internal rotation about the C(7)-C(8) bond in PBI molecule and the equilibrium configurations of the PBI conformers (B3LYP/6–31G method). The conformational energies were calculated as the total electronic energy differences between a conformer with the current value of $\Theta$ and the most stable conformer.

these curves. Configurations of PBI molecule with $\Theta$ value of 0° and 360° are identical. Minima on the curve corresponded to $\Theta$ value of 157.7° and 202.6° (B3LYP method) as well as 135.8° and $\Theta$ = 224.1° (MP2 method) are of the same energy, and are characterized by the same value of dipole moment (Table 24.1). The equilibrium structures of the PBI conformers are shown in Figure 24.2. Relative electronic energies of them are listed in Table 24.1 and indicate that conformer 1 dominate at room temperature whereas the content of conformers 2 and 3 is low. Figure 24.3 presents the visualization of HOMO and LUMO for BIP conformers (Table 24.2).

Experimentally obtained activation free energy for the interconversion between the PBI unequally populated rotamers as reported in Ref. [13] are 63.3 kJ/mol (NMR $^1$H data) 59.4 kJ/mol (NMR $^{13}$C data). Thus rotation barriers obtained within B3LYP method are in reasonable agreement with experimental data.

**TABLE 24.1**　PBI Conformers Parameters

| Parameter | B3LYP/6-31G | | | MP2/6-31G | | |
|---|---|---|---|---|---|---|
| | 1 | 2 | 3 | 1 | 2 | 3 |
| $\Delta E$, kJ mol$^{-1}$ | 0 | 47.55 | 47.55 | 0 | 41.78 | 41.78 |
| $\Theta$, deg | 0.0 | 157.7 | 202.6 | 0.0 | 135.8 | 224.1 |
| $\mu$, D | 2.451 | 5.147 | 5.147 | 2.929 | 5.533 | 5.533 |
| C(7)-C(8), Å | 1.455 | 1.465 | 1.465 | 1.464 | 1.473 | 1.473 |
| C(7)-N(1), Å | 1.332 | 1.328 | 1.328 | 1.354 | 1.349 | 1.349 |
| C(7)-N(2), Å | 1.385 | 1.402 | 1.402 | 1.395 | 1.409 | 1.409 |
| C(8)-N(3), Å | 1.359 | 1.356 | 1.356 | 1.376 | 1.376 | 1.376 |
| C(11)-N(3), Å | 1.348 | 1.346 | 1.346 | 1.369 | 1.369 | 1.369 |
| C(8)-C(9), Å | 1.403 | 1.408 | 1.408 | 1.414 | 1.416 | 1.416 |
| C(9)-C(10), Å | 1.395 | 1.397 | 1.397 | 1.407 | 1.409 | 1.410 |
| C(9)-H(6), Å | 1.083 | 1.085 | 1.085 | 1.089 | 1.092 | 1.092 |
| C(3)-N(1), Å | 1.399 | 1.399 | 1.399 | 1.417 | 1.419 | 1.420 |
| C(5)-N(2), Å | 1.387 | 1.390 | 1.390 | 1.397 | 1.401 | 1.401 |
| C(3)-C(5), Å | 1.426 | 1.422 | 1.422 | 1.435 | 1.431 | 1.431 |
| C(1)-C(3), Å | 1.401 | 1.401 | 1.401 | 1.415 | 1.414 | 1.415 |
| C(5)-C(6), Å | 1.399 | 1.399 | 1.399 | 1.415 | 1.414 | 1.414 |
| N(2)-H(5), Å | 1.008 | 1.006 | 1.006 | 1.013 | 1.012 | 1.012 |
| C(1)-H(1), Å | 1.084 | 1.084 | 1.083 | 1.090 | 1.090 | 1.090 |
| C(2)-C(4), Å | 1.415 | 1.415 | 1.414 | 1.431 | 1.430 | 1.430 |
| N(1)-C(7)-C(8) | 126.5 | 126.3 | 126.3 | 126.1 | 126.3 | 126.3 |
| C(7)-C(8)-C(9) | 121.5 | 121.4 | 121.4 | 121.5 | 120.7 | 120.7 |
| C(7)-C(8)-N(3) | 115.9 | 116.9 | 117.0 | 115.4 | 116.7 | 116.7 |
| C(8)-N(3)-C(11) | 118.2 | 118.5 | 118.4 | 117.4 | 117.3 | 117.3 |
| N(1)-C(7)-N(2) | 112.6 | 111.8 | 111.8 | 112.7 | 112.2 | 112.2 |
| C(7)-N(2)-C(5) | 107.4 | 107.4 | 107.4 | 107.5 | 107.5 | 107.5 |
| C(3)-N(1)-C(7) | 105.1 | 105.7 | 105.7 | 104.3 | 104.7 | 104.7 |
| N(1)-C(3)-C(5) | 109.9 | 110.2 | 110.2 | 110.3 | 110.6 | 110.6 |
| C(3)-C(5)-C(6) | 121.9 | 122.1 | 122.1 | 122.0 | 122.2 | 122.2 |

## 24.3.2　NMR $^1$H AND $^{13}$C CHEMICAL SHIFTS

For the identified conformers of PBI molecule $^1$H and $^{13}$C chemical shifts were estimated. Only two conformers (1 and 2) were considered. The MP2/6–31G(d,p) and MP2/6–31G(d,p)/PCM as well as

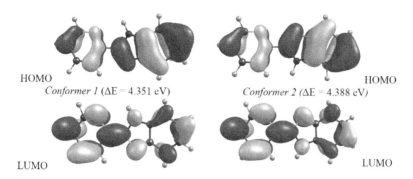

HOMO
*Conformer 1 (ΔE = 4.351 eV)*                      *Conformer 2 (ΔE = 4.388 eV)*          HOMO

LUMO                                                                        LUMO

**FIGURE 24.3**   HOMO and LUMO of the PBI conformers (B3LYP/6–311G(d,p) method).

**TABLE 24.2**   PBI Rotational Barriers Values

| Parameter | B3LYP | | MP2 | |
|---|---|---|---|---|
| | 6-31G | 6-311G(d,p) | 6-31G | 6-311G(d,p) |
| $\Delta E_{1\to 2}$, kJ mol$^{-1}$ | 57.24 | 50.14 | 44.50 | 44.86 |
| $\Delta E_{2\to 3}$, kJ mol$^{-1}$ | 0.67 | 1.30 | 7.44 | 4.24 |
| $\Delta E_{3\to 1}$, kJ mol$^{-1}$ | 9.69 | 8.24 | 2.72 | 4.51 |

B3LYP/6–31G(d,p) and B3LYP/6–31G(d,p)/PCM optimized geometries were used for magnetic shielding tensors calculation by standard GIAO method. The chemical shifts of PBI at different computational levels are listed in Table 24.3 along with corresponding experimental solution data.

Concerning the spectral patterns of protons and carbons, inspection of Table 24.3 reveals the following features. The patterns of $^1$H and $^{13}$C spectra of BIP are correctly reproduced at all used computational levels. Although both levels yield qualitatively similar results, the experimental patterns are better reproduced by B3LYP calculations.

When passing to the calculations in the PCM mode solvation accounting leads to more correct results for the MP2 and B3LYP methods. The lowest σ values are obtained with B3LYP/6–31G(d,p) basis set. Linear relationships between the experimental chemical shifts and the calculated ones have been obtained for both methods. The correlation coefficients (R) corresponding to obtained dependences are shown in Table 24.3. Joint account of σ and R values indicates possibility of B3LYP method with 6–31G(d,p) basis set using for the calculation of the BIP chemical shifts. Using of the PCM mode in calculations is preferable as compared to the isolated particle approximation.

**TABLE 24.3** Experimental and Calculated NMR $^1$H and $^{13}$C Chemical Shifts of PBI

| Atom | Conformer 1 | | | Conformer 2 | | | Experiment |
|------|------|-------|---------------|------|-------|---------------|------------|
|      | MP2  | B3LYP | B3LYP/<br>PCM | MP2  | B3LYP | B3LYP/<br>PCM |            |
| C1   | 132.15 | 127.25 | 126.08 | 133.14 | 128.63 | 126.49 | 118.93 |
| C2   | 132.41 | 127.75 | 129.14 | 132.60 | 127.90 | 129.23 | 122.42 |
| C3   | 157.25 | 152.08 | 151.42 | 157.48 | 151.57 | 150.9  | 143.79 |
| C4   | 133.19 | 129.67 | 130.91 | 133.03 | 129.99 | 131.12 | 122.42 |
| C5   | 144.55 | 139.96 | 141.53 | 144.86 | 139.81 | 141.75 | 134.8  |
| C6   | 120.33 | 114.25 | 117.40 | 119.57 | 112.64 | 117.22 | 111.83 |
| C7   | 158.48 | 154.9  | 157.16 | 156.82 | 153.4  | 156.63 | 148.62 |
| C8   | 155.98 | 155.79 | 155.05 | 160.10 | 155.23 | 154.71 | 150.27 |
| C9   | 133.46 | 127.1  | 128.07 | 133.41 | 122.52 | 127.25 | 123.78 |
| C10  | 143.94 | 141.83 | 144.96 | 142.89 | 140.7  | 144.87 | 136.6  |
| C11  | 156.77 | 154.9  | 157.16 | 159.21 | 157.5  | 158.24 | 148.79 |
| C12  | 135.24 | 128.39 | 131.77 | 134.93 | 127.3  | 131.27 | 121.22 |
| σ    | *10.31* | *6.11* | *7.46* | *10.64* | *6.10* | *7.42* | - |
| R    | *0.98* | *0.99* | *0.99* | *0.98* | *0.98* | *0.99* | - |
| H1   | 8.3114 | 8.04   | 8.244  | 8.4027 | 8.14   | 8.262  | 7.63   |
| H2   | 7.628  | 7.441  | 7.777  | 7.6677 | 7.464  | 7.783  | 7.14   |
| H3   | 7.6081 | 7.439  | 7.800  | 7.627  | 7.457  | 7.817  | 7.14   |
| H4   | 7.8177 | 7.526  | 8.063  | 7.7757 | 7.487  | 8.077  | 7.51   |
| H5   | 10.223 | 9.690  | 10.846 | 8.4752 | 8.536  | 10.415 | 12.88  |
| H6   | 9.0363 | 8.732  | 8.852  | 7.7789 | 7.536  | 8.37   | 8.47   |
| H7   | 8.0075 | 7.875  | 8.392  | 7.9712 | 7.778  | 8.389  | 7.90   |
| H8   | 8.802  | 8.791  | 9.132  | 9.1272 | 9.139  | 9.258  | 8.67   |
| H9   | 7.5145 | 7.255  | 7.819  | 7.5901 | 7.204  | 7.798  | 7.41   |
| σ    | *0.425* | *0.248* | *0.541* | *0.478* | *0.434* | *0.540* | - |
| R*   | *0.92* | *0.94* | *0.99* | *0.68* | *0.68* | *0.90* | - |

*Chemical shift of H5 was not accounted.

## 24.4 CONCLUSIONS

A comprehensive study of the 2-pyridin-2-yl-1H-benzimidazole by experimental NMR $^1$H and $^{13}$C spectroscopy and molecular modeling methods was performed. Structural parameters of the BIP conformers were obtained by MP2 and B3LYP methods. Rotation barriers obtained within B3LYP method are in reasonable agreement with experimental data. GIAO-calculated NMR chemical shifts ($^1$H and $^{13}$C) as obtained at various computational levels are reported for the 2-pyridin-2-yl-1H-benzimidazole conformers. For NMR $^1$H and $^{13}$C spectra of the BIP in DMSO-d$_6$ MP2 and B3LYP methods approximations with 6-31G(d,p) basis set allow to obtain the correct spectral pattern. A linear correlation between the calculated and experimental values of the $^1$H and $^{13}$C chemical shifts for the studied molecule were obtained. B3LYP method combined with 6–31G(d,p) basis set and PCM approximation allows to get a better agreement between the calculated and experimental data.

## KEYWORDS

- **2-pyridin-2-yl-1H-benzimidazole**
- **chemical shift**
- **GIAO**
- **intramolecular dynamics**
- **magnetic shielding constant**
- **molecular modeling**

## REFERENCES

1. Walia, R., Hedaitullah, Md., Naaz, S. F., Iqbal, Kh., & Lamba, H. S. Benzimidazole derivatives – an overview. *Int. J. Research Pharm. Chem.* 2011, 3(1), 565–574.
2. Schiffmann, R., Neugebauer, A., & Klein, C. D. Metal-Mediated Inhibition of Escherichia coli Methionine Aminopeptidase: Structure-Activity Relationships and Development of a Novel Scoring Function for Metal-Ligand Interactions. *J. Med. Chem.* 2006, 49, 511–522.
3. Tsukamoto, G., Yoshino, K., Kohono, T., Ohtaka, H., Kagaya, H., & Ito, K. 2-Substituted azole derivatives. 1. Synthesis and anti-inflammatory activity of some 2-(substituted-pyridinyl)benzimidazoles. *J. Med. Chem.* 1980, 23, 734–738.
4. Ishikawa, M., Nonoshita, K., Ogino, Y., Nagae, Y., Tsukahara, D., Hosaka, H., Maruki, H., Ohyama, S., Yoshimoto, R., Sasaki, K., Nagata, Y., Eiki, J., & Nishimura, T. Discovery of novel 2-(pyridine-2-yl)-1H-benzimidazole derivatives as potent glucokinase activators. *Bioorg. Med. Chem. Lett.* 2009, 19, 4450–4454.

5. Sontakke, V. A., Ghosh, S., & Lawande, P. P., A Simple. Efficient Synthesis of 2-Aryl Benzimidazoles Using Silica Supported Periodic Acid Catalyst and Evaluation of Anticancer Activity. *ISRN Organic Chemistry* (2013) Article ID 453682, 7 pages.

6. Chen, W., Xi, C., & Wu, Y. Highly active Pd(II) catalysts with pyridylbenzoimidazole ligands for the Heck reaction. *Journal of Organometallic Chemistry* 2007, 692, 4381–4388.

7. Das, S., Guha, S., & Banerjee, A. 2-(2-Pyridyl) benzimidazole based Co(II) complex as an efficient fluorescent probe for trace level determination of aspartic and glutamic acid in aqueous solution: A displacement approach. *Org. Biomol. Chem.* 2011, 9, 7097–7104.

8. Maiore, L., Aragoni, M. C., Deiana, C., Cinellu, M. A., Isaia, F., Lippolis, V., Pintus, A., Serratrice, M., & Arca, M. Structure–Activity Relationships in Cytotoxic AuI/AuIII Complexes Derived from 2-(2'-Pyridyl)benzimidazole. *Inorg. Chem.* 2014, 53(8), 4068–4080.

9. Guin, M., Maity, S., & Patwari, G. N. Infrared-optical double resonance spectroscopic measurements on 2-(2'-Pyridyl)benzimidazole and its hydrogen bonded complexes with water and methanol. *Phys. Chem. A.* 2010, 114, 8323–8330.

10. Iyer, S. S., Dhrubajyoti, S., Dey, A., Kundu, A., & Datta, A. 2-(2'-Pyridyl)benzimidazole as a fluorescent probe of hydration of Nafion membranes. *Indian J. Chem.* 1999, 38A, 1223–1227.

11. Iyer, E. S. S., & Datta, A. Microheterogeneity in native and cation-exchanged Nafion membranes. *J. Phys. Chem. B.* 2012, 116, 9992–9998.

12. Iyer, E. S. S., & Datta, A. Influence of external electrolyte on ion exchange in Nafion membranes. *RSC Advances* 2012, 2, 8050–8054.

13. Anchi Yeh, Chi-Yu Shih, Lieh-Li Lin, Shung-Jim Yang, Cheng-Tung Chang. Variable-temperature NMR studies of 2-(pyridin-2-yl)-1H-benzod.imidazole. *Life Science Journal* 2009, 6(4), 1–4.

14. Thakur, P., Chakravortty, V., & Dash, K. C. Synthesis and characterization of lanthanide (III) complexes of 5-methyl-2-(2'-pyridyl)benzimidazole and 2-(2'-pyridyl) benzimidazole. *Indian Journal of Chemistry* 1999, 38A, 1223–1227.

15. Frisch, M. J., Trucks, G. W., Schlegel, H. B., Scuseria, G. E., Robb, M. A., Cheeseman, J. R., et al., Gaussian 09, Revision B.01, Gaussian, Inc., Wallingford CT, 2010.

16. Wolinski, K., Hilton, J. F., & Pulay, P. Efficient implementation of the gauge-independent atomic orbital method for NMR chemical shift calculations. *J. Am. Chem. Soc.* 1990, 112, 8251–8260.

17. Mennucci, B., & Tomasi, J. Continuum solvation models: A new approach to the problem of solute's charge distribution and cavity boundaries. *J. Chem. Phys.* 1997, 106, 5151–5158.

18. Cossi, M., Scalmani, G., Rega, N., & Barone, V. New developments in the polarizable continuum model for quantum mechanical and classical calculations on molecules in solution. *J. Chem. Phys.* 2002, 117, 43–54.

# CHAPTER 25

# NMR 13C SPECTRA OF THE 1,1,3-TRIMETHYL-3-(4-METHYLPHENYL)BUTYL HYDROPEROXIDE IN VARIOUS SOLVENTS: MOLECULAR MODELING

N. A. TUROVSKIJ,[1] E. V. RAKSHA,[2] YU. V. BERESTNEVA,[2] and G. E. ZAIKOV[3]

[1]*Donetsk National University, Universitetskaya Street, 24, Donetsk, 83001, Ukraine*

[2]*L.M. Litvinenko Institute of Physical Organic and Coal Chemistry, Donetsk 83114, Ukraine*

[3]*Institute of Biochemical Physics, Russian Academy of Sciences, Kosygin Street, 4, Moscow, 117 334, Russian Federation*

*E-mail: NA.Turovskij@gmail.com; elenaraksha411@gmail.com; chembio@sky.chph.ras.ru*

## CONTENTS

## ABSTRACT

GIAO-calculated NMR [13]C chemical shifts as obtained at various computational levels are reported for the 1,1,3-trimethyl-3-(4-methylphenyl)
butyl hydroperoxide. The data are compared with experimental solution data in chloroform-d, acetonitrile-$d_3$, and DMSO-$d_6$, focusing on
the agreement with spectral patterns and spectral trends. Calculation
of magnetic shielding tensors and chemical shifts for [13]C nuclei of the
1,1,3-trimethyl-3-(4-methylphenyl)butyl hydroperoxide molecule in the
approximation of an isolated particle and considering the solvent influence
in the framework of the continuum polarization model (PCM) was carried
out. Comparative analysis of experimental and computer NMR spectroscopy results revealed that the GIAO method with MP2/6–31G(d,p) level of
theory and the PCM approach can be used to estimate the NMR [13]C chemical shifts of the 1,1,3-trimethyl-3-(4-methylphenyl)butyl hydroperoxide.

## 25.1   INTRODUCTION

Arylalkyl hydroperoxides are useful starting reagents in the synthesis of
surface-active peroxide initiators for the preparation of polymeric colloidal systems with improved stability [1]. Thermolysis of arylalkyl hydroperoxides was studied in acetonitrile [2]. NMR [1]H spectroscopy has been
already used successfully for the experimental evidence of the a complex
formation between a 1,1,3-trimethyl-3-(4-methylphenyl)butyl hydroperoxide and tetraalkylammonium bromides in acetonitrile [3–5] and chloroform solution [5]. The aim of this work is a comprehensive study of
the 1,1,3-trimethyl-3-(4-methylphenyl)butyl hydroperoxide (ROOH) by
experimental NMR [13]C spectroscopy and molecular modeling methods.

## 25.2   EXPERIMENTAL PART

The 1,1,3-trimethyl-3-(4-methylphenyl)butyl hydroperoxide (ROOH) was purified according to Ref. [1]. Its purity (99%) was controlled by iodometry method as well as by NMR spectroscopy. Experimental NMR $^{13}C$ spectra of the hydroperoxide solutions were obtained by using the Bruker Avance II 400 spectrometer (NMR $^{1}H$ – 400 MHz, NMR $^{13}C$ – 100 MHz) at 297 K. Solvents, chloroform-d, acetonitrile-$d_3$, and DMSO-$d_6$, were Sigma-Aldrich reagents and were used without additional purification but were stored above molecular sieves before using. Tetramethylsilane (TMS) was internal standard. The hydroperoxide concentration in solutions was 0.03 mol·dm$^{-3}$. Molecular geometry and electronic structure parameters, as well as harmonic vibrational frequencies of the 1,1,3-trimethyl-3-(4-methylphenyl) butyl hydroperoxide molecule were calculated after full geometry optimization in the framework of B3LYP/6–31G(d,p) and MP2/6–31G(d,p) methods. The resulting equilibrium molecular geometry was used for total electronic energy calculations by the B3LYP/6–31G(d,p) and MP2/6–31G(d,p) methods. All calculations have been carried out using the Gaussian03 [6] program.

The magnetic shielding tensors ($\chi$, ppm) for $^{13}C$ nuclei of the hydroperoxide and the reference molecule were calculated with the MP2/6–31G(d,p) and B3LYP/6–31G(d,p) equilibrium geometries by standard GIAO (Gauge-Independent Atomic Orbital) approach [7]. The calculated magnetic isotropic shielding tensors, $\chi_i$, were transformed to chemical shifts relative to TMS molecule, $\delta_i$, by $\delta_i = \chi_{ref} - \chi_i$, where both, $\chi_{ref}$ and $\chi_i$, were taken from calculations at the same computational level. Table 25.1 illustrates $\chi$ values for TMS molecule used for the hydroperoxide $^{13}C$ nuclei chemical shifts calculations.

**TABLE 25.1**   Magnetic Shielding Tensors for $^{13}C$ Nuclei of the TMS

| Solvent | MP2 | | | B3LYP | | |
|---|---|---|---|---|---|---|
| | 1 | 2 | 3 | 1 | 2 | 3 |
| - | 207.54 | 199.71 | 199.37 | 191.80 | 184.13 | 183.72 |
| Chloroform | 207.86 | 200.13 | 199.79 | 192.08 | 184.53 | 184.13 |
| Acetonitrile | 208.01 | 200.32 | 199.99 | 192.19 | 184.70 | 184.30 |
| DMSO | 208.01 | 200.33 | 200.00 | 192.30 | 184.81 | 184.40 |

*Note: 1 – 6–31G(d,p); 2 – 6–311G(d,p); 3 – 6–311++G(d,p).*

$\chi$ values were also estimated in the framework of 6–311G(d,p) and 6–311++G(d,p) basis sets on the base of MP2/6–31G(d,p) and B3LYP/6–31G(d,p) equilibrium geometries. The solvent effect was considered in the PCM approximation [8, 9]. $\chi$ values for magnetically equivalent nuclei were averaged.

Inspecting the overall agreement between experimental and theoretical spectra RMS errors ($\sigma$) were used to consider the quality of the [13]C nuclei chemical shifts calculations. Correlation coefficients (R) were calculated to estimate the agreement between spectral patterns and trends.

## 25.3  RESULTS AND DISCUSSIONS

### 25.3.1  EXPERIMENTAL NMR [13]C SPECTRA OF THE 1,1,3-TRIMETHYL-3-(4-METHYLPHENYL)BUTYL HYDROPEROXIDE

Experimental NMR [13]C spectra of the 1,1,3-trimethyl-3-(4-methylphenyl) butyl hydroperoxide (ROOH) were obtained from chloroform-d, acetonitrile-$d_3$, and DMSO-$d_6$

solutions. The hydroperoxide concentration in all samples was 0.03 mol·dm$^{-3}$. The experimental NMR [13]C spectra of the ROOH are presented in Figure 25.1.

Ten signals for the hydroperoxide carbon atoms are observed in the ROOH [13]C NMR spectrum. Signal of the carbon atom bonded with a hydroperoxide group shifts slightly to the stronger fields with the solvent polarity increasing, while the remaining signals are shifted to weak fields. Linear dependences between the [13]C chemical shifts values of the hydroperoxide are observed in the studied solvents (Figure 25.1). This is consistent with authors [11], who showed linear correlation between the chemical shifts values in chloroform-d and dimethylsulphoxide-$d_6$

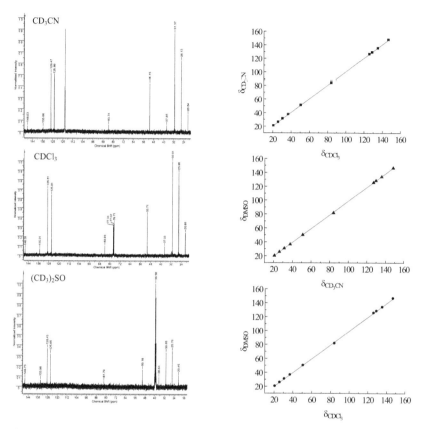

**FIGURE 25.1** The relationship between the experimental NMR $^{13}$C chemical shifts (relative to TMS) of the 1,1,3-trimethyl-3-(4-methylphenyl)butyl hydroperoxide in different solvents.

for a large number of organic compounds of different classes. Equations corresponded to the obtained relationships (Figure 25.1) are listed below.

$$\delta_{CD_3CN} = (0.02 \pm 0.23) + (1.006 \pm 0.002)\delta_{CDCl_3}$$

$$\delta_{DMSO-d_6} = (-0.46 \pm 0.41) + (0.999 \pm 0.004)\delta_{CDCl_3}$$

$$\delta_{DMSO-d_6} = (-0.48 \pm 0.24) + (0.993 \pm 0.003)\delta_{CD_3CN}$$

## 25.3.2 MOLECULAR MODELING OF THE 1,1,3-TRIMETHYL-3-(4-METHYLPHENYL)BUTYL HYDROPEROXIDE NMR $^{13}C$ SPECTRA BY MP2 AND B3LYP METHODS

The hydroperoxide molecule geometry optimization in the framework of MP2/6–31G(d,p) and B3LYP/6–31G(d,p) methods was carried out as the first step of the hydroperoxide NMR $^{13}C$ spectra modeling. Initial hydroperoxide configuration chosen for calculations was those one obtained by semiempirical AM1 method and used recently for the hydroperoxide O-O bond homolysis [2] as well as complexation with $Et_4NBr$ [4, 12] modeling. The main parameters of the hydroperoxide fragment molecular geometry obtained in the isolated particle approximation within the framework of MP2/6–31G(d,p) (Figure 25.2) and B3LYP/6–31G(d,p) levels of theory are presented in Table 25.2. Peroxide bond O-O is a reaction center in this type of chemical initiators thus the main attention was focused on the geometry of -CO-OH fragment. The calculation results were compared with known experimental values for the *tert*-butyl hydroperoxide [13], and

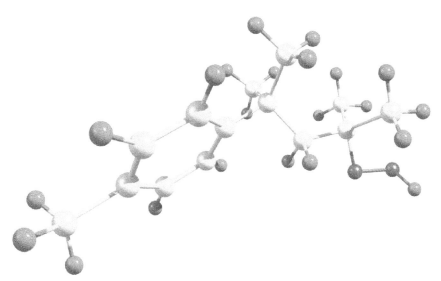

**FIGURE 25.2** The 1,1,3-trimethyl-3-(4-methylphenyl)butyl hydroperoxide structural model (MP2/6–31G(d,p) method).

**TABLE 25.2**  Molecular Geometry Parameters of the 1,1,3-Trimethyl-3-(4-Methylphenyl) Butyl Hydroperoxide -CO-OH Moiety

| Parameter | MP2/6–31G(d,p) | B3LYP/6–31G(d,p) | Experiment* |
|---|---|---|---|
| $l_{O-O}$, Å | 1.473 | 1.456 | 1.473 |
| $l_{C-O}$, Å | 1.459 | 1.465 | 1.443 |
| $l_{O-H}$, Å | 0.970 | 0.971 | 0.990 |
| C-O-O, ° | 108.6 | 110.0 | 109.6 |
| O-O-H, ° | 98.2 | 99.9 | 100.0 |
| C-O-O-H, ° | 112.4 | 109.1 | 114.0 |

*Note: experimental values are those for *tert*-butyl hydroperoxide from Ref. [13].

appropriate agreement between calculated and experimental parameters can be seen in the case of MP2/6–31G(d,p) method.

Calculation of $^{13}$C chemical shifts of the hydroperoxide was carried out by GIAO method in the approximation of an isolated particle as well as in studied solvents within the PCM model, which takes into account the non-specific solvation. Equilibrium hydroperoxide geometries obtained in the framework of MP2/6–31G(d,p) and B3LYP/6–31G(d,p) levels of theory for the isolated particle approximation were used in all cases.

The chemical shift values ($\delta$, ppm) for $^{13}$C nuclei in the hydroperoxide molecule were evaluated on the base of calculated magnetic shielding constants ($\chi$, ppm). TMS was used as standard, for which the molecular geometry optimization and $\chi$ calculation were performed using the same level of theory and basis set. Values of the $^{13}$C chemical shifts were found as the difference of the magnetic shielding tensors of the corresponding TMS and hydroperoxide nuclei (Tables 25.3 and 25.4).

The correct spectral pattern for the hydroperoxide NMR $^{13}$C spectrum was obtained for all methods and basis sets used within the isolated molecule approximation (*see* Table 25.3) as well as solvation accounting (*see* Table 25.4). Exceptions are aromatic C8 and C9 carbons, which signals are interchanged for all calculations.

The best reproduced experimental chemical shift value for the carbon atom of the CO-OH group is observed in the case of MP2/6–31G(d,p) approximation in all used solvents whereas B3LYP with the same basis set gives slightly worse values. Basis set extension

**TABLE 25.3**   NMR $^{13}$C Chemical Shifts ($\delta$, ppm) of the 1,1,3-Trimethyl-3-(4-Methylphenyl) Butyl Hydroperoxide (the Isolated Particle Approximation)

| Nuclei | MP2 | | | B3LYP | | |
|---|---|---|---|---|---|---|
| | 1 | 2 | 3 | 1 | 2 | 3 |
| C1 | 83.61 | 86.87 | 88.24 | 85.77 | 90.90 | 92.04 |
| C2 | 53.50 | 57.48 | 57.42 | 53.23 | 57.70 | 57.00 |
| C3 | 26.46 | 26.71 | 26.62 | 24.62 | 25.27 | 24.93 |
| C4 | 37.04 | 40.23 | 40.37 | 41.14 | 44.66 | 44.49 |
| C5 | 30.10 | 30.93 | 31.03 | 28.57 | 29.78 | 29.75 |
| C6 | 141.39 | 153.47 | 153.93 | 144.46 | 158.38 | 158.37 |
| C7 | 116.91 | 125.90 | 126.29 | 119.80 | 130.58 | 131.03 |
| C8 | 127.07 | 137.37 | 137.97 | 130.75 | 142.40 | 143.32 |
| C9 | 121.55 | 130.79 | 131.41 | 123.68 | 134.44 | 135.07 |
| C10 | 22.98 | 23.82 | 23.75 | 21.84 | 23.25 | 22.84 |

*Note*: $1$ – 6–31G(d,p); $2$ – 6–311G(d,p); $3$ – 6–311++G(d,p).

**TABLE 25.4**   NMR $^{13}$C Chemical Shifts ($\delta$, ppm) of the 1,1,3-Trimethyl-3-(4-Methylphenyl) Butyl Hydroperoxide in Different Solvents

| Nuclei | MP2 | | | B3LYP | | | 4 |
|---|---|---|---|---|---|---|---|
| | 1 | 2 | 3 | 1 | 2 | 3 | |
| Chloroform | | | | | | | |
| C1 | 84.29 | 87.74 | 89.27 | 86.44 | 91.86 | 93.14 | 83.93 |
| C2 | 53.69 | 57.73 | 57.64 | 53.37 | 57.89 | 57.17 | 50.71 |
| C3 | 26.62 | 26.99 | 26.89 | 24.73 | 25.52 | 25.17 | 25.98 |
| C4 | 37.46 | 40.74 | 40.92 | 41.54 | 45.20 | 45.07 | 37.03 |
| C5 | 30.18 | 31.10 | 31.20 | 28.59 | 29.90 | 29.88 | 30.91 |
| C6 | 142.14 | 154.40 | 154.90 | 145.12 | 159.23 | 159.25 | 146.55 |
| C7 | 117.33 | 126.51 | 126.91 | 120.11 | 131.14 | 131.60 | 125.81 |
| C8 | 127.92 | 138.43 | 139.03 | 131.56 | 143.43 | 144.34 | 128.81 |
| C9 | 121.90 | 131.35 | 131.91 | 123.85 | 134.81 | 135.37 | 135.01 |
| C10 | 23.01 | 23.96 | 23.88 | 21.81 | 23.36 | 22.93 | 20.86 |
| $\sigma$ | 27.86 | 25.63 | 28.66 | 20.82 | 59.16 | 63.32 | - |
| $R$ | 0.997 | 0.997 | 0.997 | 0.996 | 0.996 | 0.996 | - |

**TABLE 25.4** Continued

| Nuclei | MP2 | | | B3LYP | | | 4 |
|---|---|---|---|---|---|---|---|
| | 1 | 2 | 3 | 1 | 2 | 3 | |
| Acetonitrile | | | | | | | |
| C1 | 84.58 | 88.13 | 89.73 | 86.72 | 92.27 | 93.61 | 83.74 |
| C2 | 53.80 | 57.88 | 57.78 | 53.44 | 57.98 | 57.27 | 51.15 |
| C3 | 26.70 | 27.13 | 27.02 | 24.77 | 25.63 | 25.28 | 26.13 |
| C4 | 37.64 | 40.99 | 41.18 | 41.71 | 45.43 | 45.32 | 37.65 |
| C5 | 30.22 | 31.19 | 31.29 | 28.59 | 29.95 | 29.93 | 31.37 |
| C6 | 142.50 | 154.84 | 155.37 | 145.42 | 159.62 | 159.66 | 148.03 |
| C7 | 117.52 | 126.80 | 127.20 | 120.24 | 131.37 | 131.84 | 126.89 |
| C8 | 128.29 | 138.90 | 139.49 | 131.88 | 143.85 | 144.75 | 129.47 |
| C9 | 122.08 | 131.63 | 132.17 | 123.94 | 135.00 | 135.52 | 132.66 |
| C10 | 23.03 | 24.04 | 23.96 | 21.80 | 23.41 | 22.97 | 20.84 |
| $\sigma$ | *24.571* | *22.325* | *25.741* | *17.395* | *55.560* | *60.243* | |
| R | *0.998* | *0.998* | *0.998* | *0.997* | *0.997* | *0.996* | |
| DMSO | | | | | | | |
| C1 | 84.60 | 88.15 | 89.75 | 86.84 | 92.38 | 93.72 | 81.79 |
| C2 | 53.80 | 57.88 | 57.78 | 53.54 | 58.09 | 57.37 | 50.18 |
| C3 | 26.70 | 27.13 | 27.03 | 24.87 | 25.74 | 25.37 | 25.76 |
| C4 | 37.65 | 41.00 | 41.19 | 41.82 | 45.54 | 45.43 | 36.64 |
| C5 | 30.22 | 31.19 | 31.30 | 28.69 | 30.06 | 30.03 | 30.85 |
| C6 | 142.52 | 154.86 | 155.39 | 145.54 | 159.74 | 159.77 | 146.73 |
| C7 | 117.53 | 126.81 | 127.22 | 120.35 | 131.48 | 131.94 | 125.65 |
| C8 | 128.31 | 138.92 | 139.52 | 131.99 | 143.97 | 144.86 | 128.43 |
| C9 | 122.09 | 131.64 | 132.18 | 124.05 | 135.11 | 135.62 | 133.98 |
| C10 | 23.03 | 24.04 | 23.96 | 21.90 | 23.51 | 23.06 | 20.45 |
| $\sigma$ | *25.496* | *31.656* | *35.949* | *21.206* | *70.994* | *76.092* | |
| R | *0.997* | *0.997* | *0.997* | *0.995* | *0.996* | *0.995* | |

*Note*: *1* – 6–31G(d,p); *2* – 6–311G(d,p); *3* – 6–311++G(d,p); *4* – experimental data.

to 6–311++G(d,p) leads to a deterioration of the calculation results. Calculated value for the carbon of CO-OH group (83.61 ppm) within the isolated molecule approximation is closest to experimental one in acetonitrile (83.74 ppm). When passing to the calculations in the PCM mode solvation accounting leads to more correct results for the MP2 and B3LYP methods. The lowest σ values for all solvents are obtained with 6–31G(d,p) basis set. Linear relationships between the experimental NMR $^{13}$C chemical shifts and the calculated values $\delta_{calc}$ for the hydroperoxide $^{13}$C nuclei (see Figure 25.3) have been obtained for both methods and all basis sets. The correlation coefficients (R) corresponding to obtained dependences are shown in Table 25.4. Joint account of σ and R values indicates possibility of MP2 method with

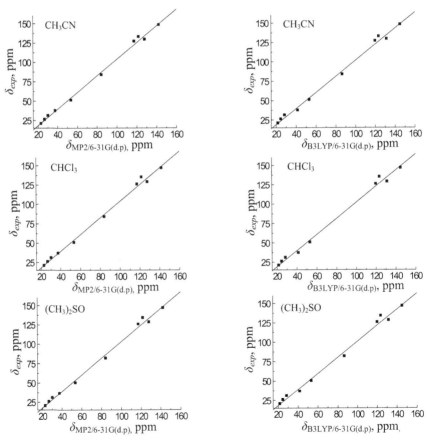

**FIGURE 25.3**    Experimental ($\delta_{exp}$) versus GIAO calculated $^{13}$C chemical shifts (relative to TMS) of the 1,1,3-trimethyl-3-(4-methylphenyl)butyl hydroperoxide.

6–31G(d,p) basis set using for the calculation of the hydroperoxide $^{13}$C nuclei chemical shifts.

## 25.4  CONCLUSIONS

A comprehensive study of the 1,1,3-trimethyl-3-(4-methyl-phenyl)butyl hydroperoxide by experimental NMR $^{13}$C spectroscopy and molecular modeling methods was performed. A comparative assessment of the $^{13}$C nuclei chemical shifts calculated by GIAO in various approximations. For NMR $^{13}$C spectra of the hydroperoxide in different solvents MP2 and B3LYP methods approximations with 6–31G(d,p), 6–311G(d,p), and 6–311++G(d,p) basis sets allow to obtain the correct spectral pattern. A linear correlations between the calculated and experimental values of the $^{13}$C chemical shifts for the studied hydroperoxide molecule were obtained for all solvents studied. In al cases, the MP method combined with 6–31G(d,p) basis set allows to get a better agreement between the calculated and experimental data as compared to the B3LYP results.

### KEYWORDS

- NMR spectroscopy
- chemical shift
- magnetic shielding constant
- GIAO
- molecular modeling
- 1,1,3-trimethyl-3-(4-methylphenyl)butyl hydroperoxide

### REFERENCES

1. Kinash, N. I., & Vostres, V. B. Synthesis of the γ-aryl containing peroxides – 2-methyl-4-pentane derivatives. *Scientific Journal of Lviv Polytechnic National University*, 2003, No 529, pp. 124–128.

2. Turovsky, M. A., Raksha, O. V., Opeida, I. O., Turovska, O. M., & Zaikov, G. E. Molecular modeling of aralkyl hydroperoxides homolysis. *Oxidation Communications*, 2007, Vol. 30, No 3, pp. 504–512.
3. Turovskij, N. A., Raksha, E. V., Berestneva Yu. V., & Zubritskij, M. Yu. Formation of 1,1,3-trimethyl-3-(4-methylphenyl)butyl hydroperoxide complex with tetrabutylammonium bromide. *Russian, J. Gen. Chem.*, 2014, Vol. 84, No. 1, pp. 16–17.
4. Berestneva, Yu. V., Raksha, E. V., Turovskij, N. A., & Zubritskij, M. Yu. Interaction of the 1,1,3-trimethyl-3-(4-methylphenyl)butyl hydroperoxide with tetraethylammonium bromide. Actual problems of magnetic resonance and its application: program lecture notes proceedings of the XVII International Youth Scientific School (Kazan, 22–27 June 2014)/edited by, M. S. Tagirov (Kazan Federal University), V. A. Zhikharev (Kazan State Technological University). – Kazan: Kazan University, 2014, 165 p., pp. 113–116.
5. Turovskij, N. A., Berestneva, Yu. V., Raksha, E. V., Opeida, J. A., & Zubritskij, M. Yu. Complex formation between hydroperoxides and Alk₄NBr on the base of NMR spectroscopy investigations. *Russian Chemical Bulletin*, 2014, No 8, P. 1717–1721.
6. Gaussian 03, Revision, B.01, Frisch, M. J., Trucks, G. W., Schlegel, H. B., Scuseria, G. E., Robb, M. A., Cheeseman, J. R., et al., Gaussian, Inc., Pittsburgh PA, 2003.
7. Wolinski, K., Hinton, J. F., & Pulay, P. Efficient implementation of the gauge-independent atomic orbital method for NMR chemical shift calculations. *J. Am. Chem. Soc.*, 1990, Vol. 112(23), P. 8251–8260.
8. Mennucci, B., & Tomasi, J. Continuum solvation models: A new approach to the problem of solute's charge distribution and cavity boundaries. *J. Chem. Phys.*, 1997, Vol. 106, pp. 5151–5158.
9. Cossi, M., Scalmani, G., Rega, N., & Barone, V. New developments in the polarizable continuum model for quantum mechanical and classical calculations on molecules in solution. *J. Chem. Phys.*, 2002, Vol. 117, pp. 43–54.
10. Raksha, E. V., Berestneva, Yu. V., Turovskij, N. A., & Zubritskij, M. Yu. Quantum chemical modeling of the 1,1,3-trimethyl-3-(4-methyl-phenyl)butyl hydroperoxide NMR ¹H and ¹³C spectra. *Scientific Publications of Donetsk National Technological University, Chemical and Chemical Technology Series*, Donetsk, Donetsk National Technological University publishing house. 2014, Iss. 1(22), pp. 150–156.
11. Abraham, R. J., Byrne, J. J., Griffiths, L., &Perez, M. ¹H chemical shifts in NMR: Part 23, the effect of dimethyl sulphoxide versus chloroform solvent on ¹H chemical shifts. *Magn. Reson. Chem.*, 2006, Vol. 44, pp. 491–509.
12. Turovskij, N. A., Raksha, E. V., Berestneva, Yu. V., Pasternak, E. N., Zubritskij, M. Yu., Opeida, I. A., & Zaikov, G. E. Supramolecular Decomposition of the AralkylHydroperoxides in the Presence of Et₄NBr. in: Polymer Products and Chemical Processes: Techniques, Analysis Applications, Eds., R. A. Pethrick, E. M. Pearce, G. E. Zaikov, Apple Academic Press, Inc., Toronto, New Jersey, 2013, 270.
13. Kosnikov, A. Yu., Antonovskii, V. L., Lindeman, S. V., Antipin, M. Yu., Struchkov, Yu. T., Turovskii, N. A., & Zyat'kov, I. P. X-ray crystallographic and quantum-chemical investigation of tert-butyl hydroperoxide. *Theoretical and Experimental Chemistry*, 1989, Vol. 25, Iss. 1, pp. 73–77.

# INDEX

Printed and bound by CPI Group (UK) Ltd, Croydon, CR0 4YY

23/10/2024

01777701-0001